FATE AND EFFECTS OF SEDIMENT-BOUND CHEMICALS IN AQUATIC SYSTEMS

Pergamon Titles of Related Interest

Bergman et al. ENVIRONMENTAL HAZARD ASSESSMENT OF EFFLUENTS

Branica & Konrad LEAD IN THE MARINE ENVIRONMENT

Cairns BIOLOGICAL MONITORING IN WATER POLLUTION

Cairns MULTISPECIES TOXICITY TESTING

Halasi-Kun POLLUTION AND WATER RESOURCES

Jenkins MICROPOLLUTANTS IN THE ENVIRONMENT

Miyamoto PESTICIDE CHEMISTRY: HUMAN WELFARE AND THE ENVIRONMENT

Stokes ECOTOXICOLOGY AND THE AQUATIC ENVIRONMENT

WHO WASTE DISCHARGE INTO THE MARINE ENVIRONMENT

Yoshida ADVANCES IN PHARMACOLOGY AND THERAPEUTICS: TOXICOLOGY AND EXPERIMENTAL MODELS, Volume 5

Related Journals*

BIOCHEMICAL SYSTEMATICS AND ECOLOGY

CHEMOSPHERE

CURRENT ADVANCES IN ECOLOGICAL SCIENCES

CURRENT ADVANCES IN PHARMACOLOGY AND TOXICOLOGY

ENVIRONMENT INTERNATIONAL

ENVIRONMENTAL TOXICOLOGY AND CHEMISTRY

MARINE POLLUTION BULLETIN

WATER RESEARCH

*Sample copies available upon request

FATE AND EFFECTS OF SEDIMENT-BOUND CHEMICALS IN AQUATIC SYSTEMS

Edited by

Kenneth L. Dickson, North Texas State University
Alan W. Maki, Exxon Corporation
William A. Brungs, USEPA

Proceedings of the Sixth Pellston Workshop,
Florissant, Colorado, August 12-17, 1984
Sponsored by USEPA, Office of Toxic Substances,
North Texas State University,
American Petroleum Institute

SETAC Special Publications Series

Series Editors

Dr. C. H. Ward
Department of Environmental Science
and Engineering, Rice University

Dr. B. T. Walton
Environmental Sciences Division,
Oak Ridge National Laboratory

Publication sponsored by the Society of
Environmental Toxicology and Chemistry

PERGAMON PRESS

NEW YORK · OXFORD · BEIJING · FRANKFURT
SÃO PAULO · SYDNEY · TOKYO · TORONTO

U.S.A.	Pergamon Press, Maxwell House, Fairview Park, Elmsford, New York 10523, U.S.A.
U.K.	Pergamon Press, Headington Hill Hall, Oxford OX3 0BW, England
PEOPLE'S REPUBLIC OF CHINA	Pergamon Press, Room 4037, Qianmen Hotel, Beijing, People's Republic of China
FEDERAL REPUBLIC OF GERMANY	Pergamon Press, Hammerweg 6, D-6242 Kronberg, Federal Republic of Germany
BRAZIL	Pergamon Editora, Rua Eça de Queiros, 346, CEP 04011, Paraiso, São Paulo, Brazil
AUSTRALIA	Pergamon Press Australia, P.O. Box 544, Potts Point, N.S.W. 2011, Australia
JAPAN	Pergamon Press, 8th Floor, Matsuoka Central Building, 1-7-1 Nishishinjuku, Shinjuku-ku, Tokyo 160, Japan
CANADA	Pergamon Press Canada, Suite No. 271, 253 College Street, Toronto, Ontario, Canada M5T 1R5

Copyright © 1987 Pergamon Books Inc.

First edition 1987

Library of Congress Cataloging-in-Publication Data
Pellston Environmental Workshop (*6th:1984:Florissant, Colo.*)
Fate and effects of sediment-bound chemicals in aquatic systems.
(SETAC special publications series).
1. Water—Pollution—Environmental aspects—Congresses.
2. Sedimentation and deposition—Congresses.
3. Biodegradation—Congresses. I. Dickson, Kenneth L.
II. Maki, Alan W., 1947- . III. Brungs, William A.
IV. United States. Environmental Protection Agency.
Office of Toxic Substances. V. American Petroleum
Institute. VI. North Texas State University.
VII. Title. VIII. Series.
QH545.W3P45 1984   574.5'222   87-16210

British Library Cataloging in Publication Data
Pellston Workshop (*6th:1984:Florissant*)
Fate and effects of sediment-bound chemicals in aquatic
systems: proceedings of the Sixth Pellston Workshop,
Florissant, Colorado, 12-17 August 1984.—(SETAC special
publications series).
1. Aquatic ecology. 2. Water—Pollution
I. Title II. Dickson, Kenneth L. III. Maki, Alan W.
IV. Brungs, William A. V. Society of Environmental
Toxicology and Chemistry VI. Series
574.5'263   QH541.5.W3

ISBN 0-08-034866-1

Printed in Great Britain by A. Wheaton & Co. Ltd., Exeter

Contents

Session 3. Biological Effects and Ecotoxicology of Sediment-Associated Chemicals

Introduction to the SETAC Special Publications Series

The SETAC Special Publications series was established by the Society of Environmental Toxicology and Chemistry to provide in-depth reviews and critical appraisals on scientific subjects relevant to understanding the impacts of chemicals and technology on the environment. The series consists of single and multiple authored/edited books on topics selected by the Board of Directors for their importance, timeliness, and their contribution to multi-disciplinary approaches to solving environmental problems. The diversity and breadth of subjects covered in this series will reflect the wide range of disciplines encompassed by environmental toxicology, environmental chemistry, and hazard assessment. Despite this diversity, the goals of these volumes are similar; they are to present the reader with authoritative coverage of the literature, as well as paradigms, methodologies, controversies, research needs, and new development specific to the featured topics. All books in the series are peer reviewed for SETAC by acknowledged experts.

The SETAC Special Publications will be useful to environmental scientists in research, research management, chemical manufacturing, regulation, and education, as well as to students considering careers in these areas, for keeping abreast of recent developments in familiar areas and for rapid introduction to principles and approaches in new subject areas.

Fate and Effects of Sediment-Bound Chemicals in Aquatic Systems, the third book published in this series, presents the proceedings of the sixth Pellston Workshop, held August 12-17, 1984, at Florissant, Colorado. Sediments are increasingly recognized to affect the transport and biological availability of chemical contaminants in fresh water and marine ecosystems.

This workshop brought together many of the leading workers to consider the scientific basis for evaluating and predicting the fate and effects of sediment-bound chemicals. These proceedings provide a valuable summary of chemical interactions with sediments and their consequences for aquatic organisms, as well as identifying critical research needs in this area.

BARBARA T. WALTON,
Oak Ridge National Laboratory
Series Editor

Preface

This publication is the product of the Sixth Pellston Environmental Workshop held on August 12-17, 1984, in Florissant, Colorado. These workshops are devoted to assessing and guiding the development of scientific inquiry of hazards to the aquatic environment. Support for the workshop series has come from a variety of sources both private and public in the form of grants. Participants of each workshop are invited from regulatory agencies, universities, and industries to share their expertise on the particular topic chosen for discussion. By serving as a forum where professionals can discuss the state of knowledge on a topic and offer direction to future research, these workshops provide stimulus to the formulation and implementation of policies pertinent to water quality management.

This workshop addressed the topic of the role of suspended and settled sediments in regulating the fate and effects of chemicals in the aquatic environment. The nature and extent of partitioning and bioavailability were identified as key elements in research efforts toward assessing the effects of sediments on water quality. As we move forward in developing regulatory and management strategies for chemicals entering public water supplies, it is essential that our knowledge of chemical interactions with solids be incorporated. This book and the personal interchange of ideas among the participants of the workshop that it represents is a compilation of the state of knowledge on the role of solids in assessing the hazards of chemicals in the aquatic environment.

Acknowledgments

The 1984 Pellston Workshop and this publication were made possible by financial support from the American Petroleum Institute, the Office of Toxic Substances Ecological Effects Branch USEPA, and North Texas State University.

The success of this workshop was enhanced by the serenity of the Rocky Mountains and the excellent conference facilities of The Nature Place in Florissant, Colorado. Mr. Dave Perry and the Sanborn Western Camps staff are thanked for their assistance and hospitality.

Workshop organizers would like to commend the participants for their dedication to the tasks defined in the opening session and for their prompt preparation of manuscripts and revisions that have made possible these proceedings. Gratitude is also expressed to Donna Argo, Jan Hansen, Kathy Henson, and Bonnie Yates for managing details of workshop organization and publication preparation.

The co-chairmen wish to thank the Society of Environmental Toxicology and Chemistry for promoting the results of this workshop as a Special Publication. Dr. Barbara T. Walton is thanked for her coordination of the peer review of this volume. Special gratitude is extended to the anonymous reviewers who devoted careful consideration and constructive appraisal of these efforts, thus ensuring a concise and scholarly product. The organizers and participants sincerely appreciate their contributions.

Introduction

A. W. Maki, K. L. Dickson and W. A. Brungs

The prediction of the environmental fate and effects of chemicals potentially reaching surface waters is a major activity of many chemists and biologists throughout academia, industry, and regulatory agencies. The increasing environmental awareness of the past 15 years has focused the need for accurate and reliable methods to predict environmental impacts of new chemicals and effluent discharges in order to provide guidance and control measures for industry and regulatory agencies. These control strategies have led to the development of integrated hazard assessment programs for new and expanded-use chemicals, effluent and discharge permitting, refinement of water quality criteria and standards, as well as several other control programs. Underlying each of these control programs is an implied assurance in our ability to predict and assess the ultimate fate of a chemical and its toxicity to aquatic life.

Historically, several assumptions have been made or implied in the development and application of environmental safety assessments and water quality criteria. The most significant factor was that introduced chemicals were generally water-soluble. This resulted in the emphasis for impact analysis being limited to the water column. This simplistic approach in turn created bias for environmental assessments that emphasized pelagic species for toxicity studies and placed the slow-to-evolve significance of environmental chemistry in a support role. The process did not consider the ultimate fate and distribution of the chemicals of concern, and when suspended and settled solids were factors, they were considered to be a safe repository of sorbed contaminants.

Two factors probably were most responsible for the insight that discredited the overly simplistic and entrenched philosophy that emphasized only dissolved chemicals and the water column. The first was the development of the Priority Pollutant List in 1976, even though the choice of chemicals was human health oriented. This list included a wide range of inorganic and organic chemicals that, upon investigation, were found to be very water-insoluble. The second factor was a result of the gradual involvement of environmental chemists in the assessment process. They expressed concern about the assumption of the irreversibility of chemicals sorbed to sediments. The words bioavailability and partitioning became the key words of environmental assessment.

The nature and extent of partitioning will ultimately influence the bioavailability of a chemical. Whether or not bioavailability is enhanced or decreased depends on the chemical, the suspended and settled sediments, and the aquatic life form of interest. As we move forward in developing regulatory and management strategies for chemicals in the aquatic environment, it is essential that our knowledge of chemical interactions with solids be incorporated.

At the present time, it is our belief that considerable knowledge exists on how chemicals interact with sediments and how these interactions affect the fate and effects of the chemicals. Research data to date have clearly shown that the fate and effects of chemicals are controlled by a combination of concurrent processes involving both the transformation and transport of the chemical. Among the most important processes for many chemicals are solubilization and adsorption to solids, since they both affect the transport of the material and influence bioavailability, toxicity, biodegradation, hydrolysis, and ultimate fate of the chemical. However, information regarding these processes is fragmentary, ad hoc, and sufficiently diverse as to preclude definition of meaningful conclusions.

This workshop represents an attempt to bring together and focus the existing knowledge on the role sediments play in regulating the fate and effects of chemicals in the aquatic environment. The United States Environmental Protection Agency's Office of Toxic Substances, the American Petroleum Institute, and North Texas State University were the sponsoring organizations. Conference facilities and accommodations were provided by The Nature Place in Florissant, Colorado.

The format and philosophy of this workshop is a continuation of what has become known as the "Pellston" Workshops. The first workshop was held in June 1977 at the University of Michigan's Biological Station in Pellston, Michigan, and twenty-four participants attended, representing industry, academia, and government. They addressed the topic of *Estimating the Hazard of Chemical Substances to Aquatic Life* and published the proceedings under the sponsorship of the American Society of Testing and Materials (ASTM) (1). Without a doubt, this workshop influenced thinking about how a hazard assessment of a chemical should be approached. It also made people aware of the state of the art for predictive fate and effects.

All participants felt the format of the workshop was an ideal mechanism for constructive synthesis of concepts. Because of the success of the first workshop and the ever-increasing need for guidance for both industrial and governmental regulatory sectors, a second workshop was planned and conducted in August 1978 at Waterville Valley, New Hampshire. This workshop expanded on the first and addressed the topic of *Analyzing the Hazard Evaluation Process*. Attended by thirty-four participants from this country, Europe, and Asia, the workshop further solidified many of the hazard assessment concepts and developed an international dialogue between the regulator and the regulated, as well as introduced several new concepts relevant to hazard evaluation and water quality criteria. The proceedings of this workshop were published by the American Fisheries Society (2). The meeting itself was funded by contributions from six industries.

Encouraged by the impact these workshops were having on the advancement of the field of aquatic hazard evaluation, it was decided that it would be productive to conduct a third workshop focusing on topics identified in the first two workshops. In August 1979, the third Pellston Workshop was held on the theme of *Biotransformation and Fate of Chemicals in the Aquatic Environment*. This workshop was funded by the Chemical Manufacturers Association (CMA) via a grant to North Texas State University. Like the first two workshops, it has proven to be helpful in the development of the state of the art in this field. The American Society of Microbiology published the proceedings in the summer of 1980 (3).

The fourth workshop on *Modeling the Fate of Chemicals in the Aquatic Environment* was held at the University of Michigan's Biological Station at Pellston and was attended by thirty-two model developers and model users. It too was an outgrowth of the earlier workshops. Being able to predict the concentration of a chemical in a particular enviromental compartment is a critical need of decision makers in industry and governmental regulatory agencies. This workshop was jointly funded by the United States Environmental Protection Agency and by the Chemical Manufacturers Association. Ann Arbor Science published the proceedings in 1982 (4).

The fifth workshop dealt with *Environmental Hazard Assessment of Effluents* and included forty-one participants involved in the evaluation of the enviromental fate and effects of complex mixtures and effluents. This workshop, held at the Valley Ranch in Cody, Wyoming, was instrumental in identifying water quality-based effluent limitations and standards, signaling a departure from technology-based limitations. Participants were successful in identifying realistic laboratory and field testing methodology useful to predictive evaluations of effluent impacts. Proceedings of the workshop are currently being published through the Society of Environmental Toxicology and Chemistry (SETAC) via Pergamon Press. The workshop was supported by contributions from the American Petroleum Institute, the Electric Power Research Institute, the Chemical Manufacturers Association, and the United States Environmental Protection Agency.

BACKGROUND

An ad hoc advisory committee met on January 19, 1983, in Dallas, Texas, to assist us in formulating the scope, and list of participants for this sixth Pellston Workshop. The following people participated in the ad hoc advisory committee meeting:

Dr. Jack Anderson
Battelle North West Laboratories

Dr. G. Fred Lee
New Jersey Institute of Technology

Dr. William Brungs
USEPA, Environmental Research Laboratory

Dr. Al Maki
Exxon Corp.

Dr. Kenneth L. Dickson
North Texas State University

Dr. Richard Peddicord
US Army Engineer Waterways Experiment Station

Dr. James H. Gilford
USEPA, Office of Toxic Substances

Dr. John H. Rodgers, Jr.
North Texas State University

Dr. James Lazorehak
USEPA

Dr. Farida Y. Saleh
North Texas State University

After spending a day in discussing the proposed topic, it was evident that a workshop was timely, sorely needed, and had the potential to be constructively influential in the formulation and implementation of policies and regulations controlling chemicals in the aquatic environment. During this ad hoc planning session, we identified session topics, as well as knowledgeable individuals in industry, academia, and regulatory agencies who would contribute to and benefit from such a workshop.

WORKSHOP CONDUCT

The results of the ad hoc advisory committee meeting were used to finalize the workshop objectives, session topics, paper authors, and list of participants. All authors and participants were notified and commitments were obtained to prepare drafts of discussion papers in advance of the actual workshop date. Approximately 1 month prior to convening the workshop, all participants were provided with a notebook containing drafts of all papers to facilitate preparation for the workshop. During the initial plenary session, the following statement of Workshop Objectives and Charge to Participants were presented to all participants:

Objectives

1.   Enhance communications and provide a forum for transfer of

information between researchers and decision makers from the regulatory, industrial, and academic communities.

2. Assess state of the science regarding the role of suspended and settled sediments in controlling the fate and effects of chemicals in aquatic systems.

3. Identify existing consensus methodology for predictive testing of sediment impacts on environmental fate and effects of chemicals in the laboratory and field.

4. Identify individual chemical properties and major chemical, physical, and biological environmentally related factors, keying the need for sediment/sorption testing in a predictive hazard assessment process.

5. Identify the water quality significance of sediment interactions/bioavailability including:
   —regulatory implications;
   —sediment action level;
   —consequences to beneficial user.

6. Identify specific research needs for sediment interactions/bioavailability assessment and document the workshop findings for general distribution.

Charge to Participants

- Participants should engage in open discussions of topics introduced in presentations.
- Participants must recognize their multiple responsibilities to the science, environment, and user groups including the public, government, and industry.
- Participants should assist in the preparation of summary statements and recommendations for their assigned sessions.
- Rapid completion of final drafts and revisions is necessary to ensure timely preparation of the workshop proceedings.

Each plenary technical session was initiated with one or more discussion papers presented by the previously-identified author(s). Following each presentation, a discussion session, open to all participants, was held under the direction of a session chairman. It was the responsibility of the chairman and his committee to capture the main points and identify the major issues for inclusion in the final workshop proceedings. Thursday of the workshop week was provided for all individual session committees to meet in separate sessions, develop consensus points from their session, and prepare a draft written summary of these findings for presentation to the entire workshop during the final plenary session. The main working sessions and their content are summarized as follows.

Session 1. Background and Perspectives

The initial session was designed to reconfirm the objectives of the workshop and present the charge to the participants. Following introductory statements from the organizers, an overview discussion of historical perspectives and water quality implications of sediment-associated contaminants was held. Additional presentations were made summarizing needs and perspectives for chemical:sediment interactions from industrial and regulatory viewpoints. The session chairman was G. Loewengart.

Session 2. Environmental Fate and Compartmentalization

The objectives of the second session were to develop an understanding of the current state of knowledge concerning methods to assess and predict the chemistry of chemical/sediment interactions. Topics such as sorption:desorption equilibria for metals and organics, physical transport of sediments, and biodegradation methodology were considered. The session chairman was D. DiToro.

Session 3. Biological Effects and Ecotoxicology of Sediment-Associated Chemicals

This session was designed to promote discussion and develop a consensus on optimal methods to assess and predict the bioavailability and biological effects of chemicals bound to sediments. Questions such as: Are chemicals bound to sediments available or toxic to benthos? Do they bioconcentrate? were addressed in the session. Chairman of the session was J. Anderson.

Session 4. Case Studies

In this session, specific case study examples of chemical/sediment interactions were presented. By highlighting what had actually been done in several recent case study examples, a discussion of problems, shortcomings, and alternative strategies was developed. Each case study was selected to represent actual examples where sediment/chemical interactions were found to be critical in determining the short- and long-term impacts to water quality and associated uses. The session chairman was K. Malueg.

Session 5. Regulatory Implications

It was the purpose of this session to examine closely the regulatory implications of sediment bound chemicals: should or can sediment criteria be established? Does the present state of knowledge allow for a quantitative assessment of sediment "action levels"? Additional questions were directed at identifying the research and data needs necessary to derive scientifically sound and ecologically relevant criteria values. The session chairman was G. Chapman.

Session 6. Workshop Summary

At the beginning of the workshop, an overall summary committee was charged with the responsibility to develop a workshop summary, including consensus conclusions from the individual sessions and an identification of recommendations and future research needs. Throughout the week, members of this committee attended the individual group working sessions and recorded the major findings. On the final day of the workshop, this committee presented their findings to a plenary session for comments and finalization. The session chairman was J. Fava, and committee members were J. Gilford, R. Kimerle, R. Parrish, T. Podoll, H. Pritchard, and N. Rubinstein.

The following chapters and sections of this book correspond to and represent the findings of each of the Workshop Sessions as outlined above. Each section, except this introduction and the final workshop summary, contains one or more papers presented to initiate discussion and a short summary report prepared by the session summary committee. Complete drafts of all texts were prepared by the final day of the workshop, and final drafts of each of the summaries were completed via subsequent mailings. However, no substantial changes were permitted after the workshop was finished. Every attempt has been made to ensure that editorial review has been carried out in the spirit of the workshop.

REFERENCES

1. **Cairns, J., Jr., K. L. Dickson and A. W. Maki** (eds.). 1978. *Estimating the Hazard of Chemical Substances to Aquatic Life.* ASTM STP 657. American Society for Testing and Materials, Philadelphia, PA.

2. **Dickson, K. L., A. W. Maki and J. Cairns, Jr.** (eds.). 1979. *Analyzing the Hazard Evaluation Process.* American Fisheries Society, Washington, DC.

3. **Maki, A. W., K. L. Dickson and J. Cairns, Jr.** (eds.). 1980. *Biotransformation and Fate of Chemicals in the Aquatic Environment.* American Society for Microbiology, Washington, DC.

4. **Dickson, K. L., A. W. Maki and J. Cairns, Jr.** (eds.). 1982. *Modeling the Fate of Chemicals in the Aquatic Environment.* Ann Arbor Science Publishers/Butterworths, Ltd., Stoneham, MA.

5. **Bergman, H.L., R. A. Kimerle and A. W. Maki** (eds.). 1985. *Environmental Hazard Assessment of Effluents.* Pergamon Press, Elmsford, NY.

WORKSHOP PARTICIPANTS

Dr. William J. Adams (32*)
Monsanto Industrial Chemical Co.
St. Louis, MO 63167

Dr. Jack Anderson (4*)
Battelle Northwest Laboratories
Sequim, WA 98382

Dr. Wesley J. Birge (34*)
University of Kentucky
Lexington, KY 40506

Dr. Ronald J. Breteler (1*)
Battelle New England Marine Research Laboratory
Duxbury, MA 02332

Dr. William Brungs (14*)
USEPA, Environmental Research Laboratory
Narragansett, RI 02882

Dr. Gary A. Chapman (29*)
USEPA, Environmental Research Laboratory
Corvallis, OR 97333

Dr. Kenneth L. Dickson (19*)
North Texas State University
Denton, TX 76203

Dr. Dominic DiToro (24*)
HydroQual Inc.
Mahwah, NJ 07430

Dr. James A. Fava (18*)
Ecological Analysts, Inc.
Sparks, MD 21152

Dr. John H. Gentile (33*)
USEPA, Environmental Research Laboratory
Narragansett, RI 02882

Dr. James H. Gilford (16*)
USEPA, Office of Toxic Substances
Washington, DC 20460

Dr. Andrea Hall (23*)
Battelle Laboratories
Washington, DC 20015

Dr. Florence Harrison (30*)
University of California
Livermore, CA 94550

Dr. Robert J. Huggett (25*)
Virginia Institute of Marine Science
Gloucester Pt., VA 23062

Dr. Everett A. Jenne (7*)
Battelle Northwest Laboratories
Richland, WA 99352

Dr. Samuel W. Karickhoff (5*)
USEPA, Environmental Research Laboratory
Athens, GA 30613

Dr. Richard A. Kimerle (28*)
Monsanto Industrial Chemical Co.
St. Louis, MO 63167

Dr. James L. Lake (11*)
USEPA Environmental Research Laboratory
Narragansett, RI 02882

Dr. G. Fred Lee (22*)
New Jersey Institute of Technology
Newark, NJ 07102

Dr. Henry Lee (12*)
USEPA, Marine Science Center
Newport, OR 97365

Dr. Wilbert Lick (26*)
University of California
Santa Barbara, CA 93106

Dr. Gordon Loewengart (10*)
Allied Corp.
Morristown, NJ 07960

Dr. Warren J. Lyman (15*)
Arthur D. Little, Inc.
Cambridge, MA 02140

Dr. Al Maki (17*)
Exxon Corp.
East Millstone, NJ 08873

Dr. Ken Malueg (8*)
USEPA, Environmental Research Laboratory
Corvallis, OR 97333

Mr. William R. Murden (**)
U.S. Army Corps of Engineers
Fort Belvoir, VA 22060

Mr. Rod Parrish (**)
USEPA, Environmental Research Laboratory
Gulf Breeze, FL 32561

Dr. Spyros P. Pavlou (**)
Envirosphere Company
Bellevue, Washington 98004

Dr. Richard K. Peddicord (20*)
U.S. Army Engineer Waterways Experiment Station
Vicksburg, MS 29180

Dr. R. T. Podoll (9*)
SRI International
Menlo Park, CA 94025

Dr. P. H. Pritchard (2*)
USEPA, Environmental Research Laboratory
Gulf Breeze, FL 32561

Dr. John H. Rodgers, Jr. (21*)
North Texas State University
Denton, TX 76203

Mr. Norman Rubinstein (**)
USEPA, Environmental Research Laboratory
Narragansett, RI 02882

Dr. Richard C. Swartz (3*)
USEPA, Marine Science Center
Newport, OR 97365

Mr. Ron Wilhelm (**)
USEPA, Office of Toxic Substances
Washington, DC 20460

Dr. Douglas A. Wolfe (6*)
NOAA, Department of Commerce
Rockville, MD 20852

Dr. Dan Woltering (31*)
Procter & Gamble Company
Cincinnati, OH 45217

Mr. Chris Zarba (13*)
U.S. Environmental Protection Agency
Washington, DC 20460

Dr. Robert Zeller (27*)
USEPA, Office of Marine and Estuarine Protection
Washington, DC 20460

Support Staff
Donna M. Argo (35*)
Rose Ann Seyerle (37*)
Bonnie Yates (36*)

 * Key to Photograph, page xxv.
** Not in Photograph.

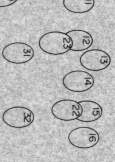

Back Row

Dr. Ronald J. Breteler Dr. P. H. Pritchard Dr. Richard Swartz Dr. Jack Anderson Dr. Samuel Karickhoff Dr. Douglas Wolfe Dr. E. A. Jenne Dr. Ken Malueg Dr. R. T. Podoll

Middle Row

Dr. Al Maki Donna M. Argo Bonnie Yates Dr. James A. Rose Ann Fava Seyerle

Dr. Gordon Loewengart Dr. James Lake Dr. Henry Lee Mr. Chris Zarba Dr. William Brungs Dr. Warren Lyman Dr. James H. Gilford

Dr. Robert Huggett Dr. Dominic DiToro Dr. Andrea Hall Dr. G. Fred Lee Dr. John H. Rodgers, Jr. Dr. Richard Peddicord

Front Row (left to right)

Dr. Wilbert Lick Dr. Robert Zeller Dr. Richard A. Kimerle Dr. Gary Chapman Dr. Florence Harrison Dr. Dan Woltering Dr. William Adams Dr. Jack Gentile Dr. Wesley J. Birge Dr. Kenneth Dickson

Session 1
Background and Perspectives

Chapter 1

Water Quality Significance of Contaminants Associated with Sediments: An Overview

G. Fred Lee and R. Anne Jones

INTRODUCTION

The evaluation of the water quality hazard that contaminants associated with aquatic sediments represent is of considerable interest in many aspects of water pollution control. While it has been known for many years that contaminants associated with particulate matter in natural water may have limited adverse impact on fish and aquatic life-related beneficial uses of water, there is considerable controversy on how best to assess the availability of sediment-associated contaminants to aquatic life. This paper presents an overview of the current state of knowledge on the water quality significance of sediment-associated contaminants, and is designed to be a discussion-initiation paper for this workshop. A number of the topics presented in this paper are discussed in great detail in the papers presented by the other workshop participants.

HISTORICAL PERSPECTIVE

There have been several major pollution control efforts undertaken that have called attention to the need for evaluating the water quality significance of contaminants associated with aquatic sediments. One of the first of these was the effort in the late 1960s to control nutrients entering the Great Lakes as a means of controlling the eutrophication of these water bodies. Inter-action of phosphorus with sediments was identified as being of potential significance in contributing phosphorus to and/or removing phosphorus from these water bodies. Another began during the late 1960s and early 1970s when it was found that some of the sediments that are dredged from the nation's waterways contained high concentrations of a wide variety of potentially toxic contaminants, such as heavy metals and pesticides. This finding stimulated one of the most intensive research efforts ever undertaken in the water pollution control field, to evaluate the water quality significance of such contaminants in the US waterway sediments that are dredged by the U.S. Army Corps of Engineers (USACE).

This paper reviews the development of our understanding of the significance of sediment-associated contaminants by focusing on the investigations into the role of lake sediment- and river particulate-associated phosphorus in eutrophication management, as well as the nature of and findings of the Corps Dredged Material Research Program (DMRP). It also reviews the work that has been done over the years on the oxygen demand of sediments as it relates to the control of biochemical oxygen demand (BOD) in rivers and estuaries.

Role of Sediment-Associated Phosphorus in Eutrophication

The water quality significance of phosphorus associated with natural water particulate matter has been an area of investigation since the early 1960s. It was at that time that the United States and Canada first began to implement eutrophication control programs based on the reduction of phosphorus loads to water bodies, specifically the Great Lakes. The focal point of phosphorus load control was usually domestic wastewater-derived phosphorus. There were many who argued in the 1960s that the removal of phosphorus from domestic wastewaters by detergent phosphate bans, or the removal of phosphorus from domestic wastewater effluent by alum, iron, or lime precipitation-coprecipitation, would have limited effect on the eutrophication-related water quality of the U.S.-Canadian Great Lakes or other water bodies because of the release of phosphorus from the sediments of these water bodies. The basis of this was the belief by some that lake sediments served as phosphate buffers such that when the input of phosphorus from external sources to the water column was reduced, phosphorus would be released from the sediments to compensate for this reduction, effectively negating the impact of the phosphorus input control program.

During the early 1960s to the early 1970s, the USEPA and its predecessor organizations focused considerable research funding on the management of eutrophication. Considerable amounts of money were devoted to developing an understanding of the aqueous environmental chemistry of phosphorus in lake and reservoir systems as it relates to the eutrophication-related water quality of these water bodies. Two approaches were followed in conducting these studies; one focused on laboratory-oriented work and the other on whole lake studies. In the former, lake sediments were taken into the laboratory and incubated under various conditions to examine the extent and rate of, as well as factors influencing, phosphorus release from the sediments. This work was patterned after the classic studies of Mortimer (1, 2) in the 1940s. The whole-lake studies involved measuring the concentrations of phosphorus in the water column over the annual cycle. Some of the most important insights provided from the work on phosphorus utilizing both of these approaches was how not to conduct studies on the water quality significance of sediment-associated contaminants. Both the laboratory and field studies led to some erroneous conclusions concerning the role of sediment phosphorus in affecting the eutrophication of natural waters.

The classical limnological studies, such as those conducted by Mortimer as well as many others, showed that large amounts of phosphorus can be released in a short time under anoxic conditions. This caused the majority of the limnologists to conclude that the release of phosphorus from the sediments to anoxic hypolimnetic waters is an important factor affecting the water quality-related trophic status of water bodies; the field, in general, tended to ignore the oxic release of phosphorus. However, the release of phosphorus from lake/reservoir sediments must be viewed in terms of the ability of the phosphorus to reach the surface waters (euphotic zone) of the water body where it could stimulate the growth of algae. As discussed by Lee et al. (3), the anoxic release of phosphorus that occurs in lakes tends to have limited impact on the eutrophication-related water quality of concern to the public, at a time of concern, usually in the summer. This is because in most water bodies having anoxic bottom waters, the anoxic conditions develop during the summer when the water body is stratified. The thermocline tends to be an effective barrier to the transport of the anoxically released phosphorus into the euphotic zone where it could stimulate algal growth. Only a small part of the phosphorus released from the sediments is transported into the surface waters. The release of this phosphorus in the hypolimnion at fall overturn does tend to stimulate late fall blooms of planktonic algae. However, these blooms are rarely of major concern to the public because of the time of the year in which they occur and the type of algae that develop.

There are some water bodies that show considerable thermocline migration. This movement promotes additional transport of the phosphorus from the hypolimnion into the epilimnion-photic zone, which could add some importance to the anoxic release of phosphorus. It should be noted, however, that although phosphorus associated with iron can be released under anoxic conditions, it would also tend to be removed once it comes in contact with oxygen in the surface waters. Stauffer and Lee (4) discussed how thermocline migration likely contributed to the periodic summer algal blooms in Lake Mendota, and other Wisconsin lakes.

The laboratory studies of Spear (3, 5) were among the first to point to the potential importance of oxic release of phosphorus from sediments associated with the mineralization of algae, in affecting the eutrophication-related water quality of water bodies. However, it was not until the whole-lake studies of the type conducted by Schindler et al. (6) and others on the Canadian Shield Experiment Lakes, as well as those of Sonzogni (3, 7), that the overall role of lake sediment phosphorus in influencing the water quality-related trophic status of water bodies began to be generally understood. These investigators used a mass balance approach to evaluate the flux of phosphorus to or from the sediments; they examined the amount of phosphorus present in the water column, the amount of phosphorus entering the water body, and the amount of phosphorus discharged from the water body, both over an annual cycle and during short time periods during the year. With this approach, it became possible to begin to quantify the overall phosphate buffering tendencies of the sediments. It was concluded from these studies that while there is both oxic and anoxic release of phosphorus from aquatic sediments at various times of the year, the net annual flux of phosphorus is to the sediments. Thus, aquatic sediments, in general, are ultimately sinks for phosphorus rather than buffers/sources.

These conclusions were supported by the work of several groups such as Sonzogni et al. (8) who utilized the concept of elemental residence times to describe whole-lake contaminant behavior, especially under altered load conditions. If lake sediments serve as phosphate buffers, then the rate of response of a lake to an alteration of its phosphorus load could be quite long; however, if the factors that cause lake sediments to become sinks for phosphorus operate to about the same degree to control phosphorus concentrations in the water column before and after phosphorus load alteration, then the rate of recovery of the water body upon reduction of phosphorus loads should be predictable based on a phosphorus residence time model. Thus, in a period of time equal to about three times the phosphorus residence time after load alteration, about 95% of the new equilibrium conditions governing the trophic status of the water body should be attained.

The concept of a phosphorus-residence time model evolved from chemical oceanography. Barth (9) developed elemental residence time models for various chemicals, and Goldberg (10) presented a discussion of the development of elemental residence times for various chemicals in the oceans. The elemental residence times are determined by dividing the total mass of the chemical in the water column by the rate of input of that chemical. It is an assessment of how rapidly contaminants added to the water column of a water body become incorporated into the sediments. Chemicals with short residence times tend to have net reactions that cause the chemicals to become a part of the sediments at a fairly rapid rate. Those chemicals that tend to be the least reactive, such as sodium, have the longest residence time in the water column. For lakes and reservoirs for which there is an outlet, the residence time is also influenced by the flux of contaminants from the water body. This, however, is automatically considered as part of the computations since it influences the concentrations in the water column.

Sonzogni et al. (8) found that the phosphorus residence times for many lakes and reservoirs are on the order of 1 year, which means that all of the phosphorus that enters a water column during a year is either transported out of the water body via the outlets or becomes incorporated into the sediments within a year. This finding has important implications for eutrophication management. It means that for many lakes, the driving force for summer planktonic algal growth is the nutrient input received during the preceding year. Jones and Lee (11) and Lee et al. (12, 13) have discussed the relationships between the external phosphorus load from a water body's watershed and the trophic state of the water body. These relationships, based on the OECD (Organization for Economic Cooperation and Development) eutrophication study, show that a water body's planktonic algal biomass (as measured by chlorophyll, water clarity-Secchi depth, hypolimnetic oxygen depletion rate, or primary productivity) is determined by the external loading of phosphorus after it is normalized by the water body's mean depth, area, and hydraulic residence time. It is sometimes alleged that the OECD eutrophication modeling approach does not consider the internal phosphorus loading. This is not the case; it is considered in the normalizing factors, with the result that the normalized phosphorus loading is theoretically equal to the mean in-lake phosphorus concentration (14, 15). Clearly, the average, in-lake phosphorus concentration does reflect the sediment/water exchange reactions. Because of the large data base developed during and after the OECD eutrophication studies (11, 16, 17), laboratory or fieldwork that does not yield results consistent with this approach should be carefully examined for reliability.

The work on phosphorus exchange with sediments provides a useful guide on how studies on the uptake and release of other contaminants should be conducted. As discussed by Lee (18), sediment/water exchange reactions are controlled in many cases by the hydrodynamics of the system rather than the concentrations of the contaminant in the sediment. Therefore, neither the typical throw-some-mud-in-a-jar-and-measure-what-is-taken-up-or-released approach, nor the bell-jar in situ study approach, provides highly reliable or predictive results. The same applies to passing water over cores of sediments. Failure to recognize this can cause substantial errors in the estimation of the amounts of available forms of contaminants that may be released as assessed in laboratory or field studies of these types. A symposium was held in the mid-1970s on the release of phosphorus from lake sediments. The proceedings of the symposium (19) should be consulted for more information on this topic.

**Impact of Dredged Sediment-associated Contaminants on
Water Quality**

The second area that has provided considerable insight into the water
quality significance of contaminants associated with sediments is the work
that was done in the 1970s on the water quality aspects of dredged-sediment
disposal. In the late 1960s, as part of the "environmental awakening" in the
United States, it was found that the sediments of a number of the U.S.
waterways, especially near urban-industrial centers, contained greatly
elevated concentrations of a wide variety of contaminants that, if available
to aquatic life, could have a significant adverse effect on the numbers, types,
health, and/or wholesomeness of organisms in the vicinity of the sediments.
There was considerable concern about these "contaminated"sediments in
conjunction with the USACE dredging operations for navigation channel
maintenance. The Corps, at that time (late 1960s), generally relied on
economic considerations for determining the best method for dredging and
dredged-sediment disposal. Often, the cheapest method for dredged-
sediment disposal was to deposit the sediment into the nearby watercourse.
Also at that time, however, the practice of open-water disposal of dredged
sediment came under attack by many environmental action groups, state and
federal pollution control agencies, as well as others, the philosophy being,
"surely the disposal of large amounts of highly contaminated sediment in
open waters must have some adverse effect on the water quality."

The concern about the elevated levels of contaminants in dredged sediment caused the Federal Water Quality Administration (a predecessor
agency of the USEPA) to develop what are frequently called the "Jensen
criteria" (20). These "criteria" were for the bulk sediment composition for
about a half-dozen selected parameters such as copper, zinc, COD, volatile
solids, etc. If the concentration of one or more of these parameters in a
sediment exceeded the criterion value, then the sediment was deemed unfit
for open-water disposal and an alternate method of dredged-sediment
disposal had to be found. As discussed by Lee and Plumb (21) and Lee et al.
(22), there were numerous problems with using the Jensen criteria, the most
important of which was the tacit assumption that the concentrations of
available forms of contaminants that could be released from sediments were
equal to or proportional to the total concentrations of the contaminants in
the sediments. An elementary knowledge of aquatic chemistry would
uncover the technical invalidity of this assumption. Contaminants "bind" to
sediments in a variety of ways; the binding strength influences the availability of contaminants to aquatic life. It is certainly inappropriate to assume
that just because a sediment has an elevated concentration of a contaminant,
the contaminant is available to affect aquatic organisms and water quality.

In the face of having to spend considerable amounts of additional money
for alternative methods of dredged-sediment disposal, the federal Congress

supported a 5-year, 32 million dollar research program devoted to investigating a variety of environmental aspects of dredged-sediment disposal. This study, known as the Dredged Material Research Program (DMRP), was conducted principally through the U.S. Army Engineer Waterways Experiment Station (WES) in Vicksburg, Mississippi. In addition to the work done through the WES, a number of Corps districts conducted substantial research programs devoted to investigating environmental aspects of dredged-sediment disposal. It is estimated by the authors that the total research effort in this area has been in excess of $90 million, about 25 to 30% of which was devoted to studies specifically designed to investigate the water quality significance of contaminants associated with dredged sediments. These studies included a variety of laboratory and detailed field investigations on the availability of contaminants associated with waterway sediments throughout the United States. Several hundred reports were developed out of the DMRP studies; a complete list of these is available from the WES. The results of the work that the authors and their associates did as part of these studies, including a summary of results of related work that had been done by others, were published in two reports, Lee et al. (22) and Jones and Lee (23). Several papers that were developed out of the DMRP studies were also published in the proceedings of an ACS Symposium on Contaminants in Sediments (24).

The primary focus of the authors' studies was to investigate the reliability of the elutriate test as a measure of the potential for rapid release of dredged sediment-associated contaminants to the water column upon their open-water disposal. The elutriate test was developed by the Corps and the EPA as an alternative to bulk sediment criteria for evaluating the potential hazard of contaminants associated with dredged sediments. Basically, it involves mixing one volume of wet sediment with four volumes of water for 30 min, settling for 1 h, and filtration of the supernatant. The concentrations of contaminants in the filtrate are then to be compared to water quality criteria-standards after appropriate allowances have been made for mixing-dilution at the dredged-sediment disposal site. The specifications for conducting the elutriate tests were established to simulate the type and duration of sediment/water contact, and mixing that would be expected to occur during hydraulic dredging, i.e., hopper-dredged disposal operations. The elutriate (the supernatant from the elutriate test) would be expected to simulate settled dredged-sediment slurry.

The authors' studies involved collecting sediment and water at about twenty locations around the country and running elutriate tests on them under the variety of physical/chemical conditions. The investigations at about ten of these sites included several detailed field studies at each, in which the water column before, during, and after disposal operations was intensely monitored. Measurement of about thirty parameters, including heavy metals, aquatic plant nutrients (nitrogen and phosphorus), chlorinated hydrocarbon pesticides and PCBs, oxygen demand, etc., were made in the field samples, as well as in the elutriates. It was thus possible to evaluate the capabilities of the elutriate test for estimating the actual release of chemicals in the water column during dredged-sediment disposal.

It is important to point out that the elutriate tests were conducted under defined redox conditions in order that the oxidation/reduction conditions would not control the contaminant release. Because the bulk of dredged sediments are anoxic, in the "oxic" tests the sediment suspensions were mixed by vigorously bubbling compressed air through the system; in the "anoxic" tests run for comparison, the suspensions were mixed using nitrogen gas, and exposure to oxygen during processing was prevented by working under a nitrogen atmosphere. Under oxic conditions expected in the water column at most dredged-sediment disposal sites, any ferrous iron present in the sediments would be rapidly oxidized to ferric iron and precipitated as ferric hydroxide. Ferric hydroxide is a highly efficient scavenger of trace contaminants such as heavy metals and phosphorus. The scavenging ability of ferric hydroxide is well known in the literature and has been discussed by Jenne (25) and Lee (26). There may be some oxic sediments such as clean sands, however, that do not contain sufficient concentrations of ferrous iron to bring about effective scavenging of contaminants upon exposure of the dredged sediment to oxygenated waters. The iron in the sediments is already present as ferric hydroxide. As discussed by Lee (26), an aged ferric hydroxide does not scavenge contaminants as effectively as a contaminant scavenger as freshly precipitated ferric hydroxide. While at the pHs of marine waters and most fresh waters (>6.5), the oxidation of ferrous iron is rapid, exceptions would be those systems receiving acid mine drainage or are heavily impacted by acid rain. If the pH of the water is low enough to significantly impact the rate of iron oxidation, however, the acidity itself would be of concern for aquatic life.

The failure of the standard elutriate test procedure to specify the redox conditions that must prevail during the test was a significant shortcoming. The focus of the authors' DMRP testing was on conducting the test under oxic conditions since this is the condition that would be present where there is concern about the release of contaminants that could have an impact on aquatic life. Since oxygen has to be present for aquatic life such as fish to survive, and since oxic conditions prevail at most disposal sites, it was felt that the oxic test was a much more realistic test than the anoxic test. Great care must be exercised in conducting the anoxic test to prevent even traces of oxygen from entering the system; otherwise, a somewhat ill-defined redox condition develops where some scavenging of the contaminants takes place. This scavenging would lower the concentrations from those typically found under true anoxic conditions, but not necessarily to the point that would occur if the systems were fully oxygenated. In addition, care must be exercised in the interpretation of the results of tests conducted under anoxic conditions. It should be noted that the vigorous mixing during elutriation may result in the volatilization of volatile organics in the test system. However, it is essential that oxic conditions be maintained in the test procedure in order to interpret the results and to conduct meaningful bioassays. Where volatile organics are of concern, the impact of the vigorous mixing on the test results should be evaluated; the potential loss of those compounds during the dredging/disposal operations should also be considered in test set-up and in data interpretation.

Ammonia and Other Toxicants. It was found that the only two compounds consistently released in the dredged-sediment elutriate test conducted by the authors were manganese and ammonia. All of the other chemicals studied generally showed little or no release. For some chemicals, such as zinc, the suspension of the sediments in the water generally resulted in a decrease in the concentration in the waters used in the tests. As expected, many of the elutriate tests conducted under anoxic conditions showed releases of a variety of contaminants. However, once the anoxic elutriates were exposed to air or oxygen-containing water, most of the soluble contaminants were incorporated into the ferric hydroxide floc that formed as a result of the oxidation of the ferrous iron to the ferric form.

The release of ammonia found in some elutriate tests was sufficient to be of concern with respect to its toxicity to aquatic life. For example, elutriates from tests conducted on sediment from the North Landing River, in the Atlantic Coast Intercoastal Waterway system in Virginia, contained more than 20 mg N/L total ammonia (27). At the temperature and pH of the receiving waters, the un-ionized ammonia in the elutriates would be sufficient to be chronically toxic to aquatic life. However, as discussed by Jones and Lee (23), it would be rare that aquatic organisms in the water column could receive a chronic exposure to chemicals released to the water column during a dredged-sediment disposal operation, much less a chronic exposure to the dredged-sediment slurry (elutriate). Indeed, in the case of the North Landing River, it was shown, using a hazard assessment approach, that the likelihood was quite small that the amount of ammonia released in the elutriate test would have any significant adverse effects on aquatic life in the river. This was a result of very rapid dilution of the dredged-sediment discharge water and the fact that there was substantial exchange of this water with low ammonia waters of this region as a result of tides and inflowing waters.

As discussed above, it is important in evaluating any release of contaminants from sediment, especially when the release is determined by tests such as the elutriate test, to give proper consideration to the physical conditions that occur at a release site. Most of the release of contaminants from sediments occurs as an episodic event associated with storms, dredging operations, etc., where elevated concentrations of the contaminants may be present for relatively short periods of time. Very high concentrations of a contaminant may be present in the supernatant of a laboratory test system in which the sediment was mixed with a limited volume of water. However, in almost all sediment/water exchange systems, there is appreciable mixing-dilution of the waters near a particular sediment with the waters of the overall system. Generally, the well-mixed laboratory systems allow consider-ably more sediment/water contact than occurs during most dredging/disposal activities, especially hopper-dredge or mechanical dredging operations. Failure to recognize field conditions in the evaluation of the potential impacts of a disposal operation could readily cause erroneous conclusions to be developed on the water quality significance of the release of contaminants from sediments.

In laboratory or field testing, it is crucial to properly account for the sediment/water ratios, mixing-hydrodynamics, redox conditions, dilution, etc., expected. Further, as discussed by Lee et al. (28), it is important in evaluating the water quality significance of any contaminant to consider concentration/duration-of-exposure relationships that exist for organisms of concern. Aquatic organisms can tolerate high concentrations of contami-nants for short periods of time. However, as the duration of their exposure is increased, the safe concentration for the organism decreases.

As evidenced in its proposed (February 1984), revised water quality criteria for certain contaminants, EPA is beginning to recognize the importance of concentration/duration-of-exposure relationships in that it is proposing to adopt single maximum criteria as well as 30-day average criteria. There is considerable difference between the two. For example, for ammonia at 20°C and pH 8.0, the single maximum total ammonia concentration is 3.3 mg/L NH_3, while a 30-d average for the same pH and temperature conditions is 0.81 mg/L NH_3. These new criteria could have a significant impact on dredging operations in that the sediment/water slurries for many hydraulically dredged sediments will have total ammonia concen-trations above these criteria. While in principle this approach to criteria-standards development is better than those previously used, there will likely be considerable difficulty encountered in implementing a single maximum value criterion in an appropriate way for dredged sediment bcause many aquatic organisms can be exposed to concentrations of total ammonia in excess of these values for short periods of times (e.g., minutes to a few hours such as may occur with dredged-sediment disposal operations) without being significantly adversely affected.

It is important to emphasize that the release of ammonia, or other contaminants for that matter, during dredged-sediment disposal is not significantly different from what occurs naturally during storms, with the tidal cycle, during elevated flow in rivers, from boat traffic, etc. It is likely that the waters in contact with any of these sediments before dredging would, following a major storm, contain elevated concentrations of ammonia. It is therefore imperative that studies be conducted to determine whether the single maximum value for ammonia proposed by EPA in February 1984 represents a value that can never be exceeded, or represents a guideline value that is indicative of potential harm and that triggers hazard assessment studies to determine whether real harm could occur.

Manganese. The other contaminant that was consistently released in the elutriate test studies of dredged sediments was manganese. Manganese is of considerably less concern than ammonia, however, since its critical concentrations tend to be considerably higher. Manganese is not particularly toxic to aquatic life; it is slowly precipitated as a dioxide as it reacts with dissolved oxygen. Manganese can cause problems for domestic water supplies, but these problems are readily controllable.

PCBs. One of the areas of intense study during the dredged-sediment disposal studies of the authors and their associates was the Houston Ship Channel, Galveston Bay, and the Galveston Bay Entrance Channel (GBEC) areas of the Texas Gulf Coast. The GBEC extends from Galveston Bay to about 10 km into the Gulf of Mexico. Studies on the PCB content and release from sediments taken from each of these areas showed that the highest concentration of PCBs occurred in the oily sediment of the Houston Ship Channel, while the lowest concentrations were found in the sand-like sediments taken from the most seaward sampling point in the GBEC. The pattern of PCB release from these sediments was the opposite of the concentration pattern; the sand-like sediments released the greatest quantity of PCBs. PCBs, as well as many other contaminants, tend to become associated with oily materials. This alone should point to the inappropriateness of using bulk sediment criteria as a basis for judging sediment-associated contaminant availability. This finding also casts considerable doubt as to the validity of hot-spot dredging as a method of minimizing the potential impact of sediment-associated contaminants on water quality. Hot-spot dredging should only be practiced where it is clearly demonstrated that the "hot spots," i.e., areas of greatly elevated contaminant concentration, are the primary sources of the contaminants causing the water quality problems in the region. It could be that the "hot spots" are the sinks and areas where the contaminants are held most tightly and therefore are not the most important sources of contaminants for the waters of the region.

Elutriate Test General Conclusions

The study approach, data, results, and conclusions drawn from the elutriate tests and field studies conducted on the use and reliability of the elutriate test for predicting contaminant release upon open-water disposal of dredged sediment are presented in Lee et al. (22). It was found that, in general, the elutriate tests predicted the direction and approximate magnitude of release of contaminants from the dredged sediments and demonstrated, as anticipated, that the bulk properties of a sediment cannot be used to estimate the release of contaminants from them. The elutriate test, on the other hand, is a useful tool for predicting the potential for release of contaminants from aquatic sediments that are stirred into the water column, especially when used in conjuction with bioassays/toxicity tests as discussed below.

Toxicity Tests. While over thirty chemical parameters were measured on each of the elutriates as well as on the samples collected in the field evaluation studies, one could not assume that there was not some other parameter that would be released into the water column after stirring the sediments into it. In order to screen for this possibility, bioassays/toxicity tests were run on the elutriates; grass shrimps (*Paleometes pugio*) were used for marine and estuarine sediments and waters, and *Daphnia magna*, for freshwater sediments. The bioassay approach involved following the elutriate test procedure in an aquarium through the l-h settling period; the test organisms were added to the system after the settling period. Most of the bioassays were run for the standard 96-h period, but a few were continued for more than 20 d to provide insight into potential long-term toxicity. (The complete results were presented by Lee et al. (22) and Jones and Lee (23). A summary of the key findings was published by Jones et al. (29).) It was found that the elutriates of almost all of the U.S. waterway sediments collected near urban or industrial centers caused 20 to 50% of the test organisms to die within a 4-d period, indicating that there was some toxicity associated with these sediments when carried through the elutriate test procedure. The specific chemicals causing the toxicity were not isolated. A comparison of the bioassay results to the bulk sediment analyses showed that only a very small part of a wide variety of contaminants was available to the test organisms. Again, these results demonstrated a lack of reliability of using bulk sediment criteria to judge potential contaminant availability to aquatic life.

The toxicity found for most of the sediments evaluated by the authors represented the toxicity under what would be considered worst-case conditions, in which the organisms were confined in a system equivalent to the dredged-sediment discharge water for 4 d or longer. In actual dredged-sediment disposal situations or where sediments are stirred into the water column for short periods by natural or other causes, the organisms would be exposed to a system considerably more dilute, and where at least the mobile organisms would likely have an opportunity and inclination to leave the area. Under these conditions, little or no toxicity would be expected at a disposal site for most of the sediments tested. There is need for work, however, on identifying the specific components of U.S. waterway sediments that cause the laboratory toxicity, since they are likely to be affecting the numbers and types of organisms inhabiting the waterway sediments. They could also affect the numbers and types of organisms that could colonize a dredged-sediment disposal area. It is important in making any investigations in a disposal area to clearly sort out the physical effects of altered sediment type from the chemical effects of contaminants associated with the sediments, and to properly account for normal seasonal and other variability in organism populations.

The Corps of Engineers USEPA technical committee developed a bioassay procedure manual in the mid-1970s to provide for standardized evaluation of the potential aquatic organism toxicity and bioaccumulation of contaminants associated with redeposited dredged sediment. The procedure involved testing separately the various elutriate components (suspended particulates, filtered elutriate, and sediment) with a variety of different types of benthic, epibenthic, and pelagic organisms. Lee et al. (30) were critical of the approach used largely because the results would be interpretable in terms of potential impacts on natural water organisms or beneficial uses of the waters. WES has recognized the potential problems with this test and is attempting to develop alternative approaches for assessing at least the bioaccumulation part of the testing procedures. Based on the discussion of this redevelopment by McFarland (31), the authors feel that there will likely be significant problems in developing an effective tool for predicting the potential for bioaccumulation of chemicals from redeposited dredged sediments. Suggested approaches that have been used by the authors for making these assessments are discussed in a subsequent section of this paper.

Overall Assessment. It is important to put dredging of a channel into proper perspective as a potential source of contaminants. There will be few situations, even where the sediments of a region contained substantial amounts of contaminants, that dredging or dredged-sediment disposal operations would have a significant effect on water quality. In most instances, the amounts of contaminants released in association with the dredged-sediment disposal would be small compared to the amounts of contaminants released from the sediments during wind or vessel-induced mixing and suspension of the sediments. An exception to this would be situations where highly contaminated sediments would be disposed of in a "clean" area. In this situation, the dredged sediments could represent a new source of contaminants that had not previously been in the region. It is important, however, not to automatically assume that the disposal of contaminated sediments in a "clean" area would represent a significant adverse impact on the aquatic life or other beneficial uses of the area. "Clean" sediments are typically identified based on bulk analysis. Dredged-sediment disposal in some areas of the United States, such as Puget Sound near Seattle, Washington, is managed according to an "anti-degradation" philosophy, i.e., the dredged sediment can contain contaminants at concentrations no greater than those of the disposal-area sediments. This approach is, thus, based on bulk analysis of the sediments without regard to contaminant release potential. A situation such as that described for PCB release in the Texas Gulf Coast waters, where the "cleaner" sediment released greater amounts of PCBs than the "contaminated" sediments, could therefore readily occur if the "anti-degradation" philosophy is followed in the manner described above. Since most of the contaminants associated with waterway sediments are unavailable to affect water quality, and since bulk sediment composition has been repeatedly demonstrated to have no relationship to contaminant release, the evaluation of a particular situation requires a careful study of the physical, chemical, and biological factors that affect contaminant release, as well as the water quality significance of such release.

From an overall point of view, the DMRP provided a substantial amount of information that helps in the understanding of the significance of contaminants associated with sediments in affecting water quality—beneficial uses of a particular water body. Most of the contaminants associated with sediments are not available to affect water quality; however, sediment-associated contaminants cannot be ignored as potential sources for water quality problems. Site-specific studies have to be conducted in order to evaluate, on a case-to-case basis, the amount of contaminants that could be released at any particular site from either deposited or slurried-suspended sediments in the water column and the effects of the release of these contaminants on water quality.

Sediment Oxygen Demand

The oxygen demand associated with sediments has received considerable attention over the years because in some instances it significantly influences the amount of BOD and ammonia that may be discharged to a water body by municipalities and industry. It influences the allowable discharge of ammonia because the oxidation of ammonia to nitrite-nitrate consumes considerable amounts of oxygen. The environmental engineering literature contains numerous papers on the oxidation of sediments. Few of these, however, have attempted to describe the origin of the demand and the factors influencing the exertion of the demand. In most cases, rather simplistic, often non-representative laboratory or field measurements of sediment oxygen demand are made and used as input in a curve-fitted, deterministic, computer model. The authors believe that this type of exercise has produced little valid information that can be used to reliably predict the sediment oxygen demand under conditions of altered BOD and/or ammonia loading.

The authors (32, 33) discussed some of the principal factors that influence the oxygen demand of sediments in rivers and lakes. They pointed out that an appreciable part of this demand is due to abiotic reaction between oxygen and ferrous iron and sulfides in the sediments. The magnitude of the rate of exertion of this demand is controlled by the mixing of the sediments and their interstitial water with the overlying waters. It is the inability of most models or study approaches to properly simulate the mixing at the sediment/water interface that severely limits their predictive capability. It is important to note that the exchange reactions that govern the exertion of sediment oxygen demand are not atypical of those that would be expected to control the release of many, if not all, other contaminants. In most situations, the mass transport of sediment-associated contaminants to overlying waters is controlled by the physical mixing of the sediments and their interstitial waters with the overlying waters (18). There is considerable need for work on the hydrodynamics/mixing processes that occur between sediments and overlying waters. Until this work is done, the modeling efforts of sediment/ water exchange reactions will have limited validity and limited predictive capability in evaluating the impact of altered contaminant loads to a water body on the uptake and release of contaminants from the sediments.

Particulate-Related Water Quality Criteria

The USEPA was issued a Congressional mandate as part of PL 92-500 to develop water quality criteria to be used as a basis for formulating state water quality standards. In response to this charge, EPA has promulgated several sets of criteria (July 1976, November 1980, and February 1984). The July 1976 and November 1980 criteria did not in general consider the availability of particulate-associated contaminants, but were developed on the basis of completely available forms of the contaminants. In the implementation of the July 1976 criteria (the only set of criteria that has been used by states as a basis for water quality standards heretofore), EPA, with few exceptions, has considered the criteria values applicable to the total concentrations of the contaminants and has followed the policy of presumptive applicability. One of the few exceptions was made for lead; the soluble lead concentration (soluble defined by passage through a 0.45 μm pore-sized filter) was used for comparison with the "standard." Lee and Jones (34, 35) discussed the importance of considering the availability of particulate-associated contaminants in assessing the water quality significance of the presence of contaminants at a particular concentration. Subsequently, Lee et al. (28) and Lee and Jones (36) discussed many of the problems associated with using the July 1976 and November 1980 water quality criteria numeric values as state quality standards and suggested approaches for their use, in a hazard assessment framework, as a basis for developing appropriate water quality standards.

The February 1984 proposed criteria were the first to specifically address contaminant availability associated with particulate forms of certain contaminants. For several of the heavy metals, such as cadmium, copper, and chromium, EPA has proposed that the concentrations of pH 4, nitric acid-soluble forms of these metals be used as an estimate of the concentrations of "available" forms for the purpose of standards application. However, at this time, EPA does not have sufficient technical justification for its selection of conditions to estimate available forms of heavy metals. The selection of pH 4, nitric acid-soluble forms was arbitrary. Further, it is to a large extent an operationally defined procedure such that the results will depend on conditions of the test that are not defined, are not necessarily controllable, or are not comparable to real world situations. For example, the duration of sample incubation at pH 4 before filtration will almost certainly affect the release of some contaminants from some sediments. EPA claims that the reason the pH 4, nitrite acid-soluble condition was adopted was that it would include the carbonate forms of many contaminants as part of "available" forms. Two references were cited in its criteria documents to attempt to justify this approach (37, 38). One of these documents, Chapman (37), is not available, i.e., it is in manuscript form and therefore has not been widely reviewed by the field. Furthermore, Chapman (personal communication, 1984) has indicated that his paper does not directly support the idea that metal carbonates are toxic. He does seem to support the concept, however, that there may be some toxicity associated with metal particulate forms that are ingested by filter feeders such as daphnids. Chapman (personal communication) indicated that he feels that the pH 4, acid-soluble fraction would likely represent a conservative estimate of the concentrations of potentially available forms that could provide for adequate protection of the environment, but that it may be in some instances overly conservative with respect to protection of beneficial uses of water. This approach is in accord with what the authors feel should be used, whereby the concentration of pH 4, acid-soluble forms should serve as a screening procedure to trigger a hazard assessment of the significance of the "excessive" concentrations. The other paper that is cited in support of the pH 4, nitric-acid soluble approach, which was published in 1966 (38), does not provide any significant support for the EPA's approach. In the authors' opinion, it appears that the pH 4, nitric acid-soluble forms may be overly stringent in quantifying available forms of heavy metals. Certainly some carbonates of heavy metals are not available to affect water quality. For example, it has been known for many years that significantly greater concentrations of copper must be added to water bodies with high alkalinity than to low-alkalinity water bodies to achieve an equivalent algae kill. It seems unlikely that copper carbonate species that are not available to algae would be available to other aquatic species.

MODELING FATE AND PERSISTENCE OF
SEDIMENT-ASSOCIATED CONTAMINANTS

It is now generally recognized that sorption reactions are often the dominant reaction governing the aqueous environmental chemistry, i.e., fate and persistence of contaminants in aquatic systems. The modeling of fate and effects of contaminants requires that a reliable description be made of the sorption reactions that a contaminant may undergo. The inability to describe sorption reactions properly has been one of the most significant impediments to developing appropriate and predictive environmental chemistry-fate models. Indeed, it is the inability to model the physical factors that makes deterministic models of contaminant exchange with aquatic sediments largely a "black box," highly system-dependent approach that has limited predictive capability. It is for this reason, also, that most, if not all, deterministic eutrophication models of eutrophication load-response relationships have limited applicability as predictive tools for eutrophication management purposes (15).

The proceedings of a workshop devoted to modeling the fate of chemicals in the aquatic environment (39) discuss the importance of coupling physical factors such as diffusion, with the chemical and biochemical factors governing uptake and release of contaminants from sediments. The paper presented by DiToro et al. (40) should be consulted for further information on this topic.

The senior author and his associates were involved in an extensive set of sorption studies in the 1960s directed toward describing the uptake and release of contaminants from aquatic sediments. Much of this work was summarized in his previously mentioned review paper (18). These studies included investigating the sorption of various pesticides, such as parathion, dieldrin, and lindane, on aquatic sediments and aquifer sands (41, 42). Consideration was given to the influence of other contaminants such as organics on the sorption of the contaminant of concern on the various types of particles. In considering sorption reactions, the liquid:solid ratio, as well as the type of sorbing solid, has been found to be of great importance.

Recent studies such as those of Voice et al. (43) and Voice and Weber (44) have confirmed the previous work of the senior author and his associates, i.e., that great caution must be exercised in extrapolating sorption data obtained for one liquid:solid ratio to another ratio. Further, the influence of other contaminants such as dissolved organics, on the sorption of a particular contaminant is not a simple relationship that can be described by dissolved organic carbon. There are situations in which one contaminant (e.g., dissolved organics) can enhance the sorption of another on a particular type of inorganic solid by coating the solid. In other situations, contaminants could be competing for the sorption sites on the inorganic material. There are some potentially significant problems with much of the work on sorption involving so-called humic materials where the humic materials have been purchased from a chemical supplier or obtained by extraction from natural waters. Many of those doing the work with humics are ignoring the work that was done during the 1960s which showed that drying of humics (dissolved organic carbon) derived from natural waters can drastically change their properties (45). It is likely that most of the work that has been done on sorption involving so-called dissolved organic carbon-humic materials, which have involved evaporative concentration at elevated temperatures, freeze drying, etc., have limited applicability to predicting the sorption of contaminants on particulates in aquatic systems. Sorption of contaminants onto organic materials is also important to consider. This can be especially significant if this sorption renders the contaminant unavailable only temporarily until the organic is decomposed. Karickhoff (46) and Karickhoff et al. (47) have recently published reviews on the sorption of hydrophobic contaminants on natural water sediments. They pointed out that in many instances, there is a high correlation between sorption tendencies for these chemicals and the organic carbon content of the solid.

Another significant deficiency with many of the sorption-sediment uptake and release studies that have been conducted and are being conducted today is the failure to properly recognize the role of iron hydroxide in influencing sorption. As discussed previously in this paper, ferric hydroxide is an efficient scavenger that can sorb large amounts of a variety of contaminants. It is, however, somewhat specific for certain types of contaminants. Studies conducted by Sridharan and Lee (48) showed that certain contaminants were readily sorbed by freshly precipitated ferric hydroxide, while others were not. The ferric hydroxide does not have to form particles, but can exist as a film on other particles. An example of this phenomenon occurred for the release of contaminants associated with taconite tailings. Taconite is an iron ore containing substantial amounts of magnetite. The processing of this ore produces tailings which are basically finely-divided iron magnesium silicates, quartz, and other minerals associated with the ore body. The studies by Plumb and Lee (49, 50) have shown that while taconite tailings contain potentially significant amounts of copper and other trace metals, which if released from the tailings could have an adverse effect on water quality, the exposure of the tailings to oxygen-containing waters did not result in any release of these materials over extended exposure periods. It was reasoned that the ferrous iron in the tailings was oxidized at the surface of the tailing particles to ferric iron, which then scavenged the copper, phosphorus, and other contaminants that would tend to migrate from the tailings particles to the water. Exposure of the tailings particles to complexing agents or reducing conditions resulted in the release of copper, phosphorus, and other contaminants to the water.

It is important for those conducting sediment/water exchange studies to review the literature critically to avoid reinventing the wheel. Also, considerable care must be given to handling the sediments properly in order to prevent the alteration of their exchange characteristics. For example, drying, freezing, or extended storage should be avoided. Finally, as pointed out in this paper, the interpretation of the environmental significance of a particular sorption/ desorption reaction must be done with a particular water body or aquatic system in mind so that the hydrodynamics (e.g., mixing, dilution, and other factors that influence the concentrations of contaminants released from the sediments) are properly considered.

Bioaccumulation of Chemical Contaminants

Thus far, the discussions have focused on the toxicity of contaminants to aquatic life or their stimulatory effect on the growth of algae or other aquatic plants. The third type of effect that must be considered is the bioaccumulation of chemicals within animal tissue to the extent that the organisms or their predators are rendered unsuitable for use as food for humans or other animals. Notable examples of chemicals that bioaccumulate are mercury, PCBs, and DDTs. While a number of attempts have been made to develop chemical testing procedures to assess bioavailability and bioaccumulation of contaminants such as these, the octanol/water partition coefficient has been shown to be fairly reliable in predicting the bioaccumulation potential of contaminants in fish. This approach, however, is limited to systems in which the organisms are in contact only with water that contains the chemical. If the water also contains sediments, then the predictions of bioaccumulation of contaminants within aquatic organisms may be highly unreliable. This is because the sediments compete with the organism for the contaminants. The amount of contaminant that accumulates within the fish is therefore a function of the relative affinity that the contaminant has for the sediments and the organisms, as well as a function of the amount of the sediment present. It is because these factors play an important role in governing bioaccumulation that microcosms have little or no predictive capabilities. The relative amounts of the various components added to the microcosms determine to some extent the amount of contaminant that is present in the aquatic organisms added to the system.

An example of potential problems in trying to use bioconcentration factors developed in waters with moderate amounts of particulate matter was provided in the use of water quality criteria for PCBs. In developing the July 1976 water quality criteria for PCBs, EPA utilized a field-developed bioconcentration factor that was based on the PCB concentrations in water and in fish in Lakes Michigan and Superior. Both of these lakes have very low concentrations of particulate matter that would tend to sorb PCBs and therefore alter PCB availability to fish and other aquatic life. Lee and Jones (23, 51) reported on the studies that they and their associates conducted on the concentrations of PCBs in the water and in a variety of aquatic organisms taken from the New York Bight near Sandy Hook, New Jersey. They found that the concentrations of PCBs in the aquatic organisms of the area were far less than they should have been based on the bioconcentration factor that was developed by EPA for Lakes Michigan and Superior. In this case, the suspended particulate matter in the water column of the New York Bight sorbed an appreciable amount of PCBs which, while measurable by the analytical procedures used, was not available to aquatic life. As discussed by Veith et al. (52) in the American Fisheries Society critique of the US EPA Red Book criteria of July 1976, as well as by Lee and Jones (36), caution should be exercised in using a bioconcentration factor developed for one type of water in another water, especially if the two waters have significantly different amounts or types of particulate matter. Lee and Jones (36) suggested an approach to detect potential problems of this type based on the water's Secchi depth and/or turbidity. If the Secchi depth of the two waters is less than 2 to 3 m, or if the turbidity is greater than about 0.1 NTU (nephelometric turbidity units), then the investigators should be alert to potential problems in translating the results of laboratory evaluations or studies on one water body to another water body because of sediment/water exchange reactions with the inorganic and/or organic particles in the water column.

Hazard Assessment

The principles of environmental hazard assessment have been described and discussed in terms of evaluating the discharge of particular chemicals and for use in assessing the potential adverse impacts of new and expanded-use chemicals on aquatic systems (53, 54). The authors and their associates have also been working on adapting the principles of hazard assessment for evaluating the impact of a variety of different types of discharges, such as domestic wastewater treatment plant discharges, and dredged-sediment discharge, on the beneficial uses of receiving waters.

Lee et al. (28) have proposed a hazard assessment approach for using the EPA criteria of July 1976, November 1980, as well as February 1984, to indicate the potential for adverse impacts on water quality—beneficial uses from a particular discharge or source of contaminants that would signal the need for continued testing through a hazard assessment scheme. Similarly, Lee et al. (55) strongly advocate the use of bioassays/toxicity tests in making this evaluation and discuss their use of in-stream bioassays in determining whether a measured concentration of a contaminant could have an adverse effect on beneficial uses of the waters being investigated. Newbry and Lee (56) described an inexpensive cage system that has been used for conducting in-stream bioassays with small fish (fathead minnows). They were able to maintain 100% survival of their control/reference fish over a 6-month period with this system.

As discussed previously in this paper, the authors used a hazard assessment approach for evaluating the water quality significance of contaminants released from redeposited Intercoastal Waterway dredged sediment (27). They have also developed a hazard assessment approach for guiding the disposal of dredged sediments as part dredging the Upper Mississippi River (57-59).

It is clear that there is need for considerable research on the availability of particulate forms of metals as well as other contaminants that may affect water quality. Particular attention should be given to the pH 4, acid-soluble forms. As for now, it is recommended that the EPA's pH 4, nitric acid-soluble leaching procedure (with a defined, e.g., 12-h, contact period) be used to estimate potential amounts of contaminants that may cause water quality problems. It should *not* be assumed that because "excessive" concentrations are found in the leachate from this procedure based on criteria or standards, a violation of state water quality standards has occurred or a significant impairment of the beneficial uses of the water would occur. Instead, this test should be used as an indicator of whether or not water quality problems may occur; finding "excessive" concentrations should trigger studies to determine whether or not any adverse impacts are occurring, in accord with hazard assessment procedures.

ASSESSING AVAILABILITY OF SEDIMENT-ASSOCIATED CONTAMINANTS

Chemical Leaching Versus Bioassays/Toxicity Tests

One of the questions most frequently asked regarding the availability of contaminants associated with sediments is, what is the best or most appropriate method for measuring the available forms. Basically, there are two approaches. One is the physical/chemical approach in which various types of leaching tests and physical separation procedures, such as filtration, centrifugation, settling, etc., are used to characterize "available" forms of specific contaminants, such as pH 4, nitric acid leaching. The other involves using aquatic organisms in a bioassay procedure in which the organisms are exposed to the sediments for a period of time, and some type of organism response is used to detect adverse impact. Most frequently, death of the organism is used as this endpoint. There are, however, a wide variety of sublethal effects that could also be used as endpoints.

There are several significant deficiencies with the physical/chemical approach. First, the physical separation processes such as centrifugation, filtration, etc., may not clearly distinguish between available and unavailable forms. This is particularly true for complexed and colloidal forms. It is possible for these forms to pass through the filters and be included as available (i.e., soluble) forms yet have little or no availability to aquatic life or impact on aquatic systems.

Whereas most of the deficiencies with the physical/chemical procedures would tend to overestimate the concentrations of available forms of contaminants associated with sediments, there are a few situations in which the concentrations of available forms would be underestimated by the physical separation procedures. Such situations would be expected to be primarily encountered in conjunction with assessing the availability of heavy metals and other contaminants "attached" to particulate organics. These organics would be removed by the filters, etc., and hence the associated heavy metals would be judged unavailable. However, bacterial transformation-degradation of the organics could cause the "insoluble-unavailable" forms to become available-toxic to aquatic life. Although this phenomenon is possible, it is likely to be uncommon and certainly should not be used as a basis for not using the pH 4, nitric acid-soluble forms, for example, as an indicator of potential water quality problems associated with particulate matter in aquatic systems.

There have been numerous attempts to develop chemical leaching tests to measure available forms of contaminants associated with particulate matter. Probably one of the most intensive investigations into relating organism-available forms to the results of chemical leaching tests has been with phosphorus. As discussed by Cowen and Lee (60) and Lee et al. (61) for the Great Lakes region of the country, the ion-exchange extractable phosphorus, as well as sodium hydroxide extractable phosphorus, in river particulates provides a reasonable estimate of the amounts of phosphorus available to algae. These relationships were developed after intensive, independent tests were run by several investigators on a variety of waters and/or suspended and deposited sediment taken from the Great Lakes region. One can not be certain, however, that this relationship would hold for sediments taken from other locations. Great caution should be exercised in using the sodium hydroxide leachable phosphorus from sediments other than those tested to estimate the phosphorus available to algae. In order to establish this relationship for a wider range of suspended sediments, it would be necessary to run a number of bioassays for each type of sediment of concern. It is about as much work to conduct the comparison testing as it is to use bioassays directly for the evaluation, however.

The same would be expected for the availability of other types of contaminants. The authors feel that chemical analyses can be useful as a worst-case screening approach for estimating the availability of sediment-associated contaminants. To do this, the total concentration or load of the contaminant(s) of concern should first be measured. If this concentration or some conservative fraction of it with respect to environmental protection is greater than a chronic safe concentration for a particular situation, then bioassays should be run to determine the fraction of the total that is available and therefore could adversely affect water quality. The bioassay would consider not only the effect of the group of chemicals of potential concern, but also synergistic and antagonistic effects of other chemicals, as well as other chemicals themselves that may have not been considered when developing the study program and that would not have been measured in a chemical testing program. It is likely that bioassays will be used to a much greater extent in the future to assess available forms of contaminants.

Analysis of Interstitial Water

An approach frequently taken to assess sediment/water exchange of chemicals is the measurement of the concentration of contaminants in interstitial waters. Caution should be exercised in interpretation of interstitial water data for several reasons. First, it is now well established that the concentrations of contaminants in interstitial waters depend on the method by which the interstitial waters were extracted. Squeezing, filtering, and centrifuging can yield significantly different concentrations of some contaminants. Second, many of the methods used for interstitial water recovery do not adequately consider the redox conditions that exist during recovery; as discussed previously in this paper, the introduction of oxygen into the sample during the recovery process could result in the oxidation of ferrous iron that would scavenge some undefinable amount of the contaminants from the interstitial waters. Further, the interpretation of the significance of contaminants in interstitial water must be viewed in light of the rate of mixing and dilution of these waters with the overlying waters and the water column in general.

AREAS OF RESEARCH NEED

Even though considerable information has been gained about the significance of contaminants associated with sediments, there is still need for a substantial amount of research to develop the information necessary to properly evaluate the water quality significance of a contaminant or a group of contaminants associated with aquatic sediments. The research needs that the authors feel are of highest priority are discussed below.

Hydrodynamics of Sediment Particles

One of the areas that is of particular importance in assessing the potential water quality significance of a contaminant associated with a deposited aquatic sediment is the hydrodynamics of the sediment particles and the waters associated with these particles. Lee (18) discussed the importance of mixing/ hydrodynamics in sediment/water contaminant exchange reactions, pointing out that mixing processes are often the dominant factor governing exchange. This type of information is also needed in order to predict the fate of a contaminant that tends to become strongly associated with sediments in the water. In either case, information is needed on the conditions that tend to suspend the particles into the water column in the area of concern. This slurried condition typically provides the greatest opportunity for exchange reactions to occur. While there is considerable ability today to describe the movement of sand/silt-sized particles in rivers as a function of river flow (62), this ability does not extend to the clay-sized particles (typically >2 μm diameter) that are by far the most important size group for sorption reactions. The sand and silt-sized particles tend to have relatively small sorption capacities compared to those of the clay-sized particles. The description of the hydrodynamics of clay-sized particles is greatly complicated by the fact that they tend to form cohesive masses that make their suspension in the water column not readily definable based on the bulk water velocity of the overlying water. For river and estuarine systems, there is need to be able to relate water velocity to scour of cohesive sediments. For estuarine, lake, and ocean systems, there is need to relate wind velocity to water velocity and to scour of cohesive sediments. In both cases, the amount of sediment that is suspended into the water column as a slurry, as well as the amount of sediment that moves along the bottom in a plastic flow and never becomes fully suspended, needs to be considered.

For sand and silt systems, consideration must be given to the rates of diffusion of waters into and out of the deposited sediments. The studies of Saleh et al. (63) on the sorption of lindane by sand particles in a microcosm have shown that the diffusion of the lindane-containing waters into the particles is a dominant factor in controlling the sorption of lindane by the sediments. The time scales of concern for these reactions would be a maximum of a few days to a year or so. Emphasis should be given to studying the sediments in the active deposition zone (18). These are the sediments in which sufficient mixing occurs so that a sediment particle at the bottom of this zone could be expected to reach the surface within a year's time. Ordinarily, the diagenetic reactions of sediments that lead to sediment consolidation, and the associated movement of contaminants in the historical sediment profile, are of limited concern from a water quality point of view.

Another area having to do with mixing of sediments that needs research is micro- and macro-organism activity. As discussed by Lee (18), various types of organisms can bring about substantial mixing of sediments. This mixing is a result of a variety of activities that include methane generation, burrowing of benthic organism, and activities of bottom-feeding fish. The feeding and mating activities of certain types of fish, such as carp, can also stir appreciable amounts of sediments into the water column and have a pronounced impact on the turbidity of a water body. There have been numerous water bodies in which the removal of carp has greatly increased the water clarity. Carp activity, therefore, can be an important factor influencing sediment/water exchange reactions.

Sorption

The mechanisms of sorption should be more thoroughly investigated. The facts that sorption reactions in general do not follow simple mass law relationships and that they show considerable hysteresis make them much more complex to investigate and model than most other types of reactions that may occur in aquatic systems. Some of the areas in which research on sorption reactions is needed have been described elsewhere in this paper. In general, they include understanding of (1) the surface characteristics of various types of aquatic particles that influence sorption/desorption reactions; (2) the extent to which sorption of various types of contaminants occurs on various types of particle surfaces; and (3) the influence of water composition on sorption/desorption reactions. The ultimate goal of this research should be the description of the extent of sorption/desorption that could occur in various types of water bodies given general information on the hydrodynamics and water quality characteristics of the water body. In the opinion of the authors, there is need to utilize a standardized sorption test such as the ASTM Committee E-47 test.

Water Quality Criteria and Standards

The attempts made by the USEPA to develop more site-specific water quality criteria and standards than have been formulated in the past (e.g., February 1984 criteria) represent a substantial advance in ensuring that funds spent for water pollution control will be more cost-effective in improving the beneficial uses of the nation's waters. To implement this approach properly, however, there will have to be a considerable expansion of EPA funding for research designed to evaluate the availability of contaminants under various aquatic system conditions. Otherwise, the criteria and standards developed will likely be seriously questioned and possibly even abandoned for a lack of technical validity. EPA and state pollution control agencies should strive to develop water quality criteria that can be readily translated to field situations such that the measured concentration of a contaminant in any particular water body is related to the numeric criteria for the contaminant in some fairly well-defined way. It is suggested that rather than trying to develop single-value numeric standards that would be applied directly to a particular water body for establishing critical concentrations of a contaminant, the EPA should try to develop fairly conservative water quality criteria that encourage the execution of a hazard assessment by the discharger, i.e., source of contaminants, to define the real hazard that the contaminant(s) of concern represents to the beneficial uses of the water of concern and to set the numeric water quality standard. This approach would be in accord with the EPA's proposed approach of comparing pH 4, nitric acid-soluble forms of heavy metals with criteria-standards. It is fairly certain that whereas this approach is overly protective if used directly in establishing allowable levels of contaminants, it would be appropriate for use in a screening step for a site-specific hazard assessment.

Site-specific studies designed to evaluate the actual impacts of chlorine and ammonia in municipal wastewater treatment plant effluents discharged to several Colorado Front Range streams were conducted by the authors and their associates in recent years. A number of papers and reports (64-70) discuss the approaches used and how the results of the hazard assessment studies were interpreted in light of the beneficial uses of the waters of concern. The results of these studies in particular were found to be valuable for developing approaches for managing chlorine and ammonia in domestic wastewaters from the cities in a cost-effective way in order to protect the beneficial uses of these waters without causing undue economic burdens to the taxpayers. It would be useful to have studies of this type conducted for discharges of contaminants that are of concern from a water quality point of view.

Available Phosphorus

The work conducted by a number of investigators on the availability of particulate-associated phosphorus that enters the U.S.-Canadian Great Lakes has shown that about 20% of particulate-associated phosphorus is available to support algal growth in these water bodies. Caution should be exercised, however, in using this rule of thumb for particulates in other geological and climatological regimes. Also, caution should be utilized in using chemical tests to assess available forms for other aquatic systems. The availability of particulate-associated phosphorus in other systems should be defined based on algal bioassay assessment of availability.

Dredged-Sediment Disposal Criteria

Much of the knowlege gained through the DMRP on the water quality significance of dredged sediment-associated contaminants has still not been implemented into public policy. Because of this, in many instances, maintaining U.S. waterway navigation depth is considerably more expensive than it need be, without a concurrent improvement in water quality. In some instances, the more expensive alternatives being touted as less environmentally degrading, such as the so-called land disposal in nearshore waters, are likely to be more environmentally damaging than the previously used, less expensive, open-water disposal methods. It is important to note that misunderstandings of this type are not restricted to the dredged-sediment disposal situation, but apply to other situations involving contaminants that tend to become strongly attached to sediments and soil particles and thereby have a significantly altered ability to affect water quality.

The problems in implementing the DMRP results into public policy stem, to a considerable extent, from the politics of the Corps of Engineers, EPA, and state water pollution control agencies. Research is needed on some of the more subtle aspects of the water quality significance of contaminants associated with dredged sediments, such as the development of a hazard assessment approach for interpretation of the results of bioassay procedures. Also, of equal importance, is research on how to reduce and possibly eliminate from public policy the political problems associated with the implementation of dredged-sediment research.

Pesticides

The sorption characteristics of many of the chlorinated hydrocarbon pesticides have been extensively studied in both the water quality and the soils fields. It is now well-known that, in general, the sorption of these contaminants on soils and sediments is related to the organic carbon content of the solids phase; many of these pesticides show little or no mobility in soils or sediment systems containing large amounts of organic carbon. On the other hand, they are fairly mobile in sandy systems. In developing pesticide use regulations, EPA has taken the approach of banning pesticides in part because of their mobility in these types of systems. Although chlorinated hydrocarbon pesticides, such as DDT, have been banned in this country because of their environmental persistence and their potential for causing cancer, the total use of these chemicals world-wide is greater today than when they were allowed to be used in the United States. This use is primarily due to the use of DDT in the control of the mosquito that transmits malaria. To many people of the world, the concern about contracting cancer from DDT is infinitesimal compared to the concern about contracting malaria.

Lee et al. (71) discussed a hazard assessment approach for evaluating the potential impact that DDT and other chlorinated hydrocarbon pesticides used in a particular manner at a particular location could have on water quality; particular attention was given to the sediment/water and soil/water interactions that govern the transport of DDT in aquatic systems. They suggested that a critical evaluation be made of the nature of the aquatic system to which DDT is applied, with particular emphasis on the ability of soils and sediments to immobilize DDT. This information should be used in a hazard assessment scheme to determine whether DDT should be used for malaria control. Utilizing this approach, it would be possible for governmental officials to determine if the DDT applied to a certain watershed would likely become available to adversely affect the utilization of fish from a region as a source of food, as well as to determine other adverse effects the

DDT may have on man and the environment. This information could then be used in a risk assessment to determine the risks of DDT-related problems relative to those of malaria or of the target organism against which it is being applied.

The key component of this hazard assessment is the development of information on the expected mobility of the chemical in contact with sediment and/or soils of the region. It is important when making this assessment to refrain from automatically equating chemical persistence in soils and sediments with the availability of the chemicals to fish and other higher aquatic trophic levels such as birds and man. In some developing countries such as Egypt, the concentrations of pesticides in the soils are very high because of continuous, intensive use of pesticides on the lands. As discussed by Lee et al. (71), however, these concentrations are only of significance if the pesticides present in soils can be transported to groundwaters or surface waters of the region or can be translocated in the crops to the extent that they contain potentially hazardous concentrations. Judging the significance of soil-associated pesticides to man and the environment based on the total concentrations of the pesticides in the soil is analogous to using bulk sediment contaminant criteria first proposed by EPA for dredged sediment. It is important to critically evaluate on a site-specific basis whether the soil-associated contaminants of potential concern are available in sufficient amounts and at sufficient rates to adversely affect beneficial uses of the waters and soils of a region.

CONCLUSIONS

For many contaminants, sediment/water exchange reactions play an important and sometimes dominant role in their aqueous environmental chemistry. Because these exchange reactions generally have a significant effect on the availability of contaminants to influence water quality, there is considerable concern today about describing or modeling sediment/water exchange reactions. The work that has been conducted on the role of phosphorus in lake sediments in affecting eutrophication, as well as the studies that have been conducted under the Corps of Engineers Dredged Material Research Program, have provided considerable information on many of the factors that govern sediment/water exchange reactions. For both suspended and deposited sediments, the character of the solid-particle surface as influenced by the composition of the water surrounding the particle governs the extent of sorption/desorption that occurs. The hydrodynamics of the sediment particles and the associated water is also a major factor that influences the water quality significance of any particular exchange reaction. It is clear that in most cases, site-specific evaluations must be made in which the overall mixing processes that occur in the particular water body or area of concern within the water body must be evaluated in order to determine the water quality impacts of a particular exchange reaction. There is need for considerable research to define the factors governing sediment/water exchange reactions with particular emphasis on their significance in influencing beneficial uses of water.

Acknowledgment—Support for development of this paper was provided by the Department of Civil and Environmental Engineering, New Jersey Institute of Technology. Much of the work on phosphorus conducted by the authors and their associates was supported by the US EPA and its predecessor organizations. The studies devoted to the uptake and release of contaminants from U.S. waterway sediments conducted by the authors and their associates was supported by the U.S. Army Engineer Waterways Experiment Station, various Corps districts, and the Chief's Office. The assistance of B. Winkler in preparation of this paper is appreciated.

REFERENCES

1. **Mortimer, C. H.** 1941. The exchange of dissolved substances between mud and water in lakes—Parts I and II. *J. Ecology* 29:280-329.
2. **Mortimer, C. H.** 1942. The exchange of dissolved substances between mud and water in lakes—Parts III and IV. *J. Ecology* 30:147-201.
3. **Lee, G. F., W. C. Sonzogni and R. D. Spear.** 1977. Significance of oxic versus anoxic conditions for Lake Mendota sediment phosphorus release. In H. Golterman, ed., *Interaction between Sediment and Fresh Water.* Junk & Purdoc, The Hague, The Netherlands, pp. 294-306.
4. **Stauffer, R. E. and G. F. Lee.** 1973. The role of thermocline migration in regulating algal blooms. In E. J. Middlebrooks, D. H. Falkenborg and T. E. Maloney, eds., *Modeling the Eutrophication Process.* Utah State University, Logan, pp. 73-82.
5. **Spear, R.** 1970. Release of phosphorus from lake sediments. Ph.D. thesis. Water Chemistry Program, University of Wisconsin, Madison.
6. **Schindler, D. W., R. Hesslein and G. Kipphut.** 1977. Interactions between sediments and overlying waters in an experimentally eutrophied precambrian shield lake. In H. Golterman, ed., *Interaction between Sediments and Fresh Water.* Junk & Purdoc, The Hague, The Netherlands, pp. 235-243.
7. **Sonzogni, W.** 1974. Effects of nutrient input reduction on the eutrophication of the Madison Lakes. Ph.D. thesis. Water Chemistry Program, University of Wisconsin, Madison.
8. **Sonzogni, W. C., P. C. Uttormark and G. F. Lee.** 1976. A phosphorus residence time model: Theory and application. *Water Res.* 10:429-435.
9. **Barth, T. W.** 1952. *Theoretical Petrology.* John Wiley & Sons, New York.
10. **Goldberg, E. D.** 1963. The oceans as a chemical system. In **The Sea**, Vol. 2, John Wiley & Sons, New York, pp. 3-25.
11. **Jones, R. A. and G. F. Lee.** 1982. Recent advances in assessing the impact of phosphorus loads on eutrophication-related water quality. *Water Res.* 16:503-515.
12. **Lee, G.F., R. A. Jones and W. Rast.** 1986. Alternative approach to trophic state classification of water quality management. I. Suitability of existing trophic state classification systems. *Submitted for publication.*
13. **Lee, G. F., R. A. Jones and W. Rast.** 1986. Alternative approach to trophic state classification for water quality management. Part II. Applicability of OECD eutrophication study results. *Submitted for publication.*
14. **Vollenweider, R. A.** 1976. Advances in defining critical loading levels for phosphorus in lake eutrophication. *Mem. 1st. Ital. Idrobio.* 33:53-83.
15. **Rast, W., R. A. Jones and G.F. Lee.** 1983. Predictive capability of U.S. OECD phosphorus loading—eutrophication response models. *J. Water Pollut. Control Fed.* 55:990-1003.

16. **Rast, W. and G. F. Lee.** 1978. Summary analysis of the North American (U.S. Portion) OECD eutrophication project: Nutrient loading—Lake response relationships and trophic state indices. EPA-600/3-78-008. USEPA, Corvallis, OR.

17. **Organization for Economic Cooperation and Development (OECD).** 1982. Eutrophication of waters: Monitoring, assessment and control. Final Report of the OECD Cooperative Programme on Monitoring of Inland Waters (Eutrophication Control), OECD, Paris.

18. **Lee, G. F.** 1970. Factors affecting the transfer of materials between water and sediments. Literature Review No. 1, Eutrophication Information Program. University of Wisconsin, Madison.

19. **Golterman, H.** (ed.) 1977. *Interaction between Sediment and Fresh Water.* Junk & Purdoc, The Hague, The Netherlands.

20. **Boyd, M. B., P. T. Saucier, J. W. Keeley, R. L. Montgomery, R. D. Brown, D. B. Mathis and C. J. Guice.** 1972. Disposal of dredge spoil. Problems identification and assessment and research program development. Technical Report H-72-8. U.S. Army Corps of Engineers, Waterways Experiment Station, Vicksburg, MS.

21. **Lee, G. F. and R. H. Plumb.** 1974. Literature review on research study for the development of dredged material disposal criteria. U.S. Army Corps of Engineers, Dredged Material Research Program, Vicksburg, MS.

22. **Lee, G. F., R. A. Jones, F. Y. Saleh, G. M. Mariani, D. H. Homer, J. S. Butler and P. Bandyopadhyay.** 1978. Evaluation of the elutriate test as a method of predicting contaminant release during open water disposal of dredged sediment and environmental impact of open water dredged materials disposal, Vol. II—Data Report. Technical Report D-78-45. U.S. Army Engineer, Waterways Experiment Station Vicksburg, MS.

23. **Jones, R. A. and G. F. Lee.** 1978. Evaluation of the elutriate test as a method of predicting contaminant release during open water disposal of dredged sediment and environmental impact of open water dredged material disposal, Vol. I—Discussion. Technical Report D-78-45. U.S. Army Engineer, Waterways Experiment Station, Vicksburg, MS.

24. **Baker, R. A.** (ed.) 1980. *Contaminants and Sediments,* Vols. I and II. Ann Arbor Science Publishers, Ann Arbor, MI.

25. **Jenne, E. A.** 1969. Controls on Mn, Fe, Co, Ni, Cu and Zn concentrations in soils and water: The significant role of hydrous Mn and Fe oxides. *Adv. Chem.* **73**:337-387.

26. **Lee, G. F.** 1975. Role of hydrous metal oxides in the transport of heavy metals in the environment. *Progress in Water Technology* **17**:137-147.

27. **Lee, G. F. and Jones, R. A.** 1981. Application of hazard assessment approach for evaluation of potential environmental significance of contaminants present in North Landing River sediments upon open water disposal of dredged sediment. Paper presented at Old Dominion University/Norfolk District Corps of Engineers Symposium, August.

28. **Lee, G. F., R. A. Jones and B. W. Newbry.** 1982. Water quality standards and water quality. *J. Water Pollut. Control Fed.* **54**:1131-1138.

29. **Jones, R. A., G. M. Mariani and G. F. Lee.** 1981. Evaluation of the significance of sediment-associated contaminants to water quality. In *Utilizing Scientific Information in Environmental Quality Planning.* American Water Resources Association, Minneapolis, MN, pp. 34-45.

30. **Lee, G. F., R. A. Jones and G. M. Mariani.** 1977. Comments on U.S. EPA Corps of Engineers dredged sediments bioassay procedures. Occasional Paper No. 26. Department of Civil and Environmental Engineering, New Jersey Institute of Technology, Newark, NJ.

31. **McFarland, V. A.** 1983. Estimating bioaccumulation potential of chemicals in sediment. In *Environmental Effects of Dredging.* Report # D-83-4:1-31,11. U.S. Army Engineer, Waterways Experiment Station, Vicksburg, MS.

32. **Lee, G. F. and R. A. Jones.** 1984. Oxygen demand of U.S. waterway sediments. *J. Water Pollut. Control. Fed.* (in press).

33. **Lee, G. F. and R. A. Jones.** 1986. Mechanisms of the deoxygenation of the hypolimnia of lakes. *Submitted for publication.*

34. **Lee, G. F. and R. A. Jones.** 1978. The importance of focusing pollution control programs on available forms of contaminants. Occasional Paper No. 32. Department of Civil and Environmental Engineering, New Jersey Institute of Technology, Newark, NJ.

35. **Lee, G. F. and R. A. Jones.** 1979. Interpretation of chemical water quality data. In L. L. Marking and R. A. Kimerle, eds., *Aquatic Toxicology.* ASTM STP 667. American Society for Testing and Materials, Philadelphia, PA, pp. 302-321.

36. **Lee, G. F. and R. A. Jones.** 1983. Translation of laboratory results to field conditions: The role of aquatic chemistry in assessing toxicity. In W. E. Bishop, R. D. Cardwell and B. B. Heidolph, eds., *Aquatic Toxicology and Hazard Assessment: Sixth Symposium.* ASTM STP 802. American Society for Testing and Materials, Philadelphia, PA, pp. 328-349.

37. **Chapman, G. A.** (principal investigator). 1984. The effects of water hardness on the toxicity of metals to *Daphnia magna.* Draft report to USEPA, Corvallis, OR.

38. **Mount, D. A.** 1966. The effect of total hardness and pH on acute toxicity of zinc to fish. *Air Water Pollut. Int. J.* **10**:49.

39. **Dickson, K. L., A. W. Maki and J. Cairns** (eds.) 1982. *Modeling the Fate of Chemicals in the Aquatic Environment.* Ann Arbor Science Pubishers, Ann Arbor, MI.

40. **DiToro, D. M., D. J. O'Connor, R. V. Thomann and J. P. St. John.** 1982. Simplified model of the fate of partitioning chemicals in lakes and streams. In K. L. Dickson, A. W. Maki and J. Cairns, eds., *Modeling the Fate of Chemicals in the Aquatic Environment.* Ann Arbor Science Publishers, Ann Arbor, MI, pp. 165-190.

41. **Boucher, F. R. and G. F. Lee.** 1972. Adsorption of lindane and dieldrin pesticides on unconsolidated aquifer sands. *Environ. Sci. Technol.* **6**:538-543.

42. **Wang, W. C., G. F. Lee and D. Spyridakis.** 1972. Adsorption of parathion in a multi-component solution. *Water Res.* **6**:1219-1228.

43. **Voice, T. C., C. P. Rice and W. J. Weber, Jr.** 1983. Effects of solids concentration on the sorptive partitioning of hydrophobic pollutants in aquatic systems. *Environ. Sci. Technol.* **17**:513-517.

44. **Voice, T. C. and W. J. Weber, Jr.** 1983. Sorption of hydrophobic compounds by sediments, soils and suspended solids. I. Theory and Background. *Water Res.* **17**:1433-1441.

45. **Hall, K. and G. F. Lee.** 1974. Molecular size and spectral characterization of organic matter in a meromictic lake. *Water Res.* **8**:239-251.

46. **Karickhoff, S. W.** 1982. Sorption kinetics of hydrophobic pollutants in natural sediments. In R. A. Baker, ed., *Contaminants and Sediments,* Vol. 2. Ann Arbor Science Publishers, Ann Arbor, MI, pp. 193-206.

47. **Karickhoff, S. W., D. S. Brown and T. A. Scott.** 1979. Sorption of hydrophobic pollutants on natural sediments. *Water Res.* **13**:241-248.

48. **Sridharan, N. and G. F. Lee.** 1972. Coprecipitation of organic compounds from lake water by iron salts. *Environ. Sci. Technol.* **6**:1031-1033.

49 **Plumb, R.** 1973. Effects of taconite tailings on water quality. Ph.D. thesis. University of Wisconsin, Madison.

50. **Plumb, R. H. and G. F. Lee.** 1977. The impact of taconite tailings on dissolved solid concentrations in Lake Superior. In H. Golterman, ed., *Interaction between Sediment and Fresh Water.* Junk & Purdoc, The Hague, The Netherlands, pp. 423-434.

51. **Lee, G. F. and R. A. Jones.** 1977. An assessment of the environmental significance of chemical contaminants present in dredged sediments dumped in the New York Bight. Occasional paper No. 28. Department of Civil and Environmental Engineering, New Jersey Institute of Technology, Newark, NJ.

52. **Veith, G. D. (coordinator), T. C. Carver, Jr., C. M. Fetterolf, G. F. Lee, D. L. Swanson, W. A. Willford and M. G. Zeeman.** 1979. Polychlorinated biphenyls. In V. Thurston et al., eds., *A Review of the EPA Red Book: Quality Criteria for Water.* American Fisheries Society, Bethesda, MD, pp. 239-246.

53. **Cairns, J., K. L. Dickson and A. W. Maki** (eds.) 1978. *Estimating the Hazard of Chemical Substances to Aquatic Life.* ASTM STP 657. American Society for Testing and Materials, Philadelphia, PA.

54. **Dickson, K., A. Maki and J. Cairns** (eds.) 1979. *Analyzing the Hazard Evaluation Process.* American Fisheries Society, Bethesda, MD.

55. **Lee, G. F., R. A. Jones and B. W. Newbry.** 1982. Alternative approach to assessing water quality impact of wastewater effluents. *J. Water Pollut. Control Fed.* **54**:165-174.
56. **Newbry, B. W. and G. F. Lee.** 1984. A simple apparatus for conducting in-stream toxicity tests. *J. Testing and Evaluation ASTM* **12**:51-53.
57. **Lee, G. F. and R. A. Jones.** 1981. A hazard assessment approach for assessing the environmental impact of dredging and dredged sediment disposal for the upper Mississippi River. Occasional Paper No. 68. Department of Civil and Environmental Engineering, New Jersey Institute of Technology, Newark, NJ.
58. **Lee, G. F. and R. A. Jones.** 1981. Guidelines for conducting environmental study programs for assessing the water quality impact of dredged sediment disposal in the upper Mississippi River system. Occasional Paper No. 69. Department of Civil and Environmental Engineering, New Jersey Institute of Technology, Newark, NJ.
59. **Lee, G. F. and R. A. Jones.** 1981. Relationship between the hazard assessment approach for dredged sediment disposal in the upper Mississippi River system and the recommendations of the GREAT I and II studies. Occasional Paper No. 72. Department of Civil and Environmental Engineering, New Jersey Institute of Technology, Newark, NJ.
60. **Cowen, W. F. and G. F. Lee.** 1976. Phosphorus availability in particulate materials transported by urban runoff. *J. Water Pollut. Control Fed.* **48**:580-591.
61. **Lee, G. F., R. A. Jones and W. Rast.** 1980. Availability of phosphorus to phytoplankton and its implication for phosphorus management strategies. In R. Loehr, C. S. Martin and W. Rast, eds., *Phosphorus Management Strategies for Lakes*. Ann Arbor Science Publishers, pp. 259-308.
62. **Shen, H. W.** (ed.) 1979. *Modeling of Rivers*. John Wiley & Sons, New York.
63. **Saleh, F. Y., K. L. Dickson and J. H. Rodgers.** 1982. Fate of lindane in the aquatic environment: Rate constants of physical and chemical processes. *Environ. Toxicol. Chem.* **1**:289-297.
64. **Heinemann, T. J., G. F. Lee, R. A. Jones and B. W. Newbry.** 1983. Summary of studies on modeling persistence of domestic wastewater chlorine in Colorado Front Range rivers. In R. L. Jolley, et al., eds., *Water Chlorination-Environmental Impact and Health Effects*, Vol. 4. Ann Arbor Science Publishers, Ann Arbor, MI, pp. 97-112.
65. **Newbry, B. W., G. F. Lee, R. A. Jones and T. J. Heinemann.** 1983. Studies on the water quality hazard of chlorine in domestic wastewater treatment plant effluents. In R. L. Jolley, et al., eds., *Water Chlorination-Environmental Impact and Health Effects*, Vol. 4. Ann Arbor Science Publishers, Ann Arbor, MI, pp. 1423-1436.
66. **Lee, G. F. and R. A. Jones.** 1981. An assessment of the impact of the Colorado Springs domestic wastewater treatment plant discharges on Fountain Creek. Occasional Paper No. 62. Department of Civil and Environmental Engineering, New Jersey Institute of Technology, Newark, NJ.
67. **Lee, G. F., B. W. Newbry, T. J. Heinemann and R. A. Jones.** 1980. Demonstration of hazard assessment for evaluation of environmental impact of Pueblo, Colorado municipal wastewater discharges on water quality in the Arkansas River. Report to Pueblo Area COG, EnviroQual Consultants, Maplewood, NJ.
68. **Lee, G. F., B. W. Newbry and T. J. Heinemann.** 1981. Persistence and acute toxicity of chlorine used for domestic wastewater disinfection at Loveland, Colorado. Occasional Paper No. 70. Department of Civil and Environmental Engineering, New Jersey Institute of Technology, Newark, NJ.
69. **Lee, G. F., B. W. Newbry, T. J. Heinemann and R. A. Jones.** 1981. Persistence and acute toxicity of chlorine used for domestic wastewater disinfection at Fort Collins, Colorado. Occasional Paper No. 71. Department of Civil and Environmental Engineering, New Jersey Institute of Technology, Newark, NJ.
70   **Lee, G. F., R. A. Jones, B. W. Newbry and T. J. Heinemann.** 1982. Use of the hazard assessment approach for evaluating the impact of chlorine and ammonia in Pueblo, Colorado, domestic wastewaters on water quality in the Arkansas River. In J. G. Pearson, R. B. Foster and W. E. Bishop, eds., *Aquatic Toxicology and Hazard Assessment: Fifth Conference*. ASTM STP 766. American Society for Testing and Materials, Philadelphia, PA, pp. 356-380.
71. **Lee, G. F., R. A. Jones and F. Y. Saleh.** 1982. Environmental significance of pesticide residues associated with aquatic sediments. *J. Environ. Sci. Health* **B14**:409-423.

Chapter 2

Information Needs Related to Toxic Chemicals Bound to Sediments— A Regulatory Perspective

James H. Gilford and Robert W. Zeller

INTRODUCTION

The Environmental Protection Agency (EPA) has the responsibility and authority under various Federal laws to regulate toxic substances that pose an unreasonable risk of injury to public health or the environment. There are six major Federal laws under which EPA works to control hazardous chemicals. They are the Clean Water Act, the Safe Drinking Water Act, the Clean Air Act, the Resources Conservation Recovery Act (RCRA), the Federal Insecticide, Fungicide and Rodenticide Act (FIFRA), and the Toxic Substances Control Act (TSCA). The first four of these laws give EPA the authority to regulate discharges of toxic substances into waterways and drinking water, emissions of toxic substances into the air, and disposal of toxic and hazardous solid wastes. The last two laws, FIFRA and TSCA, authorize the Agency to regulate the production, use, and disposal of pesticides and chemicals introduced or intended to be introduced into commerce. TSCA and FIFRA also provide for assessment, before a chemical enters the market place, of the potential of a chemical, separately or as a component of a product, to cause injury to health or the environment. As it acts under these mandates, the Agency strives to exercise its regulatory authority in a way that achieves a compatible balance between the needs and activities of our society and the ability of natural systems to support and nurture life. To make decisions regarding the potential risk to health and the environment, the Agency relies on a variety of information: test data to assess the toxicity of a chemical; physical and chemical properties to estimate

35

persistence and partitioning of the chemical, and various modeling and monitoring information to estimate the concentration that the chemical is expected to reach in the environment.

For a chemical that is believed to present an unreasonable risk to health and the environment, the Agency may exercise various options to control the chemical. Those options include, but are not limited to, a total ban on production and sale of the chemical; restrictions on production, use, and disposal of the chemical; labeling requirements; monitoring and the recall of the chemical; or cancellation of its permitted use.

In the case of new chemicals intended for the U.S. market place, the Agency has the authority under Section 5 of TSCA, to review all available information about the chemical in order to assess the potential risk that it poses to health and the environment. As in the case with existing chemicals, the Agency can prohibit or regulate production, processing, distribution, use, or disposal of the new chemical if it finds that the chemical poses an unreasonable risk to health and the environment. In order to make its assessments of a new chemical, the Agency can require the manufacturer or importer to provide certain test data on the toxicity of the chemical and its chemical and physical characteristics to enable the Agency to make an adequate assessment of the risk associated with its production, distribution, use, and disposal. In order to require that such testing be performed before the chemical enters the market place, the Agency must determine that the chemical may present an unreasonable risk or that there will be a substantial release of the chemical to the environment. In either case, the Agency must decide if existing data on the toxicity of the chemical are sufficient for determining the nature and magnitude of the risk. Test data usually are not available on the environmental toxicity of chemicals submitted for review under Section 5 of the TSCA. Under those circumstances, the Agency is forced to use structure-activity relationships or analog data to predict the potential toxicity of the chemical in order to determine if the need exists to require the manufacturer or importer to provide the Agency with test data on the chemical upon which to base a risk assessment.

Chemicals that are determined to present an unreasonable risk to health or the environment may be referred by one EPA program office to another if it is determined that the chemical should be reviewed or regulated under another statutory authority. For example, a chemical, either a new chemical or an existing one, reviewed under TSCA and determined to be hazardous to the environment or health could be referred to the Office of Water for regulation under the Clean Water Act.

Many of the chemicals that come under the Agency's review are ones that are known to be bound to sediment or, because of physical and chemical properties, are expected to do so. For example, approximately half of the new chemicals that the Agency has evaluated under Section 5 of TSCA are known or expected to bind to sediment. Of sixty-five classes of toxic pollutants for which EPA has issued water quality criteria, two-thirds of those classes have constituents that will bind to sediments.

In order to assess and manage the risk to health and the environment arising as a result of the production, use, and disposal of chemicals that become associated with aquatic sediments, the agency needs information, sufficient for those purposes, concerning the partitioning, persistence, bioavailability, and toxicity of those chemicals. Because of the legislative constraints on the Agency and the economic constraints within the chemical industry on the premarket development of data on new chemicals, the Agency also needs scientifically sound and reliable methods for predicting the environmental partitioning, persistence, bioavailability, and toxicity of chemicals bound to sediments in order to evaluate adequately the risks from new chemicals reviewed under Section 5 of TSCA. Moreover, to use its data gathering authority under TSCA and FIFRA and its authority to regulate the discharge and disposal of toxic chemicals under the clean water, clean air, RCRA and safe drinking water legislation, the Agency needs methodology for testing the toxicity of chemicals that bind to sediments.

The application of this information by various regulatory offices in EPA will differ depending on the particular law from which an office derives its authority. The development of sediment criteria and their potential value to the Agency is one example of how this knowledge may be used. Ambient water quality criteria have been, and will remain, an important component of EPA strategy to maintain environmental quality. There is, however, a growing interest to supplement water quality criteria with some form of sediment criteria. Evidence has been presented of environmental degradation in areas where water quality criteria are not being exceeded. The majority of these adverse impacts are not manifest in organisms living in the water column, but rather, in those species that live in or on the sediment. Section 304 (a) of the Clean Water Act authorized EPA to develop and implement sediment criteria analogous to the Agency's water quality criteria.

If available and appropriate, sediment criteria would have broad application under the Clean Water Act. They could provide guidance in (1) assessing sediment quality, (2) deciding on remedial actions (as well as alternatives and priorities) to alleviate environmental degradation related to sediment contamination, and (3) establishing wasteload allocations, (4) assessing the quality of sediments in discharge impact areas, (5) setting requirements for ocean discharges, (6) evaluating disposal sites and dredge and fill permits, (7) assessing potential impacts on benthos, and (8) designing and evaluating monitoring programs.

Sediment criteria are only one of many needs within EPA for a better understanding of chemicals that bind to sediments and the processes and interactions involved. In order to assess the potential impact on non-target as well as target species from pesticide chemicals that bind to sediments, information on the fate, persistence, bioavailability, and toxicity of those chemicals must be considered as part of the assessment.

While the various laws under which EPA operates provide authority to collect information relevant to its regulatory function, the Agency may be required to proceed in its decision making without sufficient information to answer all of its concerns because of legislative time requirements, lack of appropriate methodologies, or the absence of scientifically sound data. Lacking adequate data on which to base its regulatory decisions, the Agency is faced with the difficult task of protecting public health and the environment from the potential adverse effects of a chemical without over-regulating its production, use, and disposal to such an extent that society is denied the beneficial uses of the chemical or that innovation in the chemical industry is inhibited.

It is important, then, that regulatory units within EPA have as sound an understanding as possible of the current status of knowledge of sediments, including such aspects as their composition and characteristics, the nature of the association between chemical and sediment, the processes involved in adsorption and desorption of toxic chemicals, and the persistence, toxicity, and bioavailability of chemicals that have bound to sediments. Equally important to the Agency is an understanding, to the extent possible, of what is not known about the bioavailability of sediment-bound toxicants and the limitations such knowledge gaps place on our ability to achieve rational approaches to managing the risk associated with the production, use and disposal of toxic chemicals that bind to sediments.

The variety of organic and inorganic chemicals that EPA is responsible for regulating represents many different chemical classes. The interactions of these compounds with the different types of particulate matter in the aquatic environment are probably enormous.

When a chemical that comes under review by the Agency is one that will bind to sediments and is known or suspected to be toxic to aquatic life or that may become incorporated into a food chain, the Agency has encountered a great deal of difficulty in gathering sufficient information to evaluate the risk to the environment associated with its production, use, and disposal. The many uncertainties that exist, either because of the absence of information or the lack of understanding and failure to use information from other scientific disciplines, clearly opens regulatory decisions to serious flaws.

The following are examples of some of the basic questions and uncertainties encountered when chemicals that associate with sediments come under review within the different program offices in EPA:

- What does the term "bioavailability" mean?
- What is sediment? Are all sediments the same with respect to the binding or sorption of chemicals?
- What is the relationship, if any, of the concentration of a chemical

in the water column and the concentration of the chemical bound to sediment?

● Are chemicals that are toxic to aquatic organisms toxic after binding to sediments? Is the answer the same for all situations or is it site specific, chemical specific, or sediment specific?

● Are currently available laboratory tests adequate for assaying the toxicity of chemicals bound to sediments?

● Can we predict the partitioning, bioavailability, and toxicity of chemicals that bind to sediment over periods of time, as well as for specific points in time? If so, how reliable are such predictions?

● How adequate is the information currently available for use in deriving sediment criteria that will be generally useful for regulatory purposes and for assessing the risk to the environment from toxic chemicals that bind to sediment?

● What are the major knowledge gaps with regard to our understanding of, and ability to predict partitioning, bioavailability, and toxicity of sorbed chemicals?

● What research should be done to fill these knowledge gaps? Which of the research areas, if any, have a greater priority than others?

In order for EPA to apply its regulatory authority to chemicals that attach to sediments, whether the issue is sediment criteria, restricted production, use or disposal of a chemical, or a ban on its production, the Agency needs to know if these chemicals will come in contact with aquatic life and if they will manifest their toxicity as a result of that contact. The relationships between the chemicals, sediments of various composition, the water column, and aquatic organisms over time and under the influence of different environmental conditions need to be defined.

Some of this information undoubtedly exists, but that which is known needs to be assembled; that which needs to be known, but is not, should be identified; and research that should be carried out to fill those knowledge gaps should be proposed.

Chapter 3

The National Dredging Program in Relationship to the Excavation of Suspended and Settled Sediments

William R. Murden

Aside from Mother Nature, the U.S. Army Corps of Engineers is one of the largest single movers of sediment in the world. I would quickly point out, however, that we are quite dwarfed in comparison with the volume of sediment moved by Nature. Second, and equally as important, is that the Corps, over the last decade or so, has been one of the world leaders in research on the environmental aspects of aquatic sediments and their removal, particularly in research and development-related (R&D) sediment/chemical relationships. Since 1972, the U.S. Government has spent over $100 million in dredging-related R&D, primarily on environmental aspects. The Corps funded over 90% of this work.

I would like to present an overview of our national dredging program; to discuss developing trends in the dredging field that I think will provide more than adequate justification for the appropriateness and timeliness of this workshop topic; and to present some personal throughts on required management perspectives and initiatives relative to sediment contamination.

THE NATIONAL DREDGING PROGRAM

The National Dredging Program involves the maintenance and development of over 25,000 linear miles of federal waters that serve 130 of the nation's largest cities and that reach forty-two of our fifty states. These waterways are utilized to transport over one-fourth of the nation's ton miles of cargo. Thus they are essential to the economic well-being of the country.

Much of the cargo is coal and oil, so the waterways are important in meeting our energy needs. Many of these waterways also serve a critical defense role. For example, fifty-six of our coastal ports around the country are designated as national defense ports. These ports, as well as our major inland waterways and ports, such as the Mississippi River, our intracoastal waterways and the Great Lakes, will serve a vital support role in any defense-related activity.

Most of these navigation channels require periodic dredging to provide safe and efficient operational conditions for maritime traffic. Only one estuary, Puget Sound in the state of Washington, has natural deep water adequate to accommodate the large vessels engaged in commercial and international maritime activities. Thus, because of the significant economic, energy, and defense importance of this extensive navigation system, it is not a question of whether or not to dredge, but a question of how best to dredge and dispose of dredged materials in an environmentally acceptable yet economical and efficient fashion.

The following items are basic facts about our national dredging program:

- Maintenance and development of this waterway system involves dredging and disposal of about 300 million cubic yards (MCY) of sediment annually.
- Another 100-150 MCY are dredged annually by other federal and state agencies and by the private sector. Disposal of this material is regulated by the Corps under the various federal authorities.
- Of this 400-450 MCY of sediment requiring disposal each year within the United States,
   — About 60 MCY are deposited in ocean waters each year. Nationwide, we have about 130 ocean disposal sites designated for this purpose. On an average, forty to fifty of these sites are used each year.
   — Another 100-150 MCY are dredged each year from coastal areas and disposed of by means other than ocean disposal.
   — The remainder, or approximately 200 MCY, is dredged annually from our inland waterways, primarily from the Mississippi River and its tributaries and the Great Lakes.
- The annual federal dredging workload is accomplished primarily through contractual arrangements with private industry. However, the Corps is required by law to maintain a minimum fleet of federal dredges to provide a rapid and effective response to meet defense and national emergency dredging requirements.

The amount of annual dredging the Corps has been responsible for has been fairly constant over the past several years because there have been only a few major deepening projects authorized by the Congress. However, there is reason to believe that the role of dredging in the United States will be significantly expanded in the near future. For example, navigation improvements to a number of our coastal port facilities are proposed to depths of 50 ft or more. These proposed channel deepenings would involve the removal of large quantities of dredged material and, in turn, would require careful consideration of the disposal of these materials. Some of the major proposed deepenings, such as at Baltimore, Norfolk, Mobile, Galveston, and New Orleans, will require excavation ranging from an estimated 300 MCY to 400 MCY per project. Because this would involve the removal of virgin material, problems with contamination are not anticipated. However, because of the large volumes of material, our environmental concerns will be directed toward physical impacts and the identification of appropriate disposal alternatives for minimizing these impacts.

Additionally, there are projections for significantly increased defense dredging needs. This trend is developing primarily on two fronts. First, there seems to be a real potential to not only expand but also to modernize our naval defense fleet. Several categories of naval vessels now entering the fleet will require either significantly deeper access channels, more frequent maintenance dredging, or both. Second, and closely related, are the navy plans to significantly expand the number and types of homebase ports in each of the coastal regions of the United States.

Another example of expansion concerns the probable role of dredging in a number of identified clean-up projects, either under Superfund or other authorities. Clean-up will require a careful and thorough scientific analysis of the sediments to be dredged, appropriate dredging technology for the removal of these sediments, and appropriate alternatives for disposal.

A fourth example is the rehabilitation of our nation's existing waterway and water supply infrastructure, specifically our reservoirs and associated locks and dams. Many of our existing locks and dams are fast approaching their projected 50-year life expectancy. This is the age, based on our experience, when locks and dams generally require either major rehabilitation or replacement. Certainly, a closely related problem is how to deal with the significant sediment accumulations that have built-up behind many of these structures in order to effectively restore water quality, floodwater storage, hydroelectric potential, irrigation, and other project design purposes. Dredging will most likely play a significant role in any such rehabilitation efforts. Again, sound and complete scientific information must be available to safely and economically dispose of these large volumes of sediments, particularly in cases where contaminated sediments are involved.

A final example is found in the increasing federal, state, and industrial interests in offshore mining, particularly in the deep-sea mining of strategic minerals. Innovative dredging technology and safe, efficient disposal of the dredged material must play a major role if such efforts are to prove successful.

For some time, disposal of dredged material has been our number one problem in the management of the national dredging program. Under existing laws and regulations, this will continue to be our most critical problem for the foreseeable future.

One perspective on the magnitude of this problem is that each year the Corps moves enough dredged material to bury the city of Washington, D.C. to a depth of 6 ft. However, before anyone jumps up and yells "good idea," he should be aware of the extensive and often highly complex environmental procedures the Corps must follow whether we choose an upland, diked, open water, or ocean disposal site.

There are presently over thirty federal environmental statutes and executive orders that govern the manner in which dredged material is disposed of within the United States. The Corps is required, by law, to fully address the substantive and/or procedural requirements of each of these statutes for each individual dredging project. Superimposed upon these myriad federal laws are the numerous and often conflicting environmental laws of the forty-two states in which federal waterways have been constructed. However, the substantive and procedural requirements of these laws must also be followed to the maximum extent feasible. A significant point to be made here is that in most cases where litigation has halted dredging projects, such rulings have been based on procedural deficiencies, not technical or scientific inadequacies.

A major problem in disposal management is the manner in which the individual federal environmental statutes have evolved. Each law (e.g., Clean Water Act, Ocean Dumping Act, Coastal Zone Management Act, etc.), with its implementing regulations, protects one environmental medium (e.g., the ocean) from disposal, often at the environmental expense of other available disposal media (e.g., upland or inland waters). Each implementing regulation requires application of testing and evaluation procedures (e.g., bioassays) to ensure against harm to a specific environmental medium. However, no clear procedural mechanism is provided in these regulations to provide for a weighing and balancing of environmental effects between the various media.

During the planning process for the dredging of any federal navigation project, the Corps is also required by law to select the least costly and environmentally acceptable disposal alternative. This may be termed the "federal standard" for environmental compliance. Beach nourishment provides an appropriate example of how this required navigation analysis can work. A federal navigation project may require the excavation of beach-quality sand and a definite need may exist in the immediate vicinity for beach nourishment. However, if an environmentally acceptable ocean disposal site is available and its use would result in less cost to the government (in most cases significantly so), ocean disposal would be identified as the preferred alternative or "federal standard" for the project if this type of disposal does not pose a significant adverse environmental effect. Under existing statutes, the local entity (state, city, etc.) must pay for the cost differential in order for the sand to be used for beach nourishment.

In specific regard to contaminated sediments, which is of principal concern to this group, our navigation assessment might proceed as follows:

- A standard navigation analysis would first be undertaken, involving a comprehensive public interest review, weighing the dredging-related costs versus benefits, both from an economic as well as environmental perspective. Environmentally, this requires a technical weighing and balancing of all "practicable" disposal alternatives, including beneficial uses. During this analysis, we must assure that the substantive and procedural (including testing) requirements of each appropriate federal and state environmental statute are met. If environmentally acceptable disposal options can be identified that are economically feasible and are justified in light of project and other engineering constraints, the dredging activity will proceed.

There are situations, however, involving contaminated sediments where costly dredging and disposal restrictions are imposed, as a result of either technically justified environmental concerns, or technically unjustified yet publicly perceived environmental concerns. If such costs are excessive and either cannot be justified based on our navigation analysis (cost vs. benefits) or for other reasons such as inability or unwillingness of the local sponsor to pay for required disposal restrictions, we must proceed as follows:

- First, there would be an analysis to determine if, from a federal perspective, cleanup efforts are justified to restore or maintain navigation projects authorized by the Congress. Although the Corps does have extensive and unique capabilities in dredging equipment and procedures that are applicable to cleanup operations, we do not, except for a few site-specific authorities, have an assigned mission to undertake dredging-related cleanup activities. However, we are involved if any cleanup assessment is aimed at restoring or maintaining federal navigation projects.
- If this analysis indicates that a cleanup operation cannot be justified from a navigational standpoint, the question then becomes whether or not it is safer environmentally to leave the contaminated material in place or to remove it to a less environmentally sensitive area.

PERSPECTIVES ON TECHNICAL AND
MANAGEMENT NEEDS

I was asked to give a personal perspective on technical and management needs in the area of sediment/contaminant associations. It would seem to me our ultimate goal must be to develop technically and economically feasible measures to eliminate sources of contamination. Contrary to what some believe, the Corps is not a generator but an unwilling recipient of these contaminants.

A related problem is one of vocabulary and the public perception caused by some of the terms still used in our business. People often use the phrase "dredge material" and "sewage sludge" as if these materials were the same. Also, we frequently hear the term "dredge spoil" which has an unpleasant and inaccurate connotation. Actually, more than 90% of the material we excavate each year is not polluted and is suitable for construction and other beneficial uses. We need to correct this public misconception and several of the existing environmental regulations which include these incorrect terms.

There is a critical need to develop a logical, technically-valid and systematic approach and framework for evaluating the contamination potential of dredged materials. Such a framework must include a careful and thorough weighing and balancing of all potential disposal alternatives to ensure maximum protection to the overall environment. Present criteria regulating open water disposal rely heavily on bioassays and bioaccumulation assessments to define contamination potential. The Toxic Substances Control Act (TSCA) classified any sediment containing 50 μg/g or greater PCB (dry weight basis) as toxic and required elaborate and costly disposal restrictions. This numerical approach is arbitrary and does not consider site-specific sediment characteristics nor the specific environmental media available to receive this material. A similar example is with the implementing criteria to define hazardous materials under the Resource Conservation and Recovery Act (RCRA). A bulk-type sediment analysis is applied. Any sediment containing contaminant concentrations in excess of 100 times established drinking water standards is classified as "hazardous," which also requires extremely expensive disposal management procedures. Our experience is that appropriate testing and evaluative procedures are available for defining sediment contamination potential, and in turn defining appropriate disposal alternatives and any required disposal management restrictions. However, these procedures must presently be applied on a case-by-case basis, considering not only the sediment to be dredged, but also the site-specific environmental conditions associated with each available disposal option.

The Corps has taken a major step in developing such a technically sound management strategy (1). This strategy is intended (1) to be applicable to the wide range of dredged sediments that are encountered in our federal waters; (2) to provide guidance on appropriate testing and evaluation procedures for sediment contamination potential within all potential disposal options; and (3) to incorporate the state of the art knowledge on appropriate disposal management strategies for a wide range of sediments. We are presently refining this management strategy and hope to conduct selected demonstration projects in several regions of the country in the near future.

A definite need exists to develop and demonstrate new and innovative approaches for the cost-effective disposal of highly contaminated sediments.

Our average national unit cost for each cubic yard of sediment presently dredged is $1.47, up some 475% from 1970 when NEPA was enacted and when the unit cost was $0.30/cubic yard. In areas such as the Great Lakes where material must often be transported long distances to diked disposal areas constructed to meet statutory requirements, the average unit cost for these areas is $3.60/cubic yard with some projects in the $30.00 to $40.00 unit cost range. This is a very expensive operational procedure and deserves reflection especially since we now know that the diked disposal method is not the environmental panacea it was once thought to be. One such alternative disposal procedure, which we are now carefully evaluating, involves aquatic or subaqueous disposal (borrow pits, dredged trenches, etc.) and capping these areas with clean material to isolate contaminants from the water column. Under certain situations, this alternative offers considerable potential as a cost-effective yet environmentally sound option for disposal of highly contaminated sediments. Tests of this type of disposal have been conducted for several years, and I am confident our tests will lead to economical yet environmentally safe disposal alternatives.

Before closing, I would like to offer some personal thoughts from my perspective as an engineer and manager. As discussed earlier, over 90% of the sediments we excavate in our annual national dredging program are non-contaminated and non-toxic to plants and animals. In many cases, this material is predominantly sand, suitable for construction aggregate, beach nourishment, or other engineering purposes. Much of the remainder, the fine-grained sediment, is basically topsoil, which is in great demand.

This beneficial-use potential has been obvious to me for many years. As such, I have strongly encouraged the Corps personnel to consider in all project planning efforts the beneficial uses of this material.

Over the last 5 years or so with the very capable assistance and innovative thinking of our Waterways Experiment Station staff and the many other scientists involved in our R&D programs, we have begun to make significant advances in beneficial use applications:

— Marsh creation and habitat development are now proven technologies that are being routinely used in many areas of the nation, as well as by other countries.

— Beach nourishment, a critical need in many areas of the coastal zone, is now a much more cost-effective disposal alternative because of direct pumpout dredging capabilities and the shallow-draft, split-hut designs that are now being used on many of our newly constructed hopper dredges.

Many other beneficial uses have been identified, demonstrated, and are now in use on a limited basis. We are continuing to encourage innovative new approaches:

— Extensive marsh creation and barren land restoration offer considerable promise along the Gulf Coast as a means of disposing of large quantities of sediments that will be dredged when the port deepening projects are authorized by Congress. This type of disposal offers a beneficial effect on the environment and at the same time can be economically feasible.

— Underwater berms constructed of dredged material generally parallel to the coastline offer the potential for significantly mitigating wave energy and erosion along our beaches and resort areas. Our pilot studies have indicated the engineering feasibility of this approach, and contrary to the views of many, no significant erosion of the berm has occurred in the high energy coastal zone in which we constructed the berm.

Clearly there is room for considerable innovation in beneficial uses in the dredging field. Public misconceptions, logistical considerations, cost effectiveness, and legal and regulatory restrictions are still major barriers that must be overcome. However, I am confident that in the not too distant future such uses will be routinely considered and applied. Then we can appropriately and effectively concentrate our research and development efforts on the small percentage of dredged material that is contaminated.

REFERENCE

1. **Francinques, N. R. Jr., M. R. Palermo, C. R. Lee and R. K. Peddicord.** 1985. Management strategy for disposal of dredged material: Contaminant testing and controls. Misc. Paper D-85-1, U.S. Army Waterways Experiment Station, Vicksburg, MS.

Chapter 4

A Chemical Industry's Perspective on Issues Related to Chemicals Bound to Sediments

Richard A. Kimerle

INTRODUCTION

The first Pellston Conference in 1977 played a significant role in advancing the science of aquatic hazard evaluation (1). Important questions were asked and new insights were gained on how to estimate chemical exposure and biological effects and how to use these data in making decisions on chemical safety. It was also recognized at that time that a deficiency existed in our ability to understand chemical binding with solids and what this phenomenon meant to interpreting chemical exposure and hazards in aquatic environments. Now, seven years later, the subject of chemicals in sediments has been elevated from the position of being mentioned in a number of the paragraphs in the first Pellston book to one commanding the undivided attention of forty expert scientists for five full days, and resulting in this publication. What has happened in the past seven years to move us and the science from where it was then to where it is now? A perspective on the issues and needs of today lies partially in the answer to that question.

Some of the important accomplishments deserve mention. For example, methods of testing for chemical effects with single species have become more standardized and accepted, and methods of predicting and measuring chemical exposure have improved. Our ability to use the effects and exposure data has also improved. Furthermore, regulatory agencies have systematized their procedures for collection and interpretation of environmental data, and many industries have developed environmental expertise and established new data bases. The environmental sciences have "come of age" with the integration of disciplines and establishment of new scientific societies and outlets for publication.

However, for the aquatic environment, it must be said that the above accomplishments are most applicable to the water column and not the sediments. Our newly acquired knowledge in dealing with one part of the aquatic environment has made our deficiencies in the other part more obvious. Simply stated, the topic of this workshop is timely and needed to enable us to bring more science to the environmental compartment that has been to a large extent neglected, namely the sediments.

When we do have a better understanding of chemicals and sediments, we will also advance our understanding of chemical exposure in all environmental compartments where solids are present. The bottom line benefit of these studies should be that the environment is appropriately protected and that any forthcoming regulations involving sediments are founded with defensible science.

At Monsanto, as is the case in many industries, a fairly active program in environmental toxicology has progressed for a number of years. It has become obvious to our group of environmental scientists that as more and more data are acquired on chemical exposure and toxicological effects of organic chemicals that the nation's ambient waters do not generally contain concentrations of many organic chemicals very near the concentrations that would be expected to cause any toxic conditions or residue problems [2, 3]. What we do not know is how to interpret the observed concentrations of chemicals on sediments found in some localized areas near high population densities and heavy chemical use. Therefore, we at Monsanto have concluded that chemicals in sediments is an important environmental subject.

Another point that has come to be more appreciated is that the universe of chemicals has very broad physical and chemical properties, and these properties control how each chemical partitions in the environment [4]. Chemicals that volatilize should receive more attention from an air standard viewpoint than chemicals that stay in water. Our experience suggests that chemicals possessing the physical and chemical properties to partition from water to sediments should receive special attention from a sediment viewpoint. What should not happen is that we begin to require that all chemicals be intensively studied for sediment effects. By using the best science available on chemical partitioning, the subject of chemical hazard in sediments should be limited to those chemicals that tend to strongly partition to sediments and those that are produced in high volumes.

SEDIMENT-RELATED ISSUES

Several scientific and regulatory issues need to be addressed by the scientific and regulatory communities to fill the data gaps:

Definition of the Chemical/Sediment "Problem"

From our viewpoint, it looks like sediments tend to be sinks for newer chemicals and reservoirs for chemicals that are declining in use or have been banned. However, the exact extent of the situation nationally is largely unknown and could use additional study.

Further Refinement of Methods for Testing Sediment Toxic Effects

A simple laboratory test needs to be adopted for screening the toxicity of chemicals associated with sediments. We need to be able to go into the field and determine if the sediments are acutely and chronically toxic to aquatic organisms. It is also important to be able to relate the laboratory data to observed field data and real world impacts. A test is also needed to determine the potential of sediment-bound chemicals to accumulate in biological tissues.

A Basis for Interpreting Exposure from Sediments

The issue of understanding exposure from sediment-sorbed chemicals is absolutely critical to establishing a scientific basis for almost all decisions concerning the safety or hazard of chemicals in sediments. Multidisciplinary research is needed to provide some conclusions on the fate and availability of chemical sorption-desorption as influenced by such factors as chemical structure types, organic carbon, clay particles, and other important factors. Issues such as bioavailability, food chain biomagnification, and species variability as they relate to benthic organisms also need to be addressed.

Progress in these areas would be beneficial to the environmental program of a company such as Monsanto in a number of ways. First of all, in evaluating new and existing chemicals for their safety in the environment, we would be including a segment of the environment that has been somewhat neglected because the science was not in place. Second, our effluent aquatic safety evaluation program would be enhanced by the additional capability of being able to test for effluent impact on benthic communities and to evaluate instream conditions. And lastly, the application of some new science to evaluating hazardous waste sites and the potential for groundwater contamination has enormous implication in deciding the extent of remedial practices needed to reduce exposure to acceptable limits.

A Basis for Establishing Sediment Criteria

Accepting the fact that chemicals do find their way into sediments in the nation's waters, is there a need for chemical-specific sediment criteria as has been established for individual chemicals in waters? Probably not, but what is needed is a method to define exposure, or the true biological availability of the chemical from the many types of sediments found in the nation's waterways. The actual water quality criterion established using the existing method of the Environmental Protection Agency probably provides a good estimate of the chronically safe concentration of a chemical. If this method adequately accounts for species variability as it in fact seems to, then the method should adequately cover the species that happen to live in or near the sediment. The key is being able to measure true exposure to the organisms. If a method is established to define chemical exposure from sediment in the spirit of a "national sediment criterion," then it would have to be site specific. Our experience suggests that for non-polar organic chemicals, total sediment concentration has little to do with the biologically available chemical concentration (5). Organic carbon is a very important factor, and it can vary by two orders of magnitude around a national average of 2.0% (6). Once you know, or can predict, the available concentration, the traditional water quality criterion should be useful to protect benthic communities. Something that we might consider is to include a sensitive benthic species in the array of species tested to establish the water quality criterion.

We have a unique opportunity, in dealing with chemicals in sediments and the subject of national sediment criteria, to let the science drive the establishment of regulation. I am confident that this is the approach that best serves the environment and the interest of all parties concerned.

REFERENCES

1. **Cairns, J. Jr., K. L. Dickson and A. W. Maki** (eds.) 1978. *Estimating the Hazard of Chemical Substances to Aquatic Life.* ASTM STP 657. American Society for Testing and Materials, Philadelphia, PA.

2. **Staples, C. A., A. F. Werner and T. H. Hoogheem.** 1985. Assessment of ambient priority pollutant concentrations using Storet-I. Occurrence in United States waterways. *Environ. Toxicol. Chem.* **4**:131-142.

3. **Staples, C. A. and A. F. Werner.** Priority pollutant assessment in the U.S.A.: Scientific and regulatory implications. *Toxic Substances Journal (in press).*

4. **Mackay, D.** 1979. Finding fugacity feasible. *Environ. Sci. Technol.* **13**(10):1218-1223.

5. **Adams, W. J., Kimerle, R. A., and Mosher, R. G.** 1985. Aquatic safety assessment of chemicals sorbed to sediments. In R. D. Cardwell, R. Purdy, and R. C. Bahner, eds., *Aquatic Toxicology and Hazard Assessment: Seventh Symposium.* ASTM STP 854. American Society for Testing and Materials, Philadelphia, PA, pp. 429-453.

6. **Adams, W. J.** 1986. Levels of organic carbon in aquatic sediments of the United States. Monsanto Internal Report MSL-5413.

Chapter 5
Synopsis of Discussion Session 1

Perspectives and Needs in the
Hazard Assessment of
Sediment-Bound Chemicals

G. Loewengart *(Chairman),*
A. Hall, G. F. Lee, C. Zarba and R. Zeller

HISTORICAL PERSPECTIVE

The interest in the water quality significance of chemical contaminants associated with aquatic sediments has evolved in several areas. There has been interest in the role of sediment-associated phosphorus in influencing the degree of eutrophication of lakes and reservoirs since the early 1960s. It was not until the late 1960s that it became generally known in the field of water quality management that sediments of some U.S. waterways were highly contaminated with a variety of potentially hazardous chemicals. The environmental quality awakening of the early 1970s resulted in federal and state water quality regulatory agencies to examine the chemical characteristics of aquatic sediments. It was found that aquatic sediments near urban centers frequently contained significantly elevated concentrations of heavy metals, nutrients (nitrogen and phosphorus), chlorinated hydrocarbon pesticides, and PCBs, as well as a wide variety of other toxicants, which if biologically available, could have significant adverse effects on aquatic life. This finding led to the establishment of water quality regulatory agencies in the early 1970s such as the Federal Water Quality Administration (FWQA) (predecessor to the U.S. Environmental Protection Agency) to attempt to develop sediment-based water quality criteria based on the bulk-total concentration of a contaminant or group of contaminants in the sediments.

The overall purpose of these criteria was to regulate the U.S. Army Corps of Engineers' (USACE) dredging of the nation's waterways. The focal point of concern was the open-water disposal of dredged sediment.

The implementation of the FWQA sediment water quality criteria led to significant increase in the cost of the Corps of Engineers' maintenance of the navigational depth of many U.S. waterways and harbors. The Corps of Engineers concluded that they were likely spending far more money for disposal of contaminated sediments than was justifiable, based on the technical information available. Further, it appeared that the alternative, more expensive dredged-sediment disposal methods that were adopted in response to the FWQA sediment criteria would be more damaging to aquatic life than the previously used open-water disposal.

This situation caused the Corps to obtain from Congress about $32 million to develop a 5-year Dredged Material Research Program (DMRP) devoted, to a considerable extent, to assessing the water quality significance of contaminants associated with dredged sediment.

The DMRP was completed in 1978. It confirmed that some waterway sediments near U.S. urban and industrial centers were highly contaminated. It also confirmed that the bulk sediment criteria originally developed by FWQA, and which were being used by some of the U.S. Environmental Protection Agency (EPA) regions, were technically invalid. As an alternative approach, EPA and USACE develped the elutriate test and dredged-sediment bioassay procedures to better evaluate the potential water quality problems associated with open-water disposal of dredged sediments. Even though these procedures were developed in the mid-1970s, there are still significant problems, largely political in origin, that have hampered the appropriate implementation of these or revised test procedures into cost-effective, environmentally protective approaches to regulation of dredged-sediment disposal.

Another impetus for assessing the water quality significance of sediment-associated contaminants is derived from the chemical industry's need to screen new or expanded use of commercial chemicals for potential environmental impact prior to manufacture, use, and disposal. This assessment would address both regulatory requirements such as Toxic Substances Control Act (TSCA) and National Pollution Discharge Elimination System (NPDES) permits and internal environmental safety assessments.

The studies of the mid-1970s conducted by, or on behalf of, the Corps, EPA, National Oceanic and Atmospheric Administration (NOAA), and other agencies found a number of areas of the United States where highly contaminated sediments exist. Several post-DMRP studies have focused on these problem areas. One of the areas of greatest concern is Commencement Bay in Puget Sound, at Tacoma, Washington. Again as in the early 1970s, the regulatory agencies of the region, EPA Region X and the State of Washington, have chosen to adopt bulk sediment criteria as part of their program to regulate dredged-sediment disposal. This approach, while administratively simple, suffers from the same deficiencies as the FWQA sediment criteria. This situation reflects the difficulty encountered by water quality regulatory agencies in utilizing the results of the DMRP in regulating water quality aspects of dredged-sediment disposal.

Over the last several years, EPA has found sediment chemical characteristics to be very important for regulatory decision making in the ocean dumping, ocean discharge, and dredge and fill programs. Recently, the EPA Office of Water has started a major effort towards developing sediment water quality criteria. Further, other offices of EPA such as those implementing TSCA are beginning to address the water quality significance of new chemicals that tend to become associated with aquatic sediments. Several EPA research laboratories (Corvallis, Gulf Breeze, and Narragansett), major chemical companies, NOAA, USGS, USACE, and several universities and research organizations are actively involved in research and field studies designed to provide information that is pertinent to assessing the water quality significance of sediment-associated contaminants. Thus, there is widespread interest today in developing approaches for assessing the water quality significance of chemical contaminants associated with aquatic sediments.

INFORMATION NEEDS

With the background of the above historical perspective, a statement of information needs was provided by representatives of the Corps of Engineers, the Environmental Protection Agency, and industry to be addressed during this workshop. It was apparent from the first day's discussions that a gap exists between research that has been done, available methods, implementation of test results, and current information needs. In essence, a challenge was presented to the workshop to address what were perceived to be important questions relative to chemicals associated with sediments. These are summarized below.

Scope of the Problem

Some chemicals become associated with sediments. This association will likely alter the availability of the chemicals to aquatic life. Basic questions discussed in this session include:

● What does it mean to society to have biological effects and bioaccumulation in aquatic organisms associated with sediment-bound chemicals? Stated another way, are we as scientists, engineers, and regulators cognizant of and responsible to the public's perception of the problem and their priorities?

- Under what conditions do sediments serve as sinks, and when are they reservoirs for chemicals that tend to cause enviromental problems?

- Is the problem a widespread national concern or is it a local problem associated with point source discharges from municipal or industrial effluents or dredged-sediment disposal?

- Is the problem limited to relatively few chemicals that tend to partition strongly to sediments?

- There are basic questions still needing to be addressed, such as what is sediment and are all sediments the same with respect to the binding or sorption of chemicals? Are there fundamental differences between sediments or are the differences between sediments just a matter of degree?

Environmental Fate and Compartmentalization

Sentinel to all discussions was that only a fraction of sediment-bound chemicals are available for chemical and biological processes. A key operational question then becomes whether there is agreement on what bioavailability means:

- How is the fraction of the total sediment-bound chemical that is biologically available to aquatic life determined? That is, how is bioavailability assessed?

- What are the factors (e.g., sediment characteristics) that influence bioavailability, and what are the mechanisms through which such factors influence bioavailability?

A second key question is what is the relationship, if any, of the concentration of a chemical in the water column and the concentration of the chemical bound to the sediment? This includes suspended and settled sediment particles as well as the interstitial water of the sediment bed. Related to this information need are additional queries:

- To what extent do we need to increase our understanding of the hydrodynamics of sediment/water systems in order to predict chemical partitioning and bioavailability reliably? The current inadequacy of hydrodynamic information results in the inability to physically model sediment transport and mixing and thus calculate chemical exchanges and availability.

- Can we predict the partitioning and bioavailability of chemicals (between the water column, the interstitial water, and sediment) over long periods of time, as well as for specific points in time? How reliable are such predictions?

- How much impact does micro- and macro-organism activity (in the bioturbation zone) have on sediment mixing and, if significant, how should such impacts be measured?

Biological Effects

Are chemicals that are toxic to aquatic organisms toxic after binding to sediments? Would the result be the same for all situations or is it site and chemical specific? In addition, the following questions relative to biological effects were posed to the workshop:

● How does one best determine the concentration-duration of exposure relationships associated with sediment-bound chemicals? Two extreme cases considered were (1) episodic events, such as dredged-sediment disposal or storm-associated sediment suspension, which are likely to produce short-term elevated water column concentrations of sediment and sediment-derived contaminants; and (2) long range events, where sorption of chemicals on sediments may represent a true sink or, at least change the assessment to one of lower exposure concentrations for extended periods of time.

Test Methods/Field Studies

In order to address the water quality significance of sediment-bound chemicals, the usefulness of available and newly developed test methods should be assessed. Workshop questions regarding test methods and the need for field studies to validate these methods follow:

● What are the uses and limitations of currently available tests such as (1) the elutriate test for predicting the release of chemicals from aquatic sediments, (2) the bioaccumulation test for assessing contaminant accumulation in animal tissue, and (3) sediment bioassay tests for assessing toxicity?

● Are other laboratory tests currently available, or needed, which adequately assay the significance of chemicals bound to the sediments? Such tests should be appropriately designed to consider physical and chemical factors associated with sediment. The tests should account for test-species variability and should attempt to relate the laboratory data to environmental effects. In addition, a simple laboratory test is needed to screen for toxicity of chemicals associated with sediments and indicate the need for more definitive bioassays.

● What constitutes "proper" handling of sediments for different mechanistic studies and/or for regulatory and hazard assessment decision purposes? Consideration should be given to how much sediment alteration is acceptable and what compromises in the handling of sediment are reasonable.

- Are laboratory sorption/desorption tests available that can be used for aquatic hazard assessment? These tests should consider factors that influence sorption and the role of hydrous metal oxides and organic chemicals on sorption. Would a standardized, though imperfect, test procedure assist in understanding the mechanisms of sorption/desorption or in making regulatory decisions?
- Are field studies needed to validate laboratory tests used to assess the toxicity and bioaccumulation of sediment-bound chemicals? Should field studies be conducted to determine if contaminated sediments are acutely and chronically toxic to aquatic life?
- Are existing database management systems adequate to ensure timely availability of information for sediment-bound chemical hazard assessment and regulatory decision making?

Regulatory and Control Implications

- Are sediment-based water quality criteria needed for chemicals? If sediment criteria are needed, what would be their objective, and what form should they take (e.g., generic tests, criteria numbers)?
- How would they best be used to meet identified needs?
- How adequate is the information currently available for sediment criteria development? In what areas are data needed to assess more accurately the level of environmental risk from sediment-bound chemicals? How does this relate to or impact on criteria development? In what areas should future sediment criteria development efforts be focused? Are bulk concentrations of chemicals in sediment reliable indices of potential water quality harm?
- To what extent should site-specific studies be conducted? Should they determine water column concentrations of chemicals released from sediment, and the impact of released chemicals on water quality on a case-by-case basis?
- Are predictive approaches available to assess aquatic hazards of chemicals bound to sediments? Can a systematic data collection and interpretation procedure be developed? Can tiered screening and assessment procedures and tests be developed?

CONCLUSIONS

Based on the papers presented in Session 1 and the discussions which were stimulated, it is evident that many questions exist regarding the fate and effects of sediment-bound chemicals. Understanding sediment/chemical interactions is essential in the development of a chemical hazard assessment. The questions posed by the presenters in Session 1 and summarized in the synopsis provide a clear overview of the needs and issues.

Session 2
Environmental Fate and Compartmentalization

Chapter 6

The Transport of Sediments in Aquatic Systems

Wilbert Lick

INTRODUCTION

Sediments are significant contaminants in themselves, but they are also of importance since many contaminants are readily adsorbed onto sediments and are transported with them. Because of this, if we understand the transport of sediments, we can then more readily understand the transport and fate of sediment-associated contaminants.

The primary sources of suspended sediments in aquatic systems are river inflows and shore erosion and, to a lesser extent, the disposal of wastes by man. After being introduced into the water, these sediments are dispersed by wave action and currents, and then deposited on the bottom, and later resuspended and again dispersed by wave action and currents; a series of events that continues until the sediments are transported out of the system or until they become a permanent part of the bottom sediments through diagenesis. During resuspension, the finer particles may aggregate to form larger particles (flocculation), and/or larger particles may disaggregate into smaller particles with the rates of aggregation and disaggregation dependent on the turbulence, sediment concentration, and other factors. Transport of sediments may occur as suspended load (for finer sediments) or as bed load (for coarser sediments). The emphasis here is on fine-grained sediments (viz., suspended load) because of their greater surface area and hence adsorptive capacity as compared with coarser sediments.

In the present context of water pollution, the goals of research on sediments are not only to obtain a basic understanding of the processes that govern sediment transport, but also to obtain a quantitative, predictive model of sediment transport that embodies this understanding. Such a model is necessarily based on the conservation of mass equation for sediments, which is

$$\frac{\partial C}{\partial t} + \frac{\partial (uC)}{\partial x} + \frac{\partial (vC)}{\partial y} + \frac{\partial ((w + w_s)C)}{\partial z} = \frac{\partial}{\partial x}\left(D_H \frac{\partial C}{\partial x}\right) + \frac{\partial}{\partial y}\left(D_H \frac{\partial C}{\partial y}\right) + \frac{\partial}{\partial z}\left(D_v \frac{\partial C}{\partial z}\right) + S$$

(1)

where C is the mass concentration of the sediments, t is time, u, v, and w are fluid velocities in the x, y, and z directions respectively (where z is vertical and positive upwards); w_s is the settling speed of the sediments relative to the fluid; D_H is the horizontal diffusivity; D_v is the vertical diffusivity; and S is a source term.

In most work on sediment transport, sediments have been described in terms of average quantities, e.g., an average or mean settling speed, w_s. Although this is a valid first approximation, a quantitative analysis of sediment transport must consider the fact that sediments consist of a mixture of particles with widely varying sizes and composition. This distribution of properties can be taken into account by separating the sediments into components with each component described by its own average quantities. The above equation is then valid for each component. In this case, the source term $S(x, y, z, t)$ results from changes in particle sizes caused by aggregation and disaggregation.

In order to solve the above equation for the sediment concentration, C, and hence produce a quantitative, predictive model of sediment transport, several parameters must be known or be determined. These are (1) the settling speeds and related quantities such as particle sizes and the rates at which particles aggregate and disaggregate; (2) the net flux of sediment at the sediment-water interface; and (3) the hydrodynamics as described by the fluid velocities and the diffusivities.

In the following sections, recent work on these processes is briefly reviewed. The emphasis is on the present state of the art and the knowledge still needed to produce a quantitative, predictive model and to develop an understanding of sediment transport.

PARTICLE SIZES, SETTLING SPEEDS, AND FLOCCULATION

It is important to realize that in real sediments the particle sizes and settling speeds vary over a wide range. For a typical sediment, this range may be over two orders of magnitude for particle diameters and four orders of magnitude for settling speeds. As examples, the distribution for two typical samples of disaggregated sediments are shown in Figure 6.1. The first sample consists of primarily fine-grained particles with the median settling speed being about

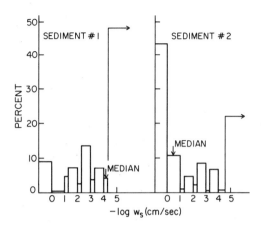

Fig. 6.1
Settling speed distribution for two
disaggregated sediments from Lake Erie.

7×10^{-5} cm/sec. The second sample is a bimodal mixture of coarse and fine-grained particles with a median settling speed of 4×10^{-1} cm/sec. The first sample is typical of off-shore sediments, whereas the second sample is more typical of near-shore sediments.

A knowledge of particle sizes, settling speeds, and the effects of flocculation on these parameters is important in determining (a) settling rates, (b) deposition rates at the sediment-water interface, and (c) the release of contaminants from sediments to the surrounding waters. Through these processes, the concentrations of sediments and contaminants throughout the water column but especially at and near the air-water interface, the thermocline region, and the sediment-water interface are greatly influenced.

In general, it is necessary to know the distributions of both the sizes and settling speeds. If the particle is spherical with known density, the two are related by the well-known Stokes' law, which is

$$w_s = \frac{gd^2}{18\mu}(p_s - p) \tag{2}$$

where w_s is the settling speed of the particles; g is the acceleration due to gravity; d is the diameter of the particle; p_s is the density of the particle; μ is the density of water; and μ is the molecular viscosity of water. The formula is valid for settling speeds such that the Reynolds' number ($Re = p\, w_s\, d/\mu$) is less than about 0.5, i.e., for sedimentary particles (with p_s about 2.65 g/cm^3) with a diameter less than about 100 μm.

Fig. 6.2
Schematic diagram
of a typical floc

However, most particles are not spherical and, more than that, individual particles tend to aggregate together into flocs, groups of particles more or less tightly bound together. A schematic diagram of a typical floc is shown in Figure 6.2. It can be seen that individual particles (about 1 μm in diameter) are relatively closely and therefore strongly bound together into clumps. Several of these clumps are loosely and therefore relatively weakly bound to each other to form a floc. Because of this aggregation, the effective diameter and density of a particle changes with time, and the diameter and settling speed of a particle are not related by Stokes' law. In fact, the diameter and settling speed are not even uniquely related to each other so that each parameter must be determined independently.

The aggregation of particles is a dynamic process with the state of aggregation at any particular location and time depending on the rates of aggregation and disaggregation, as well as the state variables such as sediment concentration, shear, etc. The rate of aggregation depends on the rate at which particles collide and also on the probability that they adhere to each other after collision. The rate of particle collisions depends primarily on Brownian motion, differential settling, and fluid shear. This collision rate can be written as (1, 2)

$$\frac{dN\,(i,j)}{dt} = N_i N_j \beta_{ij} \qquad (3)$$

where $N(i, j)$ is the number concentration (m^{-3}) of collisions between particles of size i and size j; N_i is the number concentration of i particles; N_j is the number concentration of j particles; and β_{ij} is the collision function (m^3/s), which is different for each collision mechanism.

The variation of the βs for a typical case where p_s is 2.65 g/cm^3; the temperature is 20°C; and the shear stress is 2 dynes/cm^2 is shown in Figure 6.3.

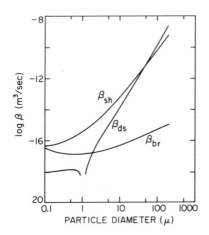

Fig. 6.3
Collision functions β_{ij}
for a 1-μm particle colliding
with particles of other sizes.
Parameters are $T = 20°C$,
shear stress is 2 dynes/cm^2,
and $p_s = 2.65$ g/cm^3.

It can be seen that, under these conditions, the collision rate will depend primarily on the fluid shear. For collisions with large particles, differential settling becomes important, whereas for collisions with small particles, Brownian motion may be significant. Whether particles will cohere after collision is a separate problem. The probability of cohesion cannot be determined theoretically at present but only through experiment. The disaggregation of particles may also be significant and is due to fluid shear and collisions between particles with sufficient relative translational energy. The rates of disaggregation of fine-grained particles are not well known and need to be determined through experiment.

Laboratory experiments have recently been conducted concerning the flocculation of fine-grained sediments in fresh and estuarine waters [3]. The purpose of these experiments was to assist in understanding flocculation and to determine the aggregation and disaggregation rates of sediments under different conditions. The experiments were conducted using a modified viscometer. The particles were initially disaggregated. However, in the viscometer, they soon aggregated primarily as a result of collisions caused by fluid shear and eventually reached an equilibrium particle size distribution in time on the order of hours.

The results of a typical experiment (3) are shown in Figure 6.4. A steady state has been reached, at least for the smaller floc size ranges. The time it takes to reach a steady state increases as the floc size increases. The experiments demonstrated that for large enough flocs, the number of flocs in a particular size range decreases as the floc size increases. This number decreases as the shear stress increases but is generally non-zero, i.e., there is no maximum floc diameter at any particular shear stress as has been assumed in some modeling of floc dynamics. The number of large flocs (per unit mass of sediments) decreases as the concentration increases.

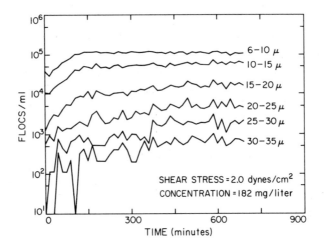

Fig. 6.4
Floc densities as a function of time
for various size ranges.

During these experiments, the use of a microscope to examine the flocs indicated that the effective densities of these flocs decrease as aggregation progresses. If an empirical relation between the effective density p_s and the particle diameter d is assumed, such as $p_s = ad^m$, the constants a and m can be determined from the particle size data and the overall mass concentration. From this it can be shown that the floc density decreases as the floc diameter increases, as the shear stress decreases, and also as the concentration decreases.

The aggregation of particles may have a significant effect on the sorption and desorption of contaminants from sediments. Consider an aggregate of many individual particles (Fig. 6.2). Because of the aggregated nature of the sediments, a contaminant cannot sorb or desorb rapidly from the particle surfaces directly to the surrounding waters but must first diffuse through the interstices of the sediment aggregates, many of which are on the order of 100 μm or more in diameter. Because of this diffusion, a significant time delay for sorption occurs and can be quantified as follows: an effective time Δt for a substance to diffuse through a distance Δx is given by $\Delta t = \Delta x^2/D$, where D is the appropriate diffusion coefficient. As an example, consider the diffusion of PCBs (with diffusion coefficients on the order of 10^{-6} to 10^{-9} cm^2/day (4)) through an aggregate 100 μm in diameter. The calculated diffusion time is on the order of 10^2 to 10^5 days. This slow diffusion would affect the sorption and desorption of a contaminant from an aggregated sediment and give an explanation, if not the only explanation, of the non-labile (but reversible) desorption of contaminants from sediments (5).

At the present time, a quantitative knowledge of flocculation is lacking. In most of the water column, fluid shear is weak and therefore the collision and aggregation of particles is primarily credited to Brownian motion and differential settling. However, in certain areas of the aquatic system, such as at and near the air/water interface, the thermocline, the sediment/water interface, and shallow, near-shore areas where wave action is significant, the fluid shear is large and will have a significant effect on aggregation and especially disaggregation of particles. Much additional work is needed to quantify the effects of fluid shear on the aggregation and disaggregation of particles and the relation between settling speed and particle size.

ENTRAINMENT AND DEPOSITION

The net flux of sediment at the sediment/water interface, denoted by q_s, can be written as the difference between the entrainment rate E and the deposition rate D, or

$$q_s = E - D \qquad (4)$$

Here the assumption has been made that entrainment and deposition are independent processes, i.e., E is the sediment flux when no suspended sediment and therefore no deposition is present, and D is the sediment flux in the absence of entrainment.

For sediments consisting of noncohesive particles of uniform size and properties, the descriptions of entrainment and deposition are relatively simple. In this case, the entrainment rate E is a function only of particle size and applied stress. The deposition rate D can be expressed as βC where the parameter β depends only on particle size and also, for particles smaller than about 1 μm, on the applied stress. For real sediments, E and D are much more complicated functions and are discussed in more detail below.

Entrainment

From experimental studies, it can be shown that for real sediments the parameters on which entrainment depends are at least the following: (1) turbulent stress at the sediment/water interface; (2) water content of the deposited sediments (time after deposition); (3) the composition of the deposited sediments including particle-size distribution, mineralogy, and organic content; (4) vertical distribution of sediment properties, i.e., manner of deposition of sediments; and (5) activity of benthic organisms and bacteria.

The effects of turbulent stress, water content, and sediment composition are qualitatively known (6-9). However, these effects are not quantitatively predictable, and each sediment must be investigated independently. Experimental results for a sediment from Lake Erie are given in Figure 6.5 and demonstrate the dependence of entrainment rates on the turbulent stress and water content and also show the large variations in entrainment rate that are possible.

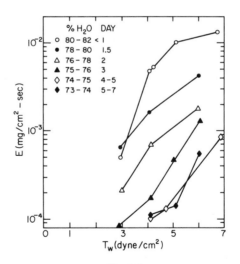

Fig. 6.5
Entrainment rate E as a
function of shear stress τ with
water content as a parameter.
Sediments from Lake Erie.

Another variable that is important but which has been investigated thoroughly is the vertical variation of deposited sediments. Here the concern is with the variations in the near-surface layer with a thickness on the order of millimeters or at most a few centimeters. In aquatic systems, this vertical variation of deposited sediments is due to the time-varying sources of sediment and to variations in entrainment and deposition rates caused by variable winds, waves, and currents. This variation is at present quantitatively unknown.

The effects of bacteria and benthic organisms are many and diverse. Both bacteria and benthic organisms secrete mucus within the sediments, at the sediment/water interface, and on burrow walls. This tends to bind the sediments. Benthic organisms also form fecal pellets that are generally larger than the non-pelletized sediment particles. But benthic organisms also burrow, stirring the sediments, disrupting the existing fabric, and changing the water content. Extensive work on these problems is presently being done both in the laboratory and in the field (10, 11). However, adequate information on the effects of organisms on entrainment and deposition processes is still not available.

For most sediments, because of cohesion of particles and/or particle size variation, the entrainment rate varies with depth such that only a finite amount of material can be entrained at a particular stress. As eqn (1) indicates, in the steady state, there is a dynamic equilibrium between entrainment and deposition of the sediments in the overlying water. This continuous entrainment and deposition of particles between the bottom sediments and the overlying water implies that, although much of the entrainable material is in the overlying water, a certain amount (dependent on the stress and the sediment concentration in the overlying water) is present on the sediment bed, i.e., has recently been deposited but not yet entrained. This recently deposited material may partially or completely cover the sediment bed and shelter the previously deposited sediments from further entrainment. The result is that the amount of material entrained is generally dependent on the concentration of the sediments in the overlying water as well as on the properties of the sediment bed.

In order to model sediment entrainment adequately, a quantitative description of the sediment bed must be constructed taking into account the above variables. At the very least, a vertically layered model must be considered where (a) each layer is described by an average water content (time after deposition) and (b) the flux of sediment from the uppermost layer is considered to be a function of the applied shear stress from wave action and currents, the water content of that layer, and the properties of the sediments. The effects of benthic organisms especially need to be investigated—not only because they can resuspend and deposit sediments by their own activities, but because of their effects on sediment properties.

Deposition

The assumption that $D = \beta C$ is consistent with recent experimental and theoretical results (6, 12). Near the sediment/water interface, the significant processes that affect the transport of particles are turbulent diffusion (which varies with distance from sediment/water interface), settling, and Brownian motion. By considering these processes, one can theoretically determine the flux at the sediment/water interface and also determine β as a function of particle size and shear stress (6). The variation of β with particle size for a shear stress of 10 dynes/cm^2 is shown in Figure 6.6. For small particles, values of β are primarily determined by diffusion (both Brownian and turbulent), whereas for larger particles, β is primarily dependent on settling and is approximately equal to the settling velocity w_s.

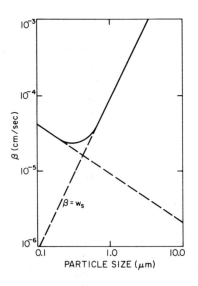

Fig. 6.6
Deposition parameter β
as a function of particle size.

Implicit in this simple model of deposition is the assumption that all particles that reach the sediment bed are captured by the bed until some later time when they are entrained. This is true of course when the stress is zero. However, for a non-zero stress, the capture efficiency should be less than one and decrease as the stress increases. This is consistent with recent experimental results on relatively uniform-size particles (13). The results of the experiment indicated that β decreased by about a factor of 4 as the shear stress increased from 5 dynes/cm^2 (where almost no entrainment occurred) to 14 dynes/cm^2 (where considerable entrainment occurred). In this case, a characteristic decay time for deposition (proportional to $1/\beta$ and therefore a constant on the basis of the simplified theory) actually decreased with increasing stress.

For real sediments that consist of particles with widely varying sizes and settling speeds, the situation is somewhat different (12). For small stresses, almost all particles are deposited. The decay time is then determined primarily by those particles that have the longest decay times, i.e., the finest particles. For large stresses, only the heaviest particles settle out, and the lighter particles stay in suspension. The decay time is then determined by the settling speed of the heavy particles and is therefore smaller than the decay time for the smaller stresses.

WAVE ACTION AND CURRENTS

Wave action and currents are significant processes when considering sediment transport since (1) they are directly responsible for the transport of sediments; (2) they cause turbulence, which disperses sediments both vertically and horizontally; and (3) they cause a shear stress (resulting primarily from turbulence) at the sediment/water interface, which is the primary cause of the resuspension of sediments. Wave action is generally dominant in shallow water, whereas currents are more significant in deeper water. Considerable research is presently being conducted on the hydrodynamics of lakes and oceans, including numerical modeling. Because of this, only a very brief outline of the present state of the art will be given here, and other references should be consulted for more details (14).

Wave Action

At present, the most widely used procedures for predicting wave parameters are semi-empirical methods, such as the PNJ method developed by Pierson, Neumann, and James (15) and the SMB method developed by Sverdrup, Munk, and Bretschneider (16-18). These procedures give the significant wave height and significant wave period as a function of wind speed, fetch, and mean depth. From these relations and by assuming inviscid flow, one can determine the horizontal periodic flow at the sediment/water interface.

Near this interface, a turbulent boundary layer exists (on the order of 10 cm thick for typical surface waves) producing a shear stress at the interface. Kajiura (19) has analyzed this problem based on the assumption of a time-independent, but spatially varying, vertical eddy viscosity. His results have been partially verified by other experiments. Recent studies have attempted to extend the analysis to the case where waves and currents are present simultaneously (20). However, no substantial field studies have been done to verify these analyses. Instrumentation is now being developed so that measurements of bottom velocities and stresses are now feasible and are being made on the continental shelves and in the Great Lakes.

The SMB and PNJ methods are semi-empirical in nature, require a great deal of data for calibration, and are therefore difficult to use in regions or conditions for which they are not calibrated. Because of this, wave models are presently being developed more extensively based on the equations of motion, i.e., the energy balance or wave transport equation (21-23). The differences in the various models are primarily due to the structure of the source function appearing in this equation, a function that accounts for wind-wave interactions, wave-wave interactions, and wave decay. The models have been partially verified and show good results for many different cases. They are being actively developed, primarily for deep waters. They need to be extended so as to be valid for shallow waters.

Currents

Extensive work has been done and is continuing on the dynamics and especially the modeling of wind-driven currents in lakes and oceans (24-26). This is a very active area of research at the present time, and the models, although quite complex, are becoming more and more capable of describing reality. The most general of these models are three-dimensional and time-dependent and include the conservation equations for mass, momentum, and energy. These models have been extensively applied to both lakes and oceans when thermal stratification is not significant and, to a lesser extent, when thermal stratification is important.

Two areas of emphasis in improving these models are (1) to make them more realistic by a better description of turbulence, and (2) to make them more numerically efficient. In the description of turbulence, many models are incorporating additional equations that describe various turbulence quantities such as a turbulence velocity, a turbulence length, or a turbulence stress-tensor. A more recent emphasis is on the direct simulation of large-scale turbulent eddies with a low-level model for small-scale turbulence.

Major difficulties with three-dimensional, time-dependent models are that they are inherently complex, consume large amounts of computer time, and are difficult to program and use. Improvements are being made by developing and using more efficient algorithms. In many cases, simple models such as two-dimensional (vertically-integrated), time-dependent models may be especially useful. Although generally not as accurate as the full three-dimensional, time-dependent models, they are useful for the general understanding of phenomena and the determination of the effects of variations in pertinent parameters. In certain situations, the simple models may give completely adequate results.

CONCLUSIONS

A great deal of progress has recently been made in understanding and being able to quantitatively model the processes that govern the transport of sediments. However, additional research needs to be done before we can do this modeling adequately. Significant problem areas are as follows.

First is our limited knowledge of flocculation and its effects on particle sizes and settling speeds. The disaggregation of flocs because of shear is poorly understood despite its significance in high-shear regions such as the benthic boundary layer.

A second area is a realistic, quantitative description of entrainment and deposition for a range of conditions as found in aquatic systems. The physical processes governing entrainment and deposition are becoming relatively well understood. However, the effects of benthic organisms are not well quantified as yet. A more accurate and realistic mathematical model that incorporates the dominant processes needs to be constructed and verified.

Our knowledge of the hydrodynamic processes affecting sediment transport is reasonably good and is improving rapidly. Mathematical models of the hydrodynamics can be improved by (a) incorporating better descriptions of the turbulence, and (b) devising more efficient and accurate numerical algorithms.

In general, we need to do extensive numerical calculations so that we can begin to understand such problems as: the conditions under which there is net resuspension or deposition of sediments; the distances that sediments travel after resuspension and before the next deposition; and the importance of a major storm relative to the effects of many days of moderate winds. And finally, of course, we need extensive verification of this mathematical modeling and laboratory work by means of field studies.

Acknowledgments—This research was supported by the United States Environmental Protection Agency. Mr. Dave Dolan, Dr. John Paul and Dr. Louis S. Swaby were the project officers.

REFERENCES

1. **Smoluchowski, M.** 1917. Versuch einer Mathematischen Theorie der Koagulations-Kinetik Kolloid Losungen, *Zeitschrift für Physikalische Chemie* **92**:129-168.
2. **Ives, K. J.** 1978. Rate theories. In K. J. Ives, ed., *The Scientific Basis of Flocculation.* Sijthoff and Noordhoff International Publishers, B.V., Alphensan den Rijn, The Netherlands, pp. 37-61.
3. **Iacobellis, S.** 1984. The flocculation of fine-grained lake sediments subjected to a uniform shear stress. M.S. thesis. University of California, Santa Barbara, CA.
4. **Fisher, J. B., R. L. Petty and W. Lick.** 1983. Release of polychlorinated biphenyls from contaminated lake sediments: Flux and apparent diffusivities of four individual PCBs. *Environ. Pollut.* (Series B) **5**:121-132.
5. **Karickhoff, S.** Sorption dynamics of hydrophobic pollutants in sediment suspensions. This volume.

6. **Lick, W.** 1982. Entrainment deposition, and transport of fine-grained sediments in lakes. *Hydrobiologia* **91**:31-40.

7. **Lee, D. Y., W. Lick and S. W. Kang.** 1981. The entrainment and deposition of fine-grained sediments in Lake Erie. *Journal of Great Lakes Research* **7**:264-275.

8. **Krone, R. B.** 1976. Engineering interest in the benthic boundary layer. In I. N. McCave, ed., *The Benthic Boundary Layer*. Plenum Press, New York, pp. 143-156.

9. **Partheniades, E.** 1972. Results of recent investigations on erosion and deposition of cohesive sediments. In H. W. Shen, ed., *Sedimentation*. Ft. Collins, CO, pp. 20-1 to 20-39.

10. **Rhoads, D. C., J. Y. Yingst and W. J. Ullman.** 1978. Seafloor stability in central Long Island Sound. In M. Wiley, ed., *Estuarine Interactions*. Academic Press, New York, pp. 221-224.

11. **Rhoads, D. C. and L. F. Boyer.** 1982. The effects of marine benthos on physical properties of sediments: A successional perspective. In P. L. McCall and M. J. S. Tevesz, eds., *Animal-Sediment Relationships*. Plenum Press, New York, pp. 3-52.

12. **Lick, W. and S. W. Kang.** 1984. Entrainment of sediments and dredged materials in shallow lake water. *Journal of Great Lakes Research*, in press.

13. **Massion, G.** 1982. The resuspension of uniform-sized fine-grained sediments. M. S. thesis. University of California, Santa Barbara, CA.

14. **Csanady, G. T.** 1982. Circulation in the coastal ocean. *Environmental Fluid Mechanics Monographs*. D. Reidel Publishing Co., Boston.

15. **Pierson, W. J., G. Neumann and R. W. James.** 1955. Observing and forecasting ocean waves by means of wave spectra and statistics. U.S. Dept. Navy, Hydrography Office, Publ. No. 603, Washington, DC.

16. **Sverdrup, H.U. and W. H. Munk.** 1947. Wind, sea, and swell: Theory of relations for forecasting. U.S Dept. Navy, Hydrography Office, Washington, DC.

17. **Bretschneider, C. L.** 1958. Revisions in wave forecasting: Deep and shallow water. Proceedings of 6th Conf. ASCE Council on Wave Research, U.S. Army Coastal Engineering Research Center, Ft. Belvoir, VA, pp. 30-67.

18. **Coastal Engineering Research Center.** 1973. Shore Protection Manual, Vol. 1. U.S. Army Coastal Engineering Research Center, Ft. Belvoir, VA.

19. **Kajiura, K.** 1968. A model of the bottom boundary layer in water waves. *Bulletin of Earthquake Research International* **46**:75-123.

20. **Grant, W. D. and O. S. Madsen.** 1979. Combined wave and current interaction with a rough bottom. *J. Geophys. Res.* **84**:1797-1808.

21. **Barnett, T. P.** 1968. On the generation, dissipation and prediction of ocean waves. *J. Geophys. Res.* **73**:513-530.

22. **Hasselman, K.** 1978. On the spectral energy balance and numerical prediction of ocean waves. In A. Favre and K. Hasselman, eds., *Turbulent Fluxes Through the Sea Surface, Wave Dynamics, and Prediction*. Plenum Press, New York, pp. 531-596.

23. **Gunther, H., W. Rosenthal, T. J. Weare, B. A. Worthington, K. Hasselman and J. A. Ewing.** 1979. A hybrid parametrical wave prediction model. *J. Geophys. Res.* **84**:5727-5738.

24. **Cheng, R. T., J. M. Powell and J. M. Dillon.** 1976. Numerical models of wind-driven circulation in lakes. *Applied Mathematical Modelling* **1**:141-159.

25. **Lick, W.** 1976. Numerical modeling of lake currents. *Annual Review of Earth Planetary Science* **4**:49-74.

26. **Simons, T. J.** 1979. Hydrodynamic models of lakes and shallow seas. Report, Canadian Centre for Inland Waters, Burlington, Ontario.

Chapter 7

Pollutant Sorption: Relationship to Bioavailability

Samuel W. Karickhoff and Kenneth R. Morris

INTRODUCTION

The tendency of a pollutant to accumulate in an organism depends upon the relative affinity of the chemical for the biological medium versus affinity for other competing environmental compartments. Bioaccumulation also depends upon the rates of chemical transport/transformation processes linking environmental sources of the pollutant to ultimate chemical sinks within the organism. A plausible hypothesis is that affinity relates directly to thermodynamic potential, with chemical transport/transformation in the direction of lower chemical potential. This relationship prescribes that in the absence of constraints (kinetic, structural, or physiological), chemicals will move toward a state of equichemical potential of chemical in a sink (such as organism tissue) relative to the source. Note that fugacity or absolute activity can be used in lieu of chemical potential and can be substituted throughout this paper without loss of meaning or generality. (See references of Mackay (1, 2) for fugacity modeling.)

The chemical concentrations in an aquatic organism at a state of equichemical potential with the source defines *maximal* chemical uptake, from purely thermodynamic considerations. The ultimate model for bio-uptake incorporates thermodynamic driving forces into kinetic process modeling, but this requires highly specific characterization of the exposure environment, organism physiology and life habit, and pollutant reactivity (internal and external to the organism).

This paper focuses on the role of sediment and sorption processes in bioavailability of environmental contaminants. A conceptual framework is presented for evaluating the thermodynamic potential for uptake of organic chemical by aquatic organisms, with special emphasis on sediment sources. Also, the role of desorption kinetics (or sorbed chemical accessibility) is discussed for cases where significant chemical depletion of the sediment in the exposure environment may result from chemical uptake by the organism.

CONCEPTUAL FRAMEWORK FOR ESTIMATION OF BIOCONCENTRATION POTENTIAL

The distribution of chemical between sink and source at equichemical potential is independent of the actual pathway(s) of chemical exchange:

$$\frac{C^e_{sink}}{C^e_{source}} = D^e_{source,sink} \tag{1}$$

where C denotes mass-referenced pollutant concentration; D denotes distribution coefficient, and e denotes the state of equichemical potential.

This path independence facilitates the estimation of a chemical distribution by not requiring pathway input, which is necessarily highly system specific. If one wishes to specify a pathway for computational purposes, however, the choice can be based on computational facility in lieu of real-world "realities," which are again highly system specific. The estimation of bioconcentration potential translates to the estimation of $D^e_{source, organism}$ for the particular organisms and/or sources in question.

Pollutant sources and sinks are commonly not homogeneous phases but may be comprised of highly composited mixtures of components with wide variability in their sorbing capability for the chemical(s) in question. For example, the major components of sediment particles can be grouped into minerals, organic materials, and amorphous oxides. Each is a prospective sorbent for aquatic pollutants but differs significantly in its (1) affinity for the pollutant (in a thermodynamic sense), (2) availability for sorption, and (3) nature of the sorption process (binding mechanism, site specificity, etc.).

In addressing chemical uptake, it is advantageous to subdivide such materials into independent distribution volumes, which are regions or components perceived to have fairly uniform potential for pollutant uptake (i.e., access, affinity, etc.). For a sediment particle, the aforementioned component breakdown provides a logical distribution volume factoring; for aquatic organisms, an appropriate factoring would likely discriminate tissue types (lipid, muscle, etc.). Chemical concentrations in a pollutant source are expressed as weighted summations over all distribution volumes:

$$C_{source} = \sum_i X_i C_i \tag{2}$$

where C_i are chemical concentrations in individual distribution volumes, and X_i are the component mass fractions of each distribution volume active in sorption. Similarly, eqn (3) describes chemical concentrations in a sink:

$$C_{\text{sink}} = \sum_j X_j C_j \qquad (3)$$

The chemical distribution coefficient between source and sink is given as:

$$D_{\text{source,sink}} = \frac{C_{\text{sink}}}{C_{\text{source}}} = \frac{\sum\limits_j X_j C_j}{\sum\limits_i X_i C_i} \qquad (4)$$

which can be expressed as a properly weighted summation of chemical distribution coefficients of sink/source components.

$$D_{\text{source, sink}} = \frac{\sum\limits_j X_j}{\sum\limits_i X_i C_i/C_j} = \frac{\sum\limits_j X_j}{\sum\limits_i X_i D_{ji}} \qquad (5)$$

Furthermore, at equichemical potential, the component distribution coefficients, D_{ji}^e, can be expressed as a ratio of pollutant activity coefficients in each phase:

$$D_{ji}^e = \frac{C_i^e}{C_j^e} = \frac{\gamma_j}{\gamma_i} \qquad (6)$$

These activity coefficients (denoted γ) quantify pollutant affinity for a given phase.

Activity coefficients contrast adhesive (solute-solvent) and cohesive (solvent-solvent) interactions in a given phase with cohesive (solute-solute) interactions in an appropriate reference state. (For organic chemicals the pure liquid is an appropriate reference state.) For hydrophobic organic chemicals in water, one would expect highly nonideal behavior with variability in γ^ω from solute to solute comparable to differences in aqueous phase solubility. To the contrary, in an organic host (i.e., animal lipid, sediment organic carbon octanol), solute interactions with the host would be similar in type and magnitude for all chemicals with variations comparable to differences in solute molecular volume. In going from benzene to benzopyrene, for example, one would expect a variation in γ^ω of roughly four orders of magnitude compared to less, than an order, of magnitude for the organic hosts.

Although it appears that eqns (1) through (6) have expanded a simple distribution coefficient ($D^e_{source, sink}$) into a cumbersome series of abstract quantities, this expansion facilitates a priori estimation of ($D^e_{source, sink}$) by factoring (to a large degree) source, sink, and chemical dependencies. The distribution-volume-factoring for a given type of source or sink depends on the type of chemical in question (organic, inorganic, charged, uncharged, etc.); the weighting coefficients, X_i, X_j, however, are independent of chemical. Similarly, the preference factors contained in D^e_{ji} depend primarily on the chemical and component involved and are relatively independent of the component source. That is, an activity coefficient for pollutant in association with sediment organic carbon is largely independent of the sediment source. Likewise, an activity coefficient for a pollutant in animal lipid is independent of the individual animal. Therefore, source specificity exists largely in X_i, sink specificity in X_j, and chemical specificity in D^e_{ji}. Stated differently, source-to-source extrapolation derives from X_i, sink-to-sink from X_j; and chemical-to-chemical from D^e_{ji}.

What is frequently found in actual application is that this expanded framework shrinks considerably. Source and/or sink strength for pollutant release/uptake is usually localized in one or two distribution volumes. For example, in examining the potential for bio-uptake of hydrophobic organics from sediment particles, one can assume the dominant distribution volumes to be the organic matter component of the sediment source and the lipid component of the organism. In this case, eqn (4) is simplified as:

$$D_{sediment, organism} \approx \frac{X_l}{X_{oc}} D_{oc, l} \qquad (7)$$

where $D_{oc, l}$ is the distribution coefficient between animal lipid and sediment organic carbon. At a state of equichemical potential in the sediment and organism,

$$D^e_{oc, l} = \frac{\gamma_{oc}}{\gamma_l} \qquad (8)$$

The biological uptake of hydrophobic organic pollutants from sediment sources should depend, to a first approximation, on the organic carbon content of the sediment, the lipid content of the organism, and the relative affinities of the chemical for sediment organic carbon versus animal lipid.

Furthermore, for these pollutants, one would not expect large variations in either γ_{oc} or γ_l (at most 10-fold) from chemical-to-chemical. Besides, chemical-to-chemical variation in γ_{oc} should be highly correlated with changes in γ_l, and therefore, $D^e_{oc,l}$ should be highly chemical independent. Thus, one can postulate, the thermodynamic potential for hydrophobic organic pollutant uptake by an aquatic organism from a sediment source should be highly independent of chemical, depending only on the organic carbon content of the source and the lipid content of the organism (Figure 7.1):

$$D_{\text{sediment,organism}} \frac{X_l}{X_{oc}} D^e_{oc,l} \qquad (9)$$

The numerical value of the chemical preference factor between animal lipid and sediment organic carbon must be determined experimentally but should be on the order of unity. Initial tests of this approach (3) suggest this preference factor to be approximately 2.

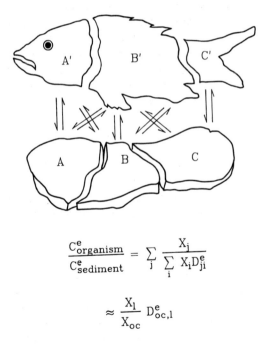

$$\frac{C^e_{\text{organism}}}{C^e_{\text{sediment}}} = \sum_j \frac{X_j}{\sum\limits_i X_i D^e_{ji}}$$

$$\approx \frac{X_l}{X_{oc}} D^e_{oc,l}$$

Fig. 7.1
Distribution of pollutant between sediments
and aquatic organisms at a state of
equichemical potential.

This postulated behavior is strikingly simplistic and needs further testing. Obvious refinements are possible, but one should be aware that the power of the postulate derives from its simplicity. Possible refinements include (1) using additional distribution volumes in either the sediment or organisms or (2) fine tuning the preference factors (d_{ji}^e) for the specific chemical, sediment, and organisms involved. Ultimately, chemical transport and physiological process kinetics can be introduced that are necessarily highly system specific (not to mention complex).

For purposes of environmental decision-making, points of advocacy for a thermodynamic approach for estimating bio-uptake potentially include the following:

- The approach can be implemented. Predictive equations can be easily formulated, calibrated, and tested.

- A conceptual framework can be developed that can be easily adapted to describe standard exposure test conditions, or extended to include kinetic considerations.

SORPTION KINETICS: RELATIONSHIP TO BIOAVAILABILITY

In cases where significant depletion of the sorbed pollutants on the sediment results from chemical uptake by an organism, accessibility of sorbed chemicals can limit bio-uptake. In most exposure environments involving sediment, this will probably not occur at the bulk system level (i.e., organisms typically constitute a minor sink for pollutants relative to sediments). Such may not be the case, however, at the microenvironment level (i.e., within organism burrows or in regions contiguous to organism membranes active in chemical uptake). In such cases, desorption kinetics may impact rates of chemical uptake by aquatic organisms. With short exposure (minutes to a few hours), a major portion of the total sorbed chemical may be inaccessible for bio-uptake, regardless of local thermodynamic gradients driving the overall process. There are no data presently available that address directly sorbed chemical accessibility for bio-uptake.

The sorption literature reveals a great deal of confusion regarding the rate (or extent) of chemical desorption with process descriptions ranging from "easily reversible" to "irreversible."

Recent studies (4-7 and DiToro, personal communication) suggest that uncharged sorbates are not irreversibly bound to sediments, but in cases where chemicals have incubated for extended periods in sediment, a significant portion of the total sorbed chemical is highly resistant to release. This nonlabile character appears to derive from physical inaccessibility; the sorbed chemical is buried within the sediment particle, perhaps in an aggregate structure or within a sorbent component such as organic matter.

The term aggregate in this context includes any of a wide variety of naturally occurring particle assemblages, from loosely agglomerated sediment particles or clay mineral tactoids to more highly structured micelles or layered lattice clay minerals (see Lick (8) for discussion of aggregation). Chemical release appears to be primarily mediated by intraparticle diffusion, although release may involve aggregate fracture. Fixed-geometry diffusion models (9) and a two-box linear model (4, 5) have been used to describe sorption kinetics. Fixed-geometry models more accurately describe the temporal sorption behavior where applicable, but they are highly restrictive in scope (i.e., range of system variable, such as time, solids concentration, etc. over which a given model is applicable).

Sediment particle size and structural variability (inherent in a given sediment or between sediments) seems to preclude the general applicability of any single fixed-geometry model. The two-box model is much more general, with no definition of particle geometry or sorbate release mechanism. The model simply discriminates a labile or rapid equilibrium sorption component (requiring at most a few hours to achieve) and a nonlabile or resistant component that may require days, weeks, or even months to achieve. Over a broad range of sediment and hydrophobic sorbates, the labile component accounted for from 1 to 60% of the total sorption. In general, the labile fraction varied inversely with the solids concentration and equilibrium sorption constant (4, 5).

At very low solids concentrations (50 mg/L or less), roughly half the total sorption was labile for all sorbates tested, but this component decreased with increased solids, becoming vanishingly small for highly sorbed chemicals. The rate constant for accessing the nonlabile component varied inversely with the equilibrium constant. For equilibrium constants of 10^2, 10^3, 10^4, 10^5, the corresponding times-to-equilibrium were roughly a day, week, month, year, respectively (5).

The potential relevance of release kinetics to bio-uptake kinetics is obvious. In regions contiguous to uptake membranes, the aqueous phase may be depleted of chemical, thus inducing desorption from sediment particles within the region. The maximal release that can be effected, however, is the labile component. In spite of this kinetic limitation, sediments can still enhance uptake rates substantially (relative to water alone) in many circumstances by increasing the total pollutant flux to uptake membranes.

A similar kinetic role of sediments applies to passive chemical excretion (i.e., across gill membranes, etc.) if the thermodynamic gradients are reversed. For example, the presence of uncontaminated sediments in a region contiguous to an excretion membrane can enhance excretion rates, but the full sorption potential is not realizable during brief contact periods. At equilibrium, the same thermodynamic equations developed previously for uptake apply.

REFERENCES

1. **Mackay, D.** 1979. Finding fugacity feasible. *Environ. Sci. Technol.* **13**:1218-1223.

2. **Mackay, D.** 1982. Fugacity revisited. *Environ. Sci. Technol.* **16**:654A-660A.

3. **Lake, J. L., N. Rubinstein and S. Pavignano.** Predicting bioaccumulation: Development of a simple partitioning model for use as a screening tool for regulating ocean disposal of wastes. This volume, pp. 151-166.

4. **Coates, J. and A. Elzerman.** 1984. Kinetics of desorption from sediment for selected PCB mixtures and individual congeners. *Abstracts: International Chemical Congress of Pacific Basin Societies,* Honolulu, HI.

5. **Karickhoff, S. and K. Morris.** 1985. Sorption dynamics of hydrophobic pollutants in sediment suspensions. *Environ. Toxicol. Chem.* (in press).

6. **DiToro, D. M., L. M. Horzempa, M. M. Casey and W. Richardson.** 1982. Reversible and resistant components of PCB adsorption-desorption:adsorbent concentration effects. *J. Great Lakes Res.* **8**:336-349.

7. **DiToro, D. M. and L. M. Horzempa.** 1982. Reversible and resistant components of PCB adsorption-desorption:isotherms. *Environ. Sci. Technol.* **16**:594-602.

8. **Lick, W.** The transport of sediments in aquatic systems. This volume, pp. 61-74.

9. **Gschwend, P. and S. Wu.** 1984. Effects of compound and sediment properties on sorption kinetics. *Abstracts: International Chemical Congress of Pacific Basin Societies,* Honolulu, HI.

Chapter 8

Factors Influencing the Sorption of Metals

Everett A. Jenne and John M. Zachara

INTRODUCTION

Three principal factors limit the quantitative description of toxic element mobility in natural waters and soil-water systems. First, scientists lack quantitative data to adequately describe the equilibrium and kinetic aspects of both adsorption and absorption onto earth materials. Second, thermodynamic data is insufficient to describe (a) complexation of the metals by natural dissolved organic compounds; (b) formation of selected aqueous species and solubility of certain solids; and (c) trace metal content of solids that are composed primarily of major elements (e.g., solid solution behavior). Finally, the kinetics of dissolution and precipitation reactions used in mass transfer calculations are inadequately known. This paper deals primarily with equilibrium and kinetic aspects of sorption.

The sorption to sediments by metals is of great importance to the health and well being of our populace because sorption is the primary line of defense between disposal sites and pathways to man. Intentional or inadvertent anthropogenic inputs of metals to our environment increase the concentration of metals in surface soils and surface waters. The sorption of these metals by soils and sediments is the most important and immediate means whereby their concentration in waters is reduced before contact and uptake by fish and other aquatic food sources. Sorption also reduces the concentration of these metals before they reach municipal water supplies and before contact with livestock or agricultural products via irrigation water. The absorption of these metals into the interior of aggregates or of accommodating primary minerals tends to maintain the sorbed metals in a state of decreased availability. That is, a significant portion of the sorbed metal is irreversibly sorbed within a time frame of two to several hours.

Research on fundamental sorption processes is urgently needed because knowledge of these processes is a major weakness in deterministic, error-term-containing mathematical models that are being developed to simulate the movement of metals through soils and porous aquifers and, ultimately, to calculate the dose to man. Virtually no predictive capability currently exists to anticipate the adsorption of metals from soil solutions and natural waters without incurring large (perhaps orders-of-magnitude) uncertainties. Current state of the art sorption modeling limits us to experimental parameterization for each soil or sediment and to its associated water quality, via laboratory sorption experiments, for each metal to be modeled. Laboratory estimation of sorption parameters may result in a major discrepancy when compared with field data [1]. This is an unfortunate state of affairs given the indicated health importance of sorption as a decontamination process. The lack of accurate data is all the more surprising in that extensive sorption studies have been carried out for five decades.

We believe that the lack of interpretative ability and capability to predict adsorption quantitatively stems from an inadequate basic understanding of the factors affecting sorption, particularly the nature of principal sorbing substrates, specific sorption sites, and the mechanisms of sorption. Investigations aimed at identifying the principal substrates and the nature of the sorption sites have been rather sporadic over the last two decades. Recent interest in this area is encouraging, and a review of the factors affecting metal sorption is timely.

DEFINITIONS

An effort is made herein to use descriptive terms as precisely as possible. *Sorption* is used as a general term, encompassing both adsorption and absorption. Precipitation reactions are specifically excluded from the definition of sorption. Homogeneous or heterogeneous precipitation in the course of sorption experiments is the result of inappropriate experimental design, and the inclusion of precipitation within the definition of sorption renders the latter term almost meaningless.

Adsorption is used to describe that portion of metal binding to particle surfaces, which is readily reversible. This is equivalent to saying that adsorption is that portion of the mass sorbed that is appropriately described by the surface complexation model of adsorption. A definition of reversibility should specify the time frames over which both adsorption and desorption occurs. Additional research is needed, however, before a defensible time frame can be established, particularly since desorption rates vary for different solids. Limited research suggests a time frame of 4 to 12 h as a useful measure of the quantity of reversible adsorbed metal (unpublished data of the authors).

Absorption is used herein to describe that portion of sorption that possesses a significant time dependency. It is presumed that the rate of absorption is limited by one or more diffusion barriers. The time dependency may result from diffusion into pores whose size may require the sorbing ions to diffuse through an electric field to reach the interior of aggregates. Solid state diffusion into the structure of the solid is also probable. This may be indicated by the replacement of structural ions from the adsorbent that is sometimes observed, ca. 50% in the case of Ag sorption by delta-MnO_2 (2). *Adsorbent* is the dissolved metal which is sorbed onto solid phase absorbents. *Absorbent* is the solid phase onto which the metal is sorbed.

Species refers to the actual form in which a molecule or its ion is present in solution. A minor exception is that waters of hydration are not specified inasmuch as they vary with ionic strength and are not accurately known.

Quantitative determination of the adsorption and absorption components of sorption can be accomplished by curve stripping if sorption data is obtained at sufficient and appropriate time intervals (unpublished data of the authors). Modeling the adsorption component via the surface coordination model may require that the total number of sites be determined by a chemical reaction or ion exchange techniques rather than by tritium adsorption ("exchange") because it seems likely that tritium will diffuse into small pores much more readily than charged metal ions. Thus, tritium adsorption may measure adsorption sites not only on external surfaces but also on the surfaces of pores in the interior of aggregates, which contribute to the time-dependent component of sorption.

The energetics of adsorption at microconcentrations ($< 1 \times 10^{-4}$ to 10^{-5} mol/L) are likely to be different than those at macroconcentrations because at macroconcentrations the metal is competing with the dominant cations for simple electrostatic sites. At microconcentrations, major competing cations have little effect on trace metal sorption because this sorption occurs at high energy sites with considerable specificity.

SORPTION SUBSTRATES

The sorption potential of a sediment for metals is primarily determined by the quantity of amorphic oxides of Fe and Mn and reactive particulate organic carbon present (3). Total organic carbon is an inappropriate parameter because of the presence of aluminosilicate or carbonate coatings which isolate part of the particulate organic carbon from the aqueous phase. Thus, "reactive" particulate organic carbon is that particulate organic carbon which has the opportunity to equilibrate with the aqueous phase. Amorphic oxides of Si and Al, as well as clay and zeolite minerals, may also contribute to the sorption potential of a sediment (4). Amorphic Al and Si oxides (e.g., amorphic aluminosilicates) are of particular importance for anionic forms of metals and metalloids. It has been suggested that the primary role of the clay minerals is as a substrate on which the amorphic oxides are precipitated (4). Although much evidence is suggestive of a prominent role of oxide coatings, the role of layer silicates as specific adsorbents remains undefined. Further, it is very difficult to quantitatively separate their primary adsorbent role from their role as a mechanical substrate on which the oxides and organic carbon is deposited. Selective extractions that utilize a reducing action are likely to reduce the valence of structural Fe and thus may change the sorptive properties of the clay. The removal of oxides and organic matter from clay mineral surfaces exposes sites not otherwise accessible to sorbing metals.

SORPTION SITES

Three aspects of sorption sites appear to be of consequence to a quantitative description of sorption. These are the (1) number, (2) selectivity, and (3) location.

Number

The number of surface coordination sites are determined by tritium adsorption (frequently called "tritium exchange") onto surface hydroxyls or via chemical reaction, including titration. Tritium exchange measurements were originally made by Berube et al. (5) and described in detail by Yates and Healy (6). Titration and chemical reaction techniques have been reviewed by James and Parks (7); these methods yield lower estimates of adsorption sites than does tritium exchange but are easier to perform (e.g., Kummert and Stumm (8)). Though numerous measurements of total surface adsorption sites have been made for metal oxides (e.g., (9)), few exist for clays or other heterogeneous earth materials. However, recent tritium exchange measurements on marine sediments (10, 11) indicate the presence of a significant number of surface coordination sites (ca. 2.5 mol/kg of sediment).

Selectively

For the most part, the major cations do not effectively compete for trace metal sorption sites since these sites possess considerable selectively as compared to the bulk of the constant charge sites on planar surfaces. The latter sites arise from isomorphous substitution in layer silicate structures. Variable charge sites occur at crystal edges due to the discontinuous nature of the structure. In the case of the layer silicates that expand along the c-axis upon solvation of the interlayer surfaces with water, ammonia, or certain organics, the constant charge sites constitute about 85% of the total sites located on both the interlayer surfaces and edges (12, p. 71). Layer silicates of the 1:1 type (e.g., kaolins) have very little isomorphic substitution.

Sorption sites are almost exclusively located at the crystal edge. These sites are described as being of variable charge since measured exchange capacity increases as the pH is increased.

The location of the higher energy sites associated with trace metal adsorption have not been definitely established; however, there is adequate evidence to suggest that the majority of the trace metals are adsorbed onto oxide (Fe, Mn, and Al) and organic particulate matter that adheres to the surface of layer silicates (4). This is consistent with the high selectivity of Mn oxides for trace metals (13) and the large surface area (14, 15), and more importantly, the number of adsorption sites associated with microcrystalline Fe oxides in soils and sediments as shown in Table 8.1.

Table 8.1   Selected Surface Properties

Solid/Electrolyte	N_2 Surface area m^2/g	Site density sites/nm^2	Reference
γ-Al$_2$O$_3$/NaCl	125	8.3	(16)
α-FeOOH/KNO$_3$	48	16.8	(17)
γ-FeOOH/NaNO$_3$	22	10	(18)
Fe$_2$O$_3\cdot$H$_2$O(am)/NaNO$_3$	182	9.8	(19, 18)
TiO$_2$/KNO$_3$	20	12	(17)
SiO$_2$/KCl	170	5	(20)

The electrochemical significance and contribution of exchangeable surface sites of the oxide coatings on layer silicates and other soil clays has only recently come under investigation (21, 22). Autoradiographs show that radioactive trace metals are sorbed at various points on the surface rather than in an approximately uniform manner. This distribution changes noticeably when reducing treatments are used to dissolve Fe and Mn oxides. However, a contradiction exists because extraction of expansible layer silicates with a reducing agent may not noticeably reduce the capacity of the layer silicate to sorb trace levels of a metal such as Co. The layer silicates have numerous structural defects, and it is possible that very selective, high energy sites occur at these sites.

Intrinsic adsorption constants determined for single phase Fe, Al, and Si oxides (23) indicate that adsorption intensity is comparable for Fe-OH sites on geothite and amorphous Fe oxyhydroxides and the Al-OH sites on α-Al$_2$O$_3$, which both exceed the Si-OH sites on silica (Table 8.2). Therefore, the relative importance of Fe-OH and Al-OH functionalities as adsorbents in earth materials is governed by their respective surface areas, which determines the total number of surface coordination sites. Techniques to estimate the quantity, binding intensity, and capacity of these constituents in sediments have recently been reviewed (24).

Table 8.2   Example of Intrinsic Binding Constants[a]
Calculated Using Triple Layer Model (9)

Oxide	$\log K^{int}_{Cd^{2+}}$	$\log K^{int}_{CdOH^+}$
TiO_2	7.2	10.1
α-Fe OOH	6.0	9.3
γ-Al_2O_3	5.9	9.7
$Fe_2O_3 \cdot H_2O$(am)	5.8	9.8
α-quartz	—	5.0

[a] Surface reactions

$$\overline{SO^-} + Cd^{2+} \xrightleftharpoons{K^{int}_{Cd^{2+}}} \overline{SO^- - Cd^{2+}}$$

$$\overline{SO^-} + CdOH^+ \xrightleftharpoons{K^{int}_{CdOH^+}} \overline{SO^- - CdOH^+}$$

The exact chemical species that are actually adsorbed are not known with certainty. Available models and experimental techniques are inadequate to determine whether the hydrolyzed species is itself adsorbed or if the uncomplexed cation is adsorbed and subsequently hydrolyzed. These are mechanistic processes that are thermodynamically inseparable. The net effect is the same: namely, the OH coordinated cation or a H-bonded oxyanion is the preferred surface species and is more strongly bound to the solid. Innovative applications of instrumental surface analysis techniques such as laser-induced Raman spectroscopy and internal reflectance spectroscopy hold promise in establishing the chemical nature of surface adsorbed species (e.g., Harrison and Berkheiser (25)).

Location

A significant fraction of trace metal sorption by "reference" clay minerals is generally not reversible at constant pH within a time equal to that in which the sorption occurred. This may be the result of (1) reactions that render the sorbed metal atom a part of the solid phase rather than its remaining at the solid-liquid interface; (2) diffusion into the pores of aggregates; or (3) a larger-than-average endothermic energy requirement. The adsorption surface may function as a template for heterogeneous precipitation. The transfer of solute species to a discrete solid phase may result from the presence of high absorbate concentrations (26). Precipitation on existing surface may occur in suspensions that are acid enough to dissolve Al from the structure of aluminosilicates and, for example, allow the precipitation of aluminum phosphates. This may maintain a lower level of dissolved Al than that supported by the initial layer silicate itself. Alternatively, precipitation on particle surfaces may develop a more-or-less contiguous coating on the oxide surface and additional solute retention may be controlled by diffusion through this coating, which then becomes the rate determining step for further uptake (27, 28).

A higher concentration gradient exists, at least initially, between the bulk solution and the pore waters within aggregates during sorption than desorption. Thus, desorption is expected to require a longer time period to reach apparent equilibrium than adsorption. Unpublished research of the authors suggests that both sorption and desorption of trace levels of metals onto iron oxides can be resolved into both a reversible surface-adsorption component and a diffusion-controlled component, which is linear on square root of time plots. Of course, the fact that one component of sorption is linearized with a square root of time function provides no evidence about the nature of sites at which sorption occurs. All or part of this component of sorption could be a reflection of sorption sites present on the surfaces of a pore system of high tortuosity where individual pores are sufficiently remote from the bulk solution that a diffusion gradient exists. Indeed, recent investigations (29) with synthetic Fe(III)-derived precipitates suggests the presence of narrow-necked, wide-bodied pores that induce significant hysteresis in N_2 adsorption-desorption isopleths. The presence of pores of similar geometry have not yet been demonstrated in natural cryptocrystalline precipitates. It would not be surprising if the diffusion-controlled component of sorption actually consists of both reversible surface adsorption sites on the surfaces of pores within aggregates as well as solid state diffusion into the bulk solid.

SOLUTION COMPOSITION

The composition of the solvent associated with an earth material (single phase solid, soil, or sediment) influences sorption in several ways. Dissolved constituents may complex metals ions, compete for the same adsorption sites, adsorb onto different sites altering the surface charge, and influence the valence of redox-sensitive metals.

pH

For surface coordination reactions of metallic cations and oxyanions on hydroxylated mineral surface, pH is the master variable. The strong pH dependency of adsorption reflects solution hydrolysis or protonation of the adsorbing ions, and more importantly, the surface charging properties of the adsorbent. Dissociation reactions are conceptualized by the following mass action relationships:

Surface charge is governed by the acidity of surface hydroxyl groups (SOH). The first and second equilibrium acidity constants may be calculated as follows:

$$\text{pH of ZPC} \sim (pK_{a_1}^{int} pK_{a_2}^{int})/2$$

where

$$K_{a_1}^{int} = \frac{(H_s^+)\,(\overline{SOH})}{(\overline{SOH_2^+})} \quad \text{and} \quad K_{a_2}^{int} = \frac{(H_s)\,(\overline{SO^-})}{(\overline{SOH})}$$

The zero point of charge (ZPC) is the pH where positive and negative charges are balanced and is determined by the intrinsic surface acidity constants $K_{a_i}^{int}$:: Surfaces are positively charged below the ZPC and negatively charged above the ZPC. The pH of the ZPC for some common hydroxylated oxide surfaces increases in the following order SiOH < MnOH < FeOH < AlOH. For most hydrolyzable metallic cations and protolyzable anions, adsorption increases from near zero to a high percent of the adsorbent initially present over a narrow pH range, defining an "adsorption edge." The pH at which 50% of a metal has been adsorbed varies with metal and its concentration and the absorbent concentration (e.g., solution:solid ratio).

Complexation

Foremost among the solution composition factors that significantly affect the extent of adsorption of microconcentrations of trace metals is complexation by dissolved ligands. Complexation may either increase or decrease sorption of a trace metal (30). Even though the exact chemical species involved in adsorption at the molecular level are not known with certainty, aqueous complexation is of major importance. In general, complexation with inorganic ligands other than the hydroxyl ion decreases adsorption. For example, Avotins (31) found that Hg could be systematically desorbed from amorphic iron hydroxide by adding Cl incrementally. Exceptions, however, may exist; adsorption of $CuCl^+$ complexes was postulated to explain increased Cu sorption on geothite in the presence of Cl^- (10^{-3} to 10^{-2} mol/L (32)). The outer sphere complexation of oxyanions by divalent and trivalent cations may change the overall charge of the complex from anionic to cationic, thus promoting adsorption. However, a larger effect on oxyanion behavior occurs with adsorption of cations, which increases the positive charge in inner-Helmholtz plane promoting anion adsorption. Complexation by organic ligands possessing more than one functional group and ambidentate inorganic ligands (e.g., SCN, S_2O_3) can increase adsorption of complexed metals because of the bonding of the other functional group to the surface of the solids (9, 33).

Competition

Competitive adsorption is the second most important way in which dissolved solutes affect the adsorption of trace metals. Inorganic constituents that mutually adsorb on the same sites will exhibit competitive effects when present simultaneously. The extent of this competition will depend upon the concentration of the competing ions and their relative adsorption intensities for the same sites. Limited data suggest that only specific combinations of trace metals compete for the same adsorption sites on amorphous iron oxyhydroxides (30). Similarly, the adsorption of anions, such as sulfate or carbonate, may decrease the adsorption of trace metals present as oxyanions through both competitive and electrostatic effects (34, unpublished data of the authors). This indicates that adsorption sites vary in their energetics and specificity. It is not clear whether these differences in site energies vary continuously or if site energies increase in discrete steps. Ions present in macroconcentrations (e.g., Ca, Mg, etc.) may compete with first transition series metals for surface coordination adsorption sites. Calcium in millimolar concentrations reduces Cd adsorption by amorphic $Fe_2O_3 \ H_2O$ (unpublished data of the authors), while Zn adsorption is unaffected by the presence of Ca (35). The failure of Ca to decrease Zn adsorption suggests that adsorption occurs at SOH_2 or SOH sites rather than at electrostatic sites arising from excess negative charge due to cation deficiency or lower valency substitution in the structure.

Redox

The redox status of a suspension may affect the valence and, hence, the adsorption strength of redox-sensitive elements. For example, Cr(IV) may be reduced to Cr(III) in sediments by reactions involving dissolved organic matter or, in some cases, S(-II). Chromium(III) is rapidly and strongly sorbed (36). In addition to the obvious effects in a homogeneous solution, complicated interactions (such as disproportionation) may occur at Mn oxide surfaces (37). These effects could cause Pb and Co to partition into different Mn oxide phases than Ni and Cu, because of the capability of Pb and Co to exist in the plus three oxidation state in Mn oxides (37).

Ionic Strength

Ionic strength influences sorption but, in general, its effects are relatively minor compared to pH, trace metal concentration, or complexation. The ionic strength, however, affects absorbate activities in solution, as well as surface charge and double-layer capacitance of the hydrated particles. Increasing ionic strength usually increases the aggregation of suspended sediment. Increased aggregation, in turn, is likely to increase the time-dependent portion of sorption, reflecting the necessity of solute diffusion into interaggregate pores. Concomitantly, both the exterior surface area and the rapidly reversible component of sorption decrease. Thus, depending upon the reaction time used, sorption could appear to decrease. An increase in solids concentration affects fractional adsorption (38). The solids concentration effect varies with the particular absorbent, absorbent combinations and the adsorbent. These effects may be due, at least in part, to the increased overlapping of electrical double layers. Additionally, adsorption of a finite amount of "inert" electrolyte constituents (e.g., Na^+, NO_3^-), particularly at concentrations $> 10^{-1}$ mole/L may alter the surface electrostatics of the adsorbent and decrease adsorption. This effect is particularly acute for weakly sorbing oxyanions.

ARTIFACTS OF EXPERIMENTAL METHODS

Numerous adsorption studies are flawed by inadequacies in their experimental methods, which render resulting data essentially unusable. Flawed experimental methods include incomplete phase separation, contamination, dehydration of the earth materials after sampling and before the sorption study, use of a temperature significantly different from the in situ temperature, and changed complexation of the trace metal caused by differences in pH and partial pressure of CO_2. Dehydration of the samples may result in decreased competition from other trace metals because of the slow redissolution of their precipitated forms.

Drying of Samples

Several problems are associated with using naturally wet earth materials in sorption studies. These problems are formidable when the material is anoxic and contains sulfides. Air drying may result in partial oxidation of metal sulfides with attendant increase in acidity and alteration of ionic composition of the containing water. Preservation of such samples requires immediate double bagging in plastic, inert gas flushing to remove oxygen, and freezing. Without freezing, it is virtually impossible to prevent oxidation of reduced Fe and Mn and subsequent formation of amorphous oxides. Oxidation may significantly alter the surface properties of potential adsorbents. An anoxic atmosphere (e.g., N_2, Ar) is a necessity for such studies. All approaches have limitations. Freezing and thawing may increase the exchangeable Mn as does drying (3). A further complication arises in the case of soils and sediments that contain significant amounts of allophane, an amorphous aluminosilicate. When these soils are air dried, weeks may be required for individual granules to rehydrate (38).

Two additional effects may result from air drying. Other trace metals, which may compete with the metal of interest in the natural environment, may precipitate and may not redissolve within the time frame of subsequent sorption experiments. Similarly, dissolved ligands (particularly organic compounds) that form complexes with the metal of interest may also precipitate or coagulate. As a result, the measured sorption may be greater than that in the natural environment through removal of natural competing ions and complexing ligands.

Phase Separation

Adequate phase separation is a problem commonly encountered in sorption studies with clays and earth materials containing readily dispersible clays. Centrifugation is notoriously inadequate when dispersible clays are present. Because trace metals are likely to be concentrated in solids by factors of 10^3 to 10^6, the presence of even a few solid particles in the aqueous phase may cause increased variability and can cause erroneous results [39], although only a fraction of the trace elements in the particulates are commonly desorbable [40]. The commonly used 0.45 μm filter membrane is generally not adequate for adsorption studies. Adequate phase separation generally requires filtration through 0.22 μm or finer membranes, with or without preceding centrifugation, to remove bacteria and earth materials from the aqueous phase.

Contamination

Contamination can arise from many sources. One of the authors once found that the process of "cleaning" plastic containers with 5% nitric acid resulted in sufficient Hg contamination to make the analyses of rainfall samples meaningless. Colored areas at the top of pH electrodes were found to be a major source of contamination in low-level metal analysis [41]. Even the tips of microliter pipet tips can introduce significant contamination [42]. Filter membranes may either release or adsorb trace metals [43] and may bleed significant levels of organic carbon. Release of metals from the membranes is rarely a problem unless acidified solutions are used. Acid leaching of cellulose-based membranes, as a cleaning procedure, is suspect because it has the possibility of forming H-saturated cation exchange sites on the membrane that could then remove the metal of interest during filtration. A more general technique is to preleach the membrane with a generous volume of distilled water followed by two aliquots of the solution to be filtered. This serves to leach soluble metals and organic compounds from the membrane and to bring the membrane surface into sorption equilibrium with the concentration of the solution being filtered.

Reaction Vessel Effects

The reaction vessels used for sorption experiments may themselves adsorb or release trace metals to the equilibrating suspension. Adsorption by the reaction containers is a frequent problem, the magnitude of which can be reduced by two approaches. An obvious first approach is to find reaction vessels that exhibit the minimum adsorption. This is partially a trial-and-error process because of the frequency with which such products go off the market. It is not generally realized that although plasticware frequently sorbs lesser amounts of trace metals than does glass, plasticware may also require more rigorous cleaning procedures to reduce their residual contamination to the same level as that of glass containers. Some glass is pretreated with Li solutions at elevated temperatures to minimize trace metal adsorption. The most effective approach is to use a given set of containers for a specified concentration range of the metal of interest. This concentration range is best restricted to plus or minus a factor of 5. Alternatively, the reaction vessels can be pre-equilibrated with the appropriate concentration before each experiment.

Microbial Effects

Microbial growth may occur during sorption studies and may lead to inconsistencies in data or erroneous observations. Perhaps the most common unrecognized adverse effect arises from the addition of bacteria present in the distilled water used for standards, blanks, and experimental solutions. Clumps of bacteria sorb metals and radionuclides present at trace concentrations with the result that the filtration may remove sizable amounts of the trace metal that is not sorbed to the sediment. Container adsorption may appear to be a more significant problem when bacterial clumps are present. Additionally, the sorption of trace metals by bacteria may greatly decrease the true concentration of the dissolved metal. Thus, a volatile element such as Hg may appear to be stabilized when in fact it has been largely removed from true solution through sorption by bacteria (44).

Oversaturated Solutions

Sorption studies have sometimes inadvertently been carried out using initial concentrations of the adsorbing metal element in excess of the solubility of its oxide, hydroxide, or carbonate (e.g., Corey (45)). This is a particular problem in the case of elements having very low solubility. For example, Rai et al. (46) have shown that recalculation of dissolved Am from several published distribution coefficients yielded values that were coherent with the solubility of amorphic Am hydroxide. In the case of elements that form compounds of low solubility, calculations should always be made to ensure that the solubility of known compounds of the element are not exceeded.

Temperature

Most sorption reactions of inorganic elements possess negative enthalpy (e.g., are endothermic) and are therefore influenced by the temperature of the medium. This is illustrated by the temperature dependency of molybdate sorption onto hermatite observed by Reyes and Jurinak (47). Temperature control should always be exercised and reported and may be systematically varied to assess certain thermodynamic properties of sorption reactions.

Solid-to-Solution Ratio

Finally, the ratio of solid-to-solution appears to influence sorption. The cause of this effect has not been exactly determined but may result from the interaction of sorption sites at loci where the electrical double layers of particles overlay. These effects may be related to coagulation processes. The important practical problem is the extent to which sorption studies performed in dilute sediment suspensions are applicable to bed sediments and soils wherein the solid-to-solution ratio is quite high.

SUMMARY

There is general agreement as to the major sorption substrates (sinks) in sediments, soils, and subsoil. The oxides of Fe, Mn, Al, and Si along with reactive particulate organic matter and clays provide the sorption sites for dissolved metals. Of these, Fe and Mn oxides, as well as organic matter and clays, are the more important substrates, although it is unclear whether the clays are of equivalent importance or if their primary role is as a surface to which oxides and organic matter adhere. Estimation of the quantity of these substrates, other than clays, requires the use of "selective" extractants.

Complexation, hence, competition among metals for the available ligands, can substantially influence the sorption of a metal. A computerized speciation model is required to calculate the activity of the dissolved species.

The presence of a significant time-dependent component of sorption (adsorption) suggests that diffusion-controlled phenomena may play an important role in metal retention over the long contact times that occur in natural systems.

Care is required in the course of sorption studies with inorganic constituents in trace concentrations to ensure reliable results. Preservation or storage techniques that allow a change in the chemical status of anoxic sediment should be avoided; drying is most significant in this respect. Efforts should be made to ensure that the phase separation is adequate and that contamination and reaction vessel effects are minor. It should also be determined whether bacteria are introducing variability into the experiments. The influence of experimental variables such as ionic strength and solid-to-solution ratio must be explicitly addressed.

Sorption measurements over a range of pH values used in conjunction with the surface coordination model are preferred for describing and quantifying sorption. This approach is rapidly becoming dominant for studies on single phase, well-characterized sorbents and is now being tested in studies of soils and sediments.

Acknowledgments—This work was performed for the U.S. Department of Energy under contract DE-AC06-76RLO 1830, with additional support from the Electric Power Research Institute, Palo Alto, California.

REFERENCES

1. **Coles, D. G. and L. D. Ramspott.** 1982. Migration of ruthenium-106 in a Nevada test site aquifer: Discrepancy between field and laboratory results. *Science* **215**:1235-1237.

2. **Anderson, B. J., E. A. Jenne and T. T. Chao.** 1973. The sorption of silver by poorly crystallized manganese oxides. *Geochim. Cosmochim. Acta* **37**:611-622.

3. **Jenne, E. A.** 1968. Controls on Mn, Fe, Co, Ni, Cu, and Zn concentrations in soils and water: The significant role of hydrous Mn and Fe oxides. *Adv. Chem.* **73**:337-387.

4. **Jenne, E. A.** 1977. Trace element sorption by sediments and soil—Sites and processes. In W. Chappell and K. Peterson, eds., *Symposium on Molybdenum in the Envionment*, Vol. 2. M. Dekker, Inc., New York, NY, pp. 415-553.

5. **Berube, Y. G., G. Y. Onoda and P. L. DeBruyn.** 1967. Proton adsorption at the ferric oxide aqueous solution interface. II. Analysis of kinetic data. *Surface Science* **8**:448-461.

6. **Yates, D. E. and T. W. Healy.** 1972. The structure of the silica/electrolyte interface. *J. Colloid Interface Sci.* **55**:9-19.

7. **James, R. O. and G. A. Parks.** 1982. Characterization of aqueous colloids by their electrical double layer and intrinsic surface chemical properties. *Surface and Colloid Sci.* **12**:119-216.

8. **Kummert, R. and W. Stumm.** 1981. The surface complexation of organic acids on hydrous gamma-Al_2O_3. *J. Colloid Interface Sci.* **75**:373-384.

9. **Davis, J. A. and J. O. Leckie.** 1978. Effect of adsorbed complexing ligands on trace metal uptake by hydrous oxides. *Environ. Sci. Technol.* **12**:1309-1315.

10. **Balistrieri, L. S. and J. W. Murray.** 1983. Metal-solid interactions in the marine environment: Estimating apparent equilibrium binding constants. *Geochim. Cosmochim. Acta* **47**:1091-1098.

11. **Balistrieri, L. S. and J. W. Murray.** 1984. Marine scavenging: Trace metal adsorption by interfacial sediment from Manop Site H. *Geochim. Cosmochim. Acta* **48**:921-929.

12. **James, D. W.** 1962. Sorption of NH_3 in a dry system in relation to the chemical properties of clay and soils. Ph.D. thesis, Dept. of Soils, Oregon State University, Corvallis, OR.

13. **McKenzie, R. M.** 1980. The adsorption of lead and other heavy metals on oxides of manganese and iron. *Aust. J. Soil Res.* **18**:61-73.

14. **Borggaard, O. K.** 1982. The influence of iron oxides on the surface area of soil. *J. Soil Sci.* **33**:443-449.

15. **Borggaard, O. K.** 1983. The effect of surface area of mineralogy of iron oxides on their surface charge and anion adsorption properties. *Clays Clay Min.* **31**:230-232.

16. **Huang, C. P. and W. Stumm.** 1973. Specific adsorption of cations on hydrous γ-Al_2O_3. *J. Colloid Interface Sci.* **43**:409.

17. **Yates, D. E.** 1975. The structure of oxide/aqueous electrolyte interface. Ph.D. thesis, University of Melbourne, Australia.

18. **Benjamin, M. M.** 1979. Effects of competing metals and complexing ligands on trace metal adsorption at the oxide/solution interface. Ph. D. thesis. Stanford University, Stanford, CA.

19. **Davis, J. S.** 1977. Adsorption of trace metals and complexing ligands at the oxide/water interface. Ph.D. thesis. Stanford University, Stanford, CA.

20. **Abendroth, R. P.** 1970. Surface and aqueous double layer properties for silicon dioxide. *J. Colloid Interface Sci.* **34**:591.

21. **Hendershot, W. H. and L. M. Lavkulich.** 1983. Effect of sequioxide coatings on surface charge of standard mineral and soil samples. *Soil Sci. Soc. Am. J.* **47**:1252-1260.

22. **Escudey, M. and G. Galindo.** 1983. Effect of iron oxide coatings on electrophoretic mobility and dispersion of allophane. *J. Colloid Interface Sci.* **93**:78-83.

23. **Davis, J. A. and J. O. Leckie.** 1978. Surface ionization and complexation at the oxide/water interface. II. Surface properties of amorphous iron oxyhydroxide and adsorption of metal ions. *J. Colloid Interface Sci.* **67**:90-107.

24. **Luoma, S. A. and J. A. Davis.** 1983. Requirements for modeling trace metal partitioning in oxidized estuarine sediments. *Mar. Chem.* **12**:159-181.

25. **Harrison, J. B. and V. E. Berkheiser.** 1982. Anion interactions with freshly prepared hydrous iron oxides. *Clays and Clay Minerals* **30**:97-102.

26. **Benjamin, M. M.** 1983. Adsorption and surface precipitation of metals on amorphous iron oxyhydroxide. *Environ. Sci. Technol.* **17**:686-692.

27. **van Riemsdijk, W. H., A. M.A. van der Linden and L. J. M. Boumans.** 1984. Phosphate sorption by soils: III. The P diffusion-precipitation model tested for three acid sandy soils. *Soil Sci. Soc. Am. J.* **48**:545-548.

28. **van Riemsdijk, W. H., L. J. M. Boumans and F. A. M. deHaan.** 1984. Phosphate sorption by soils: I. A model for phosphate reaction with metal oxides in soil. *Soil Sci. Soc. Am. Proc.* **48**:537-541.

29. **Crosby, S. A., D. R. Glasson, A. H. Cutler, I. Butler, D. R. Tuner, M. Whitfield and G. W. Millward.** 1983. Surface areas and porosities of Fe(III) and Fe(II) derived oxyhydroxides. *Environ. Sci. Technol.* **17**:709-713.

30. **Benjamin, M. M. and J. O. Leckie.** 1982. Effects of complexation by Cd, So_4, and S_2O_3 on adsorption behavior of Cd on oxide surfaces. *Environ. Sci. Technol.* **16**:162-170.

31. **Avotins, P.** 1975. Adsorption and coprecipitation studies of mercury on hydrous iron oxide. Ph.D. thesis. Stanford University, Stanford, CA.

32. **Bowden, J. W., A. M. Posner and J. P. Quirk.** 1977. Ionic adsorption on variable charge mineral surfaces: Theoretical charge development and titration curves. *Aust. J. Soil Res.* **15**:121-136.

33. **Bourg, A. C. and P. W. Schindler.** 1979. Effect of ethylenediamine-tetraacetic acid on the adsorption of copper(II) on amorphous silica. *Inorg. Nucl. Chem. Letters* **15**:225-229.

34. **Benjamin, M. M. and J. O. Leckie.** 1980. Multiple-site adsorption of Cd, Cu, Zn, and Pb on amorphous iron oxyhydroxide. *J. Coll. Interface Sci.* **79**:209-221.

35. **Dempsey, B. A. and P. C. Singer.** 1980. The effect of calcium on the adsorption of zinc by $MnO_x(s)$ and $Fe(OH)_3(am)$. In R. A. Baker, ed., *Contaminants and Sediments*, Vol. 2. Ann Arbor Science Publishers, Ann Arbor, MI.

36. **Mayer, L. M. and L. L. Schick.** 1981. Removal of hexavalent chromium from estuarine waters by model substrates and natural sediment. *Environ. Sci. Technol.* **15**:1482-1484.

37. **Hem, J. D.** 1978. Redox processes at surfaces of manganese oxide and their effects on aqueous metal ions. *Chemical Geology* **21**:199-218.

38. **Jenne, E. A.** 1960. Mineralogical, chemical, and fertility relationships of five Oregon coastal soils. Ph.D. thesis. Oregon State University, Corvallis, OR.

39. **Jenne, E. A., V. C. Kennedy, J. M. Burchard and J. W. Ball.** 1980. Sediment collection and processing for selective extractions and for total trace element analyses. In R. A. Baker, ed., *Contaminants and Sediments*, Vol. 2. Ann Arbor Science Publishers, Ann Arbor, MI, pp. 169-190.

40. **Gerritse, R. G. and W. Van Driel.** 1984. The relationship between adsorption of trace metals, organic matter, and pH in temperate soils. *J. Environ. Qual.* **13**:197-204.

41. **Ball, J. W. and E. A. Jenne.** 1976. Trace element contamination. II. Certain pH electrodes. *Bull. Environ. Contam. Toxicol.* **16**:767-769.

42. **Benjamin, M. M. and E. A. Jenne.** 1975. Trace element contamination. I. Copper from plastic microliter pipet tips. *Interface* **4**:25.

43. **Kennedy, V. C., E. A. Jenne and J. M. Burchard.** 1976. Backflushing filters for field processing of water samples prior to trace-element analyses. U.S. Geol. Survey Open-File Report 76-126, pp. 1-12.

44. **Avotins, P. and E. A. Jenne.** 1975. The time stability of dissolved mercury in water samples: II. Chemical stabilization. *J. Environ. Qual.* **4**:515-519.

45. **Corey, R. B.** 1981. Adsorption versus precipitation. In M. A. Anderson and A. J. Rubin, eds., *Adsorption of Inorganics at Solid Liquid Interfaces.* Ann Arbor Science Publishers, Ann Arbor, MI, pp. 161-182.

46. **Rai, D., R. G. Strickert, D. A. Moore and R. J. Serne.** 1981. Influence of an americium solid phase on americium concentrations in solutions. *Geochim. Cosmochim. Acta* **45**:2257-2265.

47. **Reyes, E. D. and J. J. Jurinak.** 1967. A mechanism of molybdate adsorption on alpha-Fe_2O_3. *Soil Sci. Soc. Am. Proc.* **31**:637-641.

Chapter 9

Factors to Consider in Conducting Laboratory Sorption/Desorption Tests

R. T. Podoll and W. R. Mabey

INTRODUCTION AND OVERVIEW

This paper examines several questions of fundamental importance to the use of laboratory sorption measurements in aquatic hazard assessments of organic pollutants:

- What laboratory sorption data are needed for an aquatic hazard assessment?

- How can these data be reliably measured?

- What uncertainties attend the use of sorption data in assessments?

In addressing the first question, we note that the concepts of partitioning, persistence, and toxicity of organic pollutants on sediments in the aquatic environment are intimately related. A chemical in an aquatic environment will partition among the water, sediment (and other particulates in the water-like biota), and the atmosphere. The chemical will persist in the aquatic environment if it remains dissolved or dispersed in the water, or if it sorbs on sediment. It will clearly not present a risk to aquatic life after it volatilizes from the water to the atmosphere, and its risk to aquatic life will differ depending on whether it is dissolved in water or sorbed on a sediment particle.

The partitioning of a chemical between sediment and water can be described by*

$$K_p = C_s/C_w \qquad (1)$$

where $K_p =$ sediment/water partition coefficient (mass of water per unit mass of sediment),

$C_s =$ amount of chemical sorbed per unit mass of sediment,

$C_w =$ amount of chemical in aqueous phase per unit mass of solution.

The fractional amount F of the chemical sorbed on a sediment at a loading S (mass of sediment per unit mass of water) is given by

$$F = \frac{K_p S}{1 + K_p S} \qquad (2)$$

Sorption will affect the persistence of a chemical in the aquatic environment by reducing the volatilization rate and probably by affecting the transformation rate of the chemical. The volatilization rate of a chemical is reduced by the factor $(1 + K_p S)^{-1}$ in the presence of sediment. Clearly, as the value of K_p and/or S increases, the volatilization rate will decrease. Sorption may also directly contribute to the degradation of chemicals by surface-associated chemical and microbial processes and thus affect the total concentration of chemical in the water body.

The bioavailability and related toxicity of chemicals sorbed on sediments is a subject of great interest to this conference. For example, only a fraction of the chemical sorbed on ingested sediment particles will partition into aquatic organisms. This fraction will depend on the rate of desorption of the chemical relative to the residence time of the particle in the organism. Additionally, benthic-dwelling organisms will have a greater exposure to the sediment-sorbed chemical than those organisms in the water column.

Thus, we are interested not only in equilibrium values of K_p, but also in the rate of sorption when the sediment and aqueous medium have not reached equilibrium with respect to partitioning of the chemical. Let us now consider the significance of laboratory K_p values to aquatic fate and hazard assessments.

First, K_p values are measured for specific chemicals sorbed on specific sediments. Sediments are heterogeneous materials with widely varying chemical compositions, structures, and particle size distributions; therefore, K_p values for a chemical may vary significantly on different sediments. To

---

* We note that this coefficient is operationally defined and may or may not apply for equilibrium conditions.

reduce this variability, it is useful to normalize measured K_p values with measurable sediment properties. For example, a number of researchers [1, 2, 3] have shown that equilibrium K_p values for a given hydrophobic solute sorbed on a series of sediments are proportional to the fraction organic content (f_{oc}) of the sediments. Thus,

$$K_{oc} = K_p/f_{oc} \qquad (3)$$

K_{oc} values are more generally useful because they are normalized for the amount of organic carbon on the sediment used to measure K_p and can be used to estimate K_p for other sediments with other values of f_{oc}.

Second, sorption mass transfer must be considered in using K_p values for fate and hazard assessment. In general, three mass transfer steps are considered in establishing sorption equilibrium: diffusion to the sediment, sorption on the external surface of the sediment, and movement to and sorption on the internal surface of the sediment. Diffusion of the chemical to the sediment depends on the turbulence of the water body and other mixing zone characteristics. In a well-mixed system of suspended sediment and low molecular weight (less than about 500) hydrophobic solute, diffusion to the sediment and sorption on the external surface of the sediment are very fast, with a combined half-life on the order of minutes. However, sorption on the internal surface of the sediment probably is kinetically controlled by diffusion within the pore structure of the sediment organic fraction [4, 5]. The half-life of the slow sorption process is about an order of magnitude longer than that for sorption on the external surface for a well mixed system—namely, on the order of several hours [5].

As mentioned above, the availability of sorbed chemicals for bio-uptake depends on the rate of desorption relative to exposure time. For a first approximation, it is reasonable to assume that bioavailability will depend on the fraction of chemical adsorbed on the accessible (or external) surface of the sediment and on the rate of desorption of the slow component. The relative amounts sorbed externally and internally and the rate of desorption during the slow process appear to depend on the magnitude of K_p for hydrophobic solutes [5]. As a rule of thumb, Karickhoff [5] suggests that the characteristic time (reciprocal of the pseudo first-order rate constant in minutes) for slow desorption approximates numerically the K_{oc} value. It would be very worthwhile to establish a relationship among K_p, adsorption/desorption kinetics, and bioavailability. Clearly, more work in this area is needed.

Third, it is important to note that sorption measurements require attention to a number of experimental variables and that consideration of slow and fast sorption components, etc., will be unimportant if care is not taken to obtain reliable K_p values. The next section of this paper considers important factors in measuring K_p values.

SORPTION MEASUREMENTS

The basic objective of sorption studies is to determine the change in solution concentration when a known volume of solution is equilibrated with a known amount of solids. The data are usually plotted as an adsorption isotherm. Criteria for evaluating sorption data are considered here in terms of sediment characterization, method of equilibration, and data analysis/quality assurance (QA).

Adsorption Isotherm

An equilibrium plot C_s versus C_w at constant temperature gives the adsorption isotherm. Isotherms commonly found for organic sorption on sediments are shown in Figure 9.1.

The linear isotherm shown in Figure 9.1a indicates that K_p is constant and equal to the slope of the isotherm. Linear adsorption isotherms are commonly found for low-molecular-weight organic solutes with low aqueous solubilities, at very low concentrations (less than one-half their aqueous solubility). The isotherms are linear because the solute/sediment interaction is nonspecific; that is, the hydrophobic solute does not interact strongly with specific surface sites on the sediment.

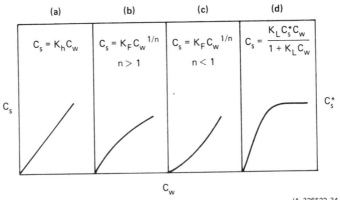

JA-326522-34

Fig. 9.1
Typical isotherms for the adsorption of organics on sediment
from dilute aqueous solution. K_F and K_h are Freundlich
constants. K_L is the Langmuir parameter and C_s^*
is the amount sorbed at monolayer capacity.

In Figure 9.1b, the slope of the isotherm decreases, and in Figure 9.1c, the isotherm slope increases with increasing concentration. These isotherms are typically fitted to the Freundlich equation

$$C_S = K_F\,C_W^{\,1/n} \tag{4}$$

and thus are defined by the Freundlich parameters K_F and n. Note that K_p increases with C_W for $n < 1$ and decreases with C_W for $n > 1$, and does not equal the slope of the isotherm. Nonlinear isotherms in the dilute concentration range indicate that the organic solute is adsorbing on specific sites on the sediment. Freundlich isotherms are common for very polar or ionizable organic solutes. The isotherm shape of Figure 9.1b is most common for these solutes and indicates that solute/sediment interaction-energies decrease for these solutes as high energy adsorption sites are filled.

The Langmuir isotherm shown in Figure 9.1d is characterized by an initial linear portion followed by a decrease in isotherm slope to a limiting plateau region of C_S. The Langmuir equation can be derived from a simple model of adsorbate molecules filling a finite number of energetically constant adsorption sites. The sorption isotherms of water-soluble polymers typically can be fitted to the Langmuir model.

Much attention has been given recently to the sorption of low molecular weight, hydrophobic solutes. Because of the nonspecificity of hydrophobic solute sorption, the adsorption isotherms are linear, and it has been possible to normalize the slope of the isotherm very simply with the organic carbon content of the sediment. The Freundlich and Langmuir isotherms, however, are both characterized by two sorption parameters, both of which may be expected to depend on sediment properties. Thus, it is inherently more difficult and uncertain to normalize the isotherms and thus predict the sorption behavior of organic solutes that exhibit nonlinear adsorption on sediments.

Sediment Characterization

A sufficient amount of sediment should be collected from a single site at a single time for all sorption measurements of interest. Careful attention should be given to storing the sediment, and in particular, the sediment should not be dried before use in a sorption experiment. Careful consideration should be given to minimizing differences (e.g., in particle size distribution) within a given set of sediment aliquots used to develop a sorption isotherm.

At a minimum, information on the particle size distribution and composition of the sediment should be known. For example, the sand fraction usually has a lower K_p value (at the same solution concentration) than the silt fraction of sediment because of the lower specific surface area and (at least for hydrophobic solutes) the lower organic carbon content of the sand fraction. Thus, adsorption should be measured with sediment samples of known and consistent particle size distributions.

The composition of the sediment should be characterized with consideration given to the types of interactions expected between sediment and solute functional groupings. For nonionic organic solutes with low aqueous solubilities, nonspecific, van der Waals' interactions of the solutes with the organic fraction of the sediment dominate sorption; therefore, the organic carbon content of the sediment should be measured. For organic solutes where specific, electrostatic interactions are expected with the sediment, the cation exchange capacity, pH, and other measures of charge distribution on the sediment should be characterized.

Because of the heterogeneity of sediments, sorption partition coefficients cannot be fully normalized with respect to a single sediment characteristic. For example, the sorption of hydrophobic solutes can be referenced to organic carbon content at the first level of approximation, but clearly K_p also depends on the particle size distribution of the suspended sediment.

Method of Equilibration

Adsorption methods can be categorized as batch or flow measurements depending on how sorption equilibrium is reached between the solid and the solution. In batch methods 6, a known mass m (g) of sediment is added to a known volume V (mL) of solution of known initial concentration C_w^o (μg/mL). The suspension is agitated, and after equilibrium is achieved, the sediment and solution are separated, usually by centrifugation. The equilibration concentration of the supernatant C_w^e is measured and the change in solution concentration ($\Delta C = C_w^o - C_w^e$) is attributed to sorption on the sediment. The specific amount sorbed at equilibrium is calculated from simple mass balance

$$C_s^e = V\Delta C/m \tag{5}$$

Measurement of C_s^e at several values of C_w^e gives the adsorption isotherm.

The uncertainty in ΔC is very high unless great care is taken to eliminate several potentially serious experimental problems. First, chemical losses from volatilization, adsorption on container walls, photochemical reactions, biodegradation, and other processes must be minimized or accounted for in blank runs without sediment present. For rigorous measurements, the amount of solute adsorbed on the sediment should be directly measured (e.g., by solvent elution analysis or scintillation counting of radio-labeled adsorbate), and this directly measured amount should be equal to the measured loss of chemical from solution.

Second, the sediment loading S and the initial solution concentration C_w^o must be carefully chosen to assure precision in the determination of ΔC and C_w^e. If the sediment loading is too low, ΔC will be small and subject to a large uncertainty. Conversely, if S is too high, the measurement of C_w^e will be imprecise. A useful rule of thumb is to estimate K_p and then calculate the value of S required to achieve a fractional amount F adsorbed equal to one-half (see eqn (2)).

It should be pointed out that several recent studies (7, 8) have indicated sharp increases in sorption at very low sediment concentrations (less than about 100 mg/L sediment to water), which may be due to disaggregation of sediment particles with a consequent increase in exposed sediment surface area. Whether this phenomenon is an experimental artifact or not is subject to continued debate (9, 10).

Third, sorption may take days to reach equilibrium, and thus care must be taken to ensure that a slow approach to equilibrium is not mistaken for an equilibrium state. Kinetic problems can become a problem for chemicals with very high values of K_p. If care is not taken during initial mixing, the sediment will be exposed to a nonuniform concentration of solute in solution. Consequently, sorbed amounts will be in excess of equilibrium values for part of the sediment. Thus, measured K_p values may represent a nonequilibrium state for a long time because desorption of strongly sorbed solutes may be very slow.

Finally, the sediment may contain elutable material, and thus the composition of the sediment surface may change during an experimental run and alter the sorption characteristics of the sediment. Thus, adsorption should be measured only after thoroughly eluting the sediment with water.

In the flow equilibration methods, solution of known concentration, C_o, is continuously passed through a column of adsorbent of mass m. The concentration of the effluent solution is monitored, and the amount sorbed is calculated by integrating the breakthrough curve as a function of time after accounting for the dead volume of the apparatus. If the curve is integrated up to the point where the effluent concentration equals C_o, the equilibrium amount adsorbed at the equilibrium solution concentration C_o is calculated. If solution flow is replaced by flow of pure water, desorption can be measured to check on adsorption/desorption hysteresis. Alternatively, flow of solution can be stopped, and the column can be thermostated at a different temperature. If flow of solution is restarted at the new temperature, the integrated area below the baseline curve at C_o is proportional to the incremental amount sorbed between the two temperatures of interest.

The advantages of the flow method are several. First, the measurements of the amount sorbed are inherently more precise. Sample handling is minimized and volatilization losses can be eliminated if the effluent is continuously monitored on-line. The solution concentration at equilibrium is known exactly, and the amount adsorbed is calculated by integrating the breakthrough curve between two known equilibrium concentrations. Second, adsorption/desorption hysteresis is simply checked. Third, the temperature dependence of sorption is easily measured. Fourth, if sorption is reversible, the same sediment sample can be used for all measurements. Fifth, the approach to sorption equilibrium is continuously monitored.

The flow method is limited, however, to the study of relatively high concentrations of chemical, and the advantage of reduced sample handling diminishes if volume fractions must be collected for analysis. Moreover, very slow sorption processes are difficult to continuously monitor and clogging of filters used to contain the sediment can be an experimental problem. Nevertheless, continuous flow methods are attractive alternatives to batch methods and should be explored more fully for investigating sorption on sediments.

Because of their simplicity, batch methods continue to be used almost exclusively to study sorption on sediments. For this reason, the last section of this paper will be limited to an examination of data analysis/QA guidelines for batch sorption methods only.

Data Analysis/QA

We have recently discussed the problems associated with undertaking a critical review of literature data with the goal of including the data in a quality environmental fate data base [11]. The criteria for evaluating K_p data for hydrophobic solutes obtained by the batch equilibrium method are given in Table 1. The criteria are organized in four major categories: (1) basis for measurement, (2) experimental information, (3) statistical information, and (4) corroborative information. We also developed a data quality indicator (DQI) system that provides a quality index for each of the four categories. Although a single DQI for each reported value (i.e., each K_p) may be desirable, we concluded that a four-DQI system was more practical because different types of judgements are required for each category and because experimental and statistical information in the literature are often inadequate.

The criteria listed in Table 9.1 should be considered an initial effort to evaluate and improve the quality of sorption data. It would be useful to arrange a workshop attended by researchers and others that develop and use sorption data so as to expand, define, and refine these criteria. One significant feature of our DQI system is that we have defined quality data as those that are accurate and fully qualified as to their use. We do not include uncertainty (or error) in the DQI evaluation, but rather consider random error as an independent measure of whether a value may be useful for a particular objective. This decision of whether or not to use the data is left to the researcher. Including criteria for a value's precision in a DQI may be misleading because a highly precise number is not useful if an invalid method is used.

Table 9.1   Checklist of Evaluation Criteria for K_p

Basis for Measurement	
Identification of method	Calculation of K_p from slope of C_s versus C_w in batch equilibration experiment
Limitation of method	None
Qualifications on value	Particular sediment character, temperature
Experimental Information	
Adherence to method	Low chemical concentration used
	Sufficient equilibration time allowed
	Sediment characterized (f_{oc}, size, etc.)
	Material balance for chemical performed[a]
Calibration/control	Chemical stability in water and sediment shown
	Experiments to verify equilibration achieved[a]
	Controls conducted for adventitious processes[a]
Characterization of sample	Standard identification procedure used[a]
	Source of chemical stated
Analytical method	Appropriate method used
	Precision of analysis of chemical in aqueous phase stated[a]
	Precision of analysis of chemical in sediment phase stated[a]
Statistical Information[a]	
Statistical analysis	Mean of replicate measurements reported
	Appropriate error analysis reported
Randomization/Replication	Randomization by varying aqueous volume, sediment loading, chemical concentration[a]
Corroborative Information	
	Independent measurement of K_p (same chemical and sediment)
	Independent measurement of K_{oc} on different sediment
	Calculation of K_p from f_{oc} and K_{oc} from $K_{oc} - K_{ow}$ correlation

[a] Should be further developed by experts.

Some comments on the particular evaluation of K_p data are appropriate. First, an equilibrium K_p value should be calculated from an adequately developed sorption isotherm. At low enough concentrations, the isotherm should be linear for solutes that bind nonspecifically with the sediment, and thus a single sorption parameter (K_p) describes the sorption isotherm. It is critical that the isotherm be developed from at least two points plus the origin for linear isotherms and many more points for nonlinear isotherms. It is recommended that a minimum of five points plus the origin be used to develop the isotherm.

Second, the solubility of the chemical should be known, and final equilibrium concentrations should be no more than about 75% (w/v) of the saturation concentration. If the solution is supersaturated, the batch depletion method will give erroneous results.

Third, it is highly desirable that the indirect calculation of the amount sorbed from the concentration change in the supernatant (after accounting for other losses with blank experiments) be confirmed with direct measurement of the amount sorbed on the sediment.

Finally, the comparability of sorption measurements would be greatly enhanced by the wider use of a set of standard, well-characterized sediment samples. The usefulness of K_p values is enhanced if several sediments with a wide range of properties, are studied.

Certainly, criteria need to be developed for other partition coefficient parameters (e.g., K_{oc}, partition coefficients for biota/water). The discussion and establishment of these criteria for K_p and the other partition coefficients not only will be useful for optimizing the correct use of available data, but also will identify critical issues for research or those specifically recognized as weaknesses in the use of data. One example of such an issue is the modeling of nonlinear sorption. Another example is the lack of knowledge about kinetics of the sorption/desorption process. The sorption/desorption of chemicals on sediment/biota may not be an instantaneous equilibrium process, and assuming that it is may lead to poor estimates of bioavailability.

REFERENCES

1. **Karickhoff, S. W., D. S. Brown and T. A. Scott.** 1979. Sorption of hydrophobic pollutants on natural sediments. *Water Res.* 13:241.

2. **Hassett, J. J., J. C. Means, W. L. Barnwart and S. G. Wood.** 1980. Sorption properties of sediments and energy related pollutants. EPA-600/3-80-041. U.S. Environmental Protection Agency, Washington, DC.

3. **Brown, D. D. and E. W. Flagg.** 1981. Empirical prediction of organic sorption in natural sediments. *J. Environ. Quality* 10:382.

4. **Freeman, D. H. and L. S. Cheung.** 1981. A gel partition model for organic desorption from a pond sediment. *Science* 214:790.

5. **Karickhoff, S. W.** 1980. Sorption kinetics of hydrophobic pollutants in natural sediments. In R. A. Baker, ed., *Contaminants and Sediments*, Vol. 2. Ann Arbor Science Publishers, Ann Arbor, MI, pp. 193-205.

6. **Mill, T., W. R. Mabey, D. C. Bomberger, T. W. Choy, D. G. Hendry and J. H. Smith.** 1982. Laboratory protocols for evaluating the fate of organic chemicals in air and water. Final Report. U.S. Environmental Protection Agency Contract No. 68-03-2227.

7. **DiToro, D. M. and L. M. Horzempa.** 1982. Reversible and resistant components of PCB adsorption-desorption isotherms. *Environ. Sci. Technol.* 16:594.

8. **O'Connor, D. J. and J. P. Connolly.** 1980. The effect of concentration of adsorbing solids on the partition coefficients. *Water Res.* 14:1517.

9. **Curl, R. L. and G. A. Keolelan.** 1984. Implicit-adsorbate model for apparent anomalies with organic adsorption on natural adsorbents. *Environ. Sci. Technol.* 18:916.

10. **Gschwen, P. M. and S. Wu.** 1985. On the constancy of sediment-water partition coefficients of hydrophobic organic pollutants. *Environ. Sci. Technol.* 19:90.

11. **Mabey, W. R., J. S. Winterle, R. T. Podoll, H. M. Jaber, D. Haynes, D. Tse, V. Barich, A. Liu and T. Mill.** 1984. Elements of a quality data base for environmental fate assessment. Final Report. U.S. Environmental Protection Agency Contract No. 68-03-2981.

Chapter 10

Assessing the Biodegradation of Sediment Associated Chemicals

P. H. Pritchard

INTRODUCTION

Monitoring studies and laboratory investigations have demonstrated that many pollutant chemicals become readily associated with sediments. Sediments are known to harbor communities of microorganisms with very diverse metabolic capabilities and consequently some of these chemicals will be biodegraded. Surprisingly little qualitative and quantitative information exists on the biodegradation of pollutants in sediments. As Bull (1) stated: "Most biodegradation studies have not, or only incidently have considered interface effects; the importance of surfaces, films and aggregations remains poorly understood." This lack of information complicates hazard assessment procedures, since the duration of exposure of sensitive plant and animal species to a chemical in sediments cannot be fully estimated.

Studying the biodegradation of xenobiotic chemicals in sediments is a fascinating and complex undertaking. For example, the degradation of chemicals in the sorbed state has only recently been addressed in natural sediments. The environmental factors affecting the biodegradation of chemicals in sediments are numerous and in some cases unique. The availability of the final electron acceptors (O_2, NO_3, SO_4 and CO_2) and the positioning of oxidation-reduction potentials (Eh), which in sediments are highly variable in time and space, will significantly affect bio-degradation rates. Processes such as diffusion, bioturbation and nutrient cycling will in turn control Eh gradients.

To adequately study these factors, laboratory systems need to be developed that will maintain the sediment as in natural a state as possible and at the same time allow quantitation of biodegradation processes. These systems should also permit examination of biochemical activities in the sediment bed with particular emphasis at the sediment-water interface. The assessment of biodegradation processes in sediments, therefore, will require some familiarity with the microbial ecology. The purpose of this paper is to summarize the relevant research which has been undertaken to detect, quantitate and characterize biodegradation processes in sediments and to relate this information, where appropriate, to hazard assessment problems.

Fate of Chemicals in Sediments

The fate of pesticides, herbicides and other chemicals in ponds, flooded soils and various laboratory sediment (soil)/water systems has been studied numerous times. However, distinguishing losses of parent compounds due to biodegradation from losses due to dilution, sorption, hydrolysis, diffusion and photolysis has not been attempted in many cases. Rapid losses of parent compound during these types of studies is often attributed to biodegradation in sediments, because losses due to other fate processes are considered unlikely. For example, Sikka and Rice (2) followed the fate of endothall in a farm pond and in laboratory microcosms. They attributed the observed decline of endothall residues in sediment to biodegradation processes therein, since it was unlikely, based on other information, that hydrolysis, photolysis or volatization would affect the fate of the endothall. Hexachloro-benzene, pentachloronitrobenzene and 4-chloroaniline, when applied to experimental ponds, were found to initially accumulate in sediments and then slowly decline (3), again suggesting possible biodegradation. Likewise, Caplan et al. (4) inferred that the fate of fenvalerate (a pyrethroid insecticide with high affinity for surfaces) in a tidal salt marsh microcosm was due to biodegradation in sediment, since they detected breakdown products in the sediments that were not typical of photolysis or abiotic degradation. A number of other fate studies have also resulted in similar deductions about the potential role of biodegradation in affecting the concentration of pollutants in sediments over time (5, 6, 7).

Other types of experiments have more directly indicated that microbial communities associated with sediments may be active in the biodegradation of pollutant chemicals. Graetz et al. (8) demonstrated that radiolabeled parathion rapidly degraded in active but not sterilized sediment-water slurries taken from a lake. Since the authors made no attempt to access biodegradation in water alone, the involvement of sediment-associated microorganisms in the biodegradation process can only be inferred. However, in general, biodegradation of parathion (and related organophosphates) in natural waters is relatively quite slow (9). Pentachlorophenol (PCP) degradation, as evidenced by rapid mineralization of radiolabeled PCP to CO_2 in sediment-water slurries taken from the Mississippi River/Gulf outlet, was observed by DeLaune et al. (10).

Mineralization was faster in sediments previously exposed to PCP, a result of a spill. Again, the extent of PCP degradation in water alone was not determined. Pignatello et al. (11) characterized PCP degradation in sediments by showing that PCP-mineralization capability of sediment-water slurries taken from a man-made stream channel continuously dosed with PCP dramatically increased at about the same time that PCP losses suddenly increased in the stream channel. In another study, toxaphene-contaminated sediments were shown to slowly lose their toxicity, presumably because of microbial activities (12). Several investigations have also detected toxaphene transformation products in sediments both under aerobic and anaerobic conditions (13, 14, 15). Pure cultures of bacteria that degraded several types of monohalogenated biphenyls (16) and the pesticide dichlobenil (17) have been isolated from sediments (although there is nothing unique about these isolations to claim that they are sediment inhabitants as opposed to water inhabitants). Thus, the microbial communities associated with sediments appear to have significant metabolic capabilities for a variety of chemical structures.

Sediment-Enhanced Biodegradation Rates

The importance of biodegradation processes in sediments is even more apparent if the quantitative aspects are examined. Several studies recently have revealed that biodegradation rates of pollutant chemicals in sediments exceed those observed in water (without sediment). This is not surprising, judging from the large numbers of studies showing more rapid turnover of simple natural organics (amino acids, glucose, organic acids) in sediments than in water (18, 19, 20). Lee and Ryan (21) were one of the first groups to carefully separate out the effects of sediments on biodegradation kinetics. Mineralization rates of radio-labeled xenobiotic chemicals were measured in estuarine water with and without natural sediments. A first order kinetic expression adequately fitted their mineralization data and allowed them to calculate half-lives of *p*-chlorophenol, trichlorophenol, chlorobenzene and trichlorophenoxy acetic acid (Table 10.1). All of the chemicals degraded slowly in water. The presence of sediment (50 g/l) decreased the half-lives by 7, 27, 2 and 41 times, respectively.

The results of these studies are significant for two reasons: (1) Several of the xenobiotic chemicals studied would be classified as persistent environmental pollutants if biodegradation information was based solely on the activities of natural microbial communities in water. Rapid degradation in the presence of sediments clearly changes this classification; (2) the chemicals tested by Lee and Ryan (21) cover a wide range of partition coefficients. This suggests that biodegradation processes in sediments may be effective with water-soluble chemicals (low affinity for surface adsorption) as they are for more insoluble chemicals (high affinity for surface adsorption). Thus, when

Table 10.1   Degradation of Radiolabeled Compounds in
Skidway River Water and Sediments[a]

Compound	Water temperature (°C)	Rate constant k (days^{-1})	Half-life $t_{1/2}$ (days)
Water			
p-Chlorophenol	21	0.035	20
2, 4, 5-Trichlorophenol	21	0.0010	690
Chlorobenzene	22	0.0045	150
1, 4, 5-Trichlorophenoxy-acetic acid	22	0.0005	1400
Water-Sediment Slurry			
p-Chlorophenol	22	0.23	3
2, 4, 5-Trichlorophenol	21	0.030	23
Chlorobenzene	21	0.0092	75
1, 4, 5-Trichlorophenoxy-acetic acid	21	0.012	35

[a] Radiolabeled compounds were added to 100 ml water samples to give a concentration of 25 μg/L. At the end of various incubation periods the $^{14}C_2$ produced was collected and counted as a measure of the degradation. The rate constant for each compound was calculated from data fitted to a first-order rate expression.

From Lee and Ryan (21)

the biodegradation of "sediment-associated" chemicals is discussed, the term "associated" should be considered as any mixing of the chemical with sediment, since it is this factor that may control biodegradation as much as the degree of sorption to sediment surfaces.

Several other investigations have shown faster degradation rates of xenobiotic chemicals in water when sediments are present. Baker et al. (22) reported sediment-enhanced degradation rates of chlorophenols in samples from a field stream. Biodegradation rates of several types of detergents (23, 24), polynuclear aromatic hydrocarbons (25), di-n-butyl-phthalate (26) and several pesticides (27, 28, 29, 30) have been shown to be stimulated by the addition of natural sediments. Spain et al. (30, 31) examined the biodegradation rates of p-nitrophenol in shake flasks containing sediment slurries and found the time to adaptation (commencement of biodegradation) to be consistently longer whenever sediments were excluded from the test systems. Exemplary decay curves for shake flask studies are shown in Figure 10.1. Commencement of p-nitrophenol degradation (adaptation) was also directly related to increases in p-nitrophenol-degrading bacteria. These degraders, although present in all of the original samples, appear to be activated more quickly by the presence of the sediment than in its absence.

Therefore, accumulating evidence supports the idea that once a chemical becomes associated with sediment by whatever mechanism, the chances for biodegradation are considerably enhanced.

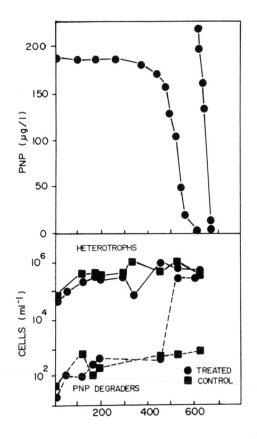

Fig. 10.1a
Degradation of paranitrophenol (PNP) and
changes in total heterotrophs and PNP
degraders in flasks containing water from a
freshwater pond. The flasks received an
additional treatment of PNP after 600 h.
All data points are means of samples from
triplicate flasks. Control flasks received no PNP.
From Spain et al. (31).

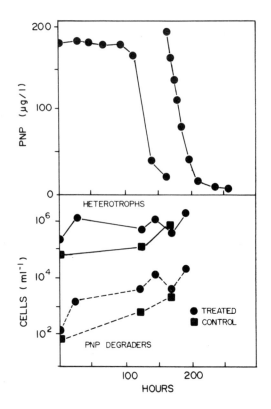

Fig 10.1b
Degradation of paranitrophenol and change in
total heterotrophs and PNP degraders in
sediment slurries (500 mg/L) from a freshwater
pond. The flasks received an additional treatment of paranitrophenol after 170 h. All data
points are means of samples from triplicate
flasks. Controls received no PNP. From
Spain et al. (31).

Potential Mechanisms for Sediment-Enhanced Biodegradation

The most likely mechanism to explain sediment-enhanced biodegradation processes is a microbial biomass effect. A biomass effect means that bacteria capable of degrading a chemical (hereafter referred to as degraders) are more numerous on the sediment surfaces than they are in the surrounding water. Adding sediments to a water sample is essentially the same as inoculating the water with a mass of microorganisms, some of which are specific degraders. Although conditions will vary considerably from one environment to another, in general the total biomass of bacteria in the top 0.2 to 0.3 cm of a sediment bed, in shallow estuarine areas, may be 100-10,000 times higher than the total bacteria in the water column above the sediment (32). This would mean a similar magnitude difference in the degrader populations (assuming they are a constant proportion of the total).

A study with *p*-nitrophenol (31) illustrates this potential biomass effect. This chemical normally requires sediment in seawater before significant degradation occurs, but if the incubation time of the chemical in seawater without sediments is long enough, biodegradation will also occur eventually. Isolation of pure cultures capable of utilizing *p*-nitrophenol as a sole source of carbon and energy showed that the same bacterium was the dominant member of the populations after degradation had commenced in systems both with and without sediment. This evidence tends to support the idea that the microbial communities in sediment and water are essentially similar in composition but there is more biomass (and, therefore, degraders) associated with the sediments.

Another mechanism for sediment-enhanced biodegradation may be that the sediment-associated bacteria which are capable of degrading a xenobiotic chemical, are, in some way, more active per cell than their counterparts suspended in the water. The possible concentration effect of sediment surfaces, whether for pollutant chemicals or for other organic or inorganic nutrients (which might promote bacterial growth or increase bacterial activity), is often suggested as an explanation for greater per cell activities. Relatively few attempts have been undertaken to experimentally verify this possibility. Novitsky (33), while demonstrating the presence of intense microbial activity at the sediment-water interface, hypothesized that the increased activity (heterotrophic activity measurements using tritiated glucose, glutamic acid and thymidine) was due to larger numbers of cells rather than to a larger percentage of active cells or a difference in per cell activity. However, Novitsky points out that some of his data are inconsistent with his own hypothesis; per cell activity with glutamate as a substrate, for example, was higher in sediments than in water. Likewise, Goulder (34) found maximum velocities (V_{max}) of glucose mineralization in water samples taken from an estuary with a normally high load of suspended particulates to be much higher for bacterial communities attached to suspended particulates than for free bacteria communities. Her results also indicated that, in general, V_{max} values per cell were approximately the same for attached and free bacteria. However, during the winter samplings, per cell activities were higher in the attached communities.

Higher per cell activities of bacteria on sediment, if they do in fact exist, could be the consequence of an increased availability, through sorption on the sediment surface of test chemical, to the degrader population. If this is true, it must first be shown that the sorbed chemical is available for degradation and that the bacteria on a surface are actually mediating degradation of the chemical. Several studies have examined the effect of sorptive surfaces on biodegradation processes.

A study by Li and Digiano (35) strongly suggests that sorption of a chemical may enhance its availability to bacteria and thereby increase rates of biodegradation. They followed the growth and biodegradation kinetics of a thin biofilm developed on three types of particles: sand, coal and granular activated carbon. The particles were incubated with secondary effluent from a sewage treatment plant and after attached cells were observed microscopically, the particles were washed to leave only tightly attached cells. The particles, with their attached biofilm, were then placed in a special reactor which allowed the utilization of substrate by the biofilm to be measured. A typical result for phenol utilization is shown in Figure 10.2. The exponential increase in phenol utilization rate suggests that unlimited growth of the biofilm was occurring at the end of the experiment; limitation by oxygen, substrate or hydrodynamic shear would give a constant utilization rate. Growth rate of the biofilm, as estimated from the substrate utilization rate, was shown by the authors to be independent of particle surface characteristics. Table 10.2 shows that growth rates with four different substrates were approximately equal for sand and coal biofilms but were considerably elevated for the activated carbon biofilm. This elevated rate was also related to particle size of the activated carbon, suggesting that the rate of resupply of the substrate to the biofilm by internal diffusion was an important rate-limiting factor. Thus the interaction between the microorganisms, the substrate and the particle surface can, in some situations, stimulate microbial activities due to biodegradation of sorbed chemical.

On the other hand, Gordon et al. (36), in a study with a periphytic marine bacterium, showed no effect of sorption on biodegradation rates of simple organic materials. They were able to regulate the attachment of the bacterium to hydroxyapatite particles by adjusting the number of bacteria per ml. Because hydroxyapatite was found to adsorb glutamic acid but not glucose from the experimental medium, experimental conditions that assessed the interaction between bacterial attachment and nutrient adsorption could be selected. Bacterial respiration activity, as measured by microcalorimetry and radiorespirometry, was not enhanced by the particles, regardless of whether the bacteria or the nutrients were sorbed to the sediment. In fact, a small

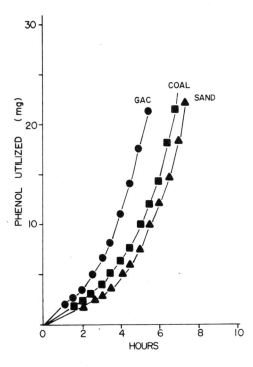

Fig. 10.2
The progression of phenol utilization with time
in an infinite batch-recycle reactor with
different solid media coated with thin biofilm.
GAC is granulated activated carbon. From
Li and Digiano (35).

Table 10.2 Calculated Specific Growth Rates (h^{-1}) Obtained from
Substrate Utilization Rates on Different Solid Media

		Solid Media[a]			
			Granular activated charcoal		
Substrate	Sand	Coal	26/25	35/45	50/60
O-cresol	0.134	0.135	0.250	0.290	0.295
Acetophenone	0.103	0.083	0.178	0.208	0.225
Phenol	0.253	0.278	0.323	0.345	0.396
Benzoic acid	0.188	0.204	0.247	0.280	0.340

[a] External surface area (cm^2) for particles is sand = 2,000, coal = 2,000,
GAC 20/25 = 614, GAC 35/40 = 1,029 and GAC 50/60 = 2,000.

From Li and Digiano (35).

decrease in respiratory activity was detected when the bacteria became attached. Their experimental design does not differentiate between a decrease due to a lower starting concentration and a decrease caused by slow desorption rates. The use of the substrate-cell-hydroxyapatite interaction (which is possibly ionic) as a model for the effects of sorption of the biodegradation of other hydrophobic chemicals (which sorb nonionically) is also uncertain. Further studies with other bacteria are needed before the idea of a surface-catalyzed effect is rejected. Also, the very rapid degradation of glucose and glutamic acid may be the key factor in developing conditions where biodegradation rates are limited by desorption rates.

Biodegradation studies in my laboratory have shown that surfaces can significantly affect the biodegradation rate of *p*-chlorophenol in seawater. *p*-chlorophenol is normally persistent in seawater samples taken from the Pensacola area unless natural sediments are present. As a consequence of this observation, experiments were designed to determine the tendency of sediment-associated bacteria to sluff off sediment particles and degrade chlorophenol in seawater in the presence and absence of surfaces. Seawater was incubated for 150 h in shake flasks under three conditions: (a) by itself, (b) with 500 mg/l sediment, (c) with 500 mg/l sediment and 200 mg/l *p*-chlorophenol. At the end of the incubation period (no *p*-chlorophenol remained in flask c), the contents of all flasks were filtered through 3.0 μm filters (to remove sediment but not all bacteria) and the filtered water was treated with *p*-chlorophenol. No degradation occurred in any of the treatments (Fig. 10.3). The addition of fresh sediment, of course, resulted in immediate degradation of the phenol. *p*-chlorophenol degraders on the sediment either are not released into the water or, if they are, they could not degrade the chemical in the seawater. Each flask was then treated with 1,000 mg/l sterile autoclaved sediment which had been thoroughly washed aseptically to remove dissolved organic and inorganic materials. Degradation was subsequently observed in all flasks. Thus, *p*-chlorophenol-degrading bacteria appeared to be in the water, but they are unable to degrade the phenol without the presence of sediment. The results are very interesting, because even bacterial communities normally associated with water (flask a) will degrade the phenol if sterile sediment is present. Preexposure of the water to active sediment (flask b) or active sediment and *p*-chlorophenol (flask c) apparently increased the number of water-borne *p*-chlorophenol degraders but degraders which require sediments for degradation. These results support the idea of chemical degradation by bacteria on sediment surfaces and they also demonstrate a possible interaction between the chemical, the bacteria and the sediment surface to enhance biodegradation. Preliminary studies indicate the interaction does not eliminate, by sorption of the chemical to the sediment surface, a toxic effect of the chlorophenol. Further studies are required before a mechanism of the sediment effect can be elucidated.

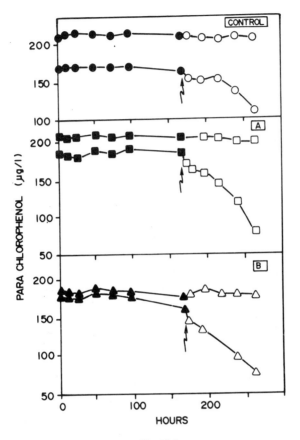

Fig. 10.3
Concentration of *p*-chlorophenol in replicate
flasks containing seawater previously exposed to
(a) nonsterile sediments (500 mg/L) and (b) nonsterile sediment (500 mg/L) plus *p*-chlorophenol
(200 g/L). Control water was not preexposed. At
approximately 170 h (arrow), sterile sediments
500 mg/L) were added to one flask of each
replicate and incubation continued.

Persistence of Chemicals in Sediments

Several studies have suggested that the association of pollutants with
sediment may, however, actually decrease chances for biodegradation. This
could be the result of several factors. The chemical structure of the
compound in question may preclude biodegradation, regardless of whether
the chemical is in water or in sediment. For example, studies with different
pesticides (6, 37) have shown that they accumulated in the sediments of a

pond dosed with the pesticide. The lack of any biotic transformation may have been the result of general recalcitrance of the chemical structure rather than its association with sediment. However, not enough data is available to ascertain the mechanism responsible for the persistence.

In other cases, the persistence of a chemical in sediment may be related to its unavailability to microorganisms. Chapman et al. (5) demonstrated that fenvalerate, a pyrethroid pesticide, degraded slower in organic soils than it did in mineral soils, suggesting that the sorption of this insecticide to organic matter possibly increased its persistence. Paraquat and diquat, dipyridinium herbicides which strongly sorb to sediments, are known to be biodegraded by soil microorganisms (38). A study using a simulated lake impoundment (39), however, showed that diquat accumulated and persisted in sediments. Sorption to the sediment surfaces through ionic interaction (a characteristic of this pesticide class) may have thus prevented biodegradation. Indeed, Weber and Coble (38) have shown that microbial decomposition of diquat was proportionally decreased as more and more clay mineral was added to the test systems. An example of their data is shown in Table 10.3.

Table 10.3a

Cumulative Release of $^{14}CO_2$ Resulting from Application of
Radiolabeled Diquat to Nutrient Solutions Containing
Montmorillonite (M) and Kaolinite (K) Clays

Diquat added μmoles)	Clay added (mg)	EVOLVED $^{14}CO_2$ cpm[a]				
		1 week	2 weeks	4 weeks	6 weeks	8 weeks
3.7	0	37,000	80,800	159,000	217,000	254,000
3.7	5 M	16,300	27,300	44,300	60,200	72,800
3.7	10 M	0	0	0	0	0
3.7	80 K	38,800	58,700	106,000	150,000	187,000
3.7	170 K	24,400	46,000	94,500	152,000	204,000

[a] Corrected for $^{14}CO_2$ derived from sterile control (< 200 cpm).

Table 10.3b

Tagged Diquat Remaining in Nutrient Solutions and in
Montmorillonite (M) and Kaolinite (K) Clay Pellets at
Termination of Experiment

Diquat added (μCi)	Clay added (mg)	C^{14}-Diquat present[a] (μCi)	
		in solution	in pellet[b]
10	0	8.4	0.4
10	5 M	5.1	4.5
10	10 M	0	9.7
10	80 K	8.1	1.0
10	170 K	5.4	3.6

[a] Corrected for background.
[b] Pellet contained microbial cells and clay.

From Weber and Coble (38).

Enough clay, in fact, could be added to completely stop degradation. Increasing sorptive surfaces by adding more natural sediment may not necessarily have the same effect as adding azooic clays, since more micro-organisms (and possibly biodegradative potential) would be added along with the sediments.

Subba-Rao and Alexander [40] reported a similar effect on the minerali-zation of benzylamine; increases in clay concentration generally decreased the percentrage of substrate mineralized. They suggest that mineralization rates were limited by the desorption rates of the benzylamine from the clay surface. Rates of methanearsenic acid demethylation to arsenic were shown by Holm et al. [41] to be inversely proportional to the concentration of the adsorbing species, such as clay minerals, and its adsorption capacity. Therefore, sorption to particulate surfaces would seem to reduce (lowered initial concentration) rather than stimulate degradation rates.

Finally, the persistence of some chemicals in natural sediments may be related to environmental factors which preclude rapid biodegradation. Anaerobic conditions, for example, could greatly reduce the rate of bio-degradation. The fate of di-2-ethylhexylphthalate (DEHP) in microcosms, designed by Perez et al. [42] to simulate a portion of Narragansett Bay (Rhode Island), may represent a situation of this type. They observed that DEHP readily biodegraded in the water column but accumulated to relatively high amounts in the sediments of the microcosms. Johnson and Lulves [43], on the other hand, have shown, using flask studies with sediment-water slurry samples from a local freshwater pond, that DEHP was readily degraded. The discrepancy here may be the result of insufficient aerobic conditions in the sediments of the microcosms, particularly as the phthalate diffuses into deeper depths of the sediment bed. Phthalates are mineralized under anaerobic conditions but at relatively slow rates.

Potential Mechanisms for Sediment-Depressed Biodegradation Rates

The persistence of chemicals in sediments has stimulated the notion that, in general, sorbed chemicals are not available for biodegradation. This has led Steen et al. [44] to develop a kinetic expression for the effect of sorption on biodegradation rates. Second-order biodegradation rate kinetics, rapid sorption relative to degradation and biodegradation of only dissolved chemical were assumed. The expression was:

$$\frac{dS}{dt} = \frac{k\,(S)\,(B)}{1 + K_p(\mathrm{p})} \tag{1}$$

where (S) = concentration of test chemical, (B) = total bacterial concentration in water and sediment (measured as ATP), k = second-order rate constant, K_p = equilibrium partition coefficient of test chemical with respect to sediment and p = the sediment-water mass ratio. An example of their data for the biodegradation of chloropropham and di-n-butyl phthalate are shown in Table 10.4.

Table 10.4

Comparison of Observed Second-order Microbial Rate Constants
(K_{obs}) vs. Second-order Constants (K_2) Affected by
Sediment Partitioning

Compound	Fraction sorbed (Hickory Hill Sediment)	K_{obs} (l org $^{-1}$ hr $^{-1}$	K_2 (l org $^{-1}$ hr $^{-1}$
Chlorpropham (CIPC)	0	$1.42 \pm 0.6 \times 10^{-13}$	—
	0.15	$4.15 \pm 0.7 \times 10^{-13}$	$3.45 \pm 0.2 \times 10^{-13}$
	0.90	$7.4 \pm 1.2 \times 10^{-15}$	$2.29 \pm 0.6 \times 10^{-13}$
Di-n-butyl phthalate	0	$3.1 \pm 0.8 \times 10^{-11}$	—
	0.15	$2.7 \pm 1.5 \times 10^{-11}$	$3.47 \pm 1.7 \times 10^{-11}$
	0.90	$6.1 \pm 0.5 \times 10^{-13}$	$2.60 \pm 0.21 \times 10^{-11}$

From Steen et al. (44).

Close agreement between second-order rate constants in the presence and absence of sediments led them to conclude that sorption of the test chemical to suspended sediments rendered the compound unavailable for degradation (no degradation in the sorbed state). Degradation rate of parent compounds also was observed to be inversely related to sediment concentration (Figure 10.4). ATP, which is a general measure of total microbial biomass, is used by Steen et al. (44) as a surrogate measure of the degrader populations (they assume degraders are always a constant fraction of the total population). Since ATP was presumably found in association with sediment, by definition, degraders should also be present on the sediment. If it is assumed, as Steen et al. suggest, that sorbed chemical is not degraded, the degraders on sediment apparently only metabolize dissolved substrate. This has been verified by the studies of Ogram et al. (45) who experimentally demonstrated that bacteria sorbed to particulate surfaces are as able to degrade a substrate as those bacteria free in solution. They measured the biodegradation kinetics (mineralization) of 2,4-dichlorophenoxy acetic acid (2,4-D) by a pure culture (capable of growing on 2,4-D as a sole source of carbon and energy) that was sorbed to the surface of montmorillinite clay. The kinetics are most readily interpreted by a mathematical model which considered sorbed chemical to be unavailable for degradation and sorbed-and solution-phase bacteria equally able to degrade solution-phase 2,4-D. The data did not fit models in which either of these two considerations were absent.

A similar type of simplified experimental system consisting of a pure culture of *Vibrio alginalyticus* and hydroxyapatite particles has been used by Gordon and Millero (46) to show that even simple organic acids when sorbed to the inorganic surface were mineralized slower than in the dissolved phase.

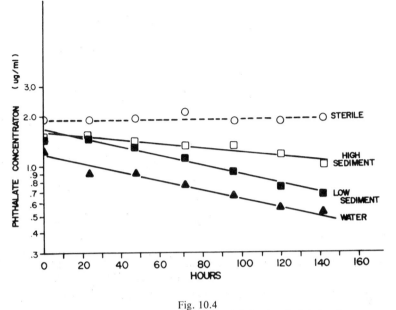

Fig. 10.4
Pseudo-order plot microbial degradation of di-n-butyl phthalate in
natural water and sediment-water systems. From Steen et al. (44).

They suggest that desorption of the substrate from the surface may be the rate-limiting step for the mineralization. Sorption of the bacteria to the surfaces did not affect their activities.

The application of results from these simplified studies to natural sediments is difficult because of the heterogeneous film of microorganisms and organic matter normally found on sediment surfaces and because of the examples where degradation seems to be enhanced. This is additionally complicated by the possibility that sorbed bacteria may behave physiologically in a different manner than free living bacteria (47). Therefore, experiments are needed to further characterize the effects of sorption on biodegradation processes carried out by microbial communities in natural sediments. Larson and Vashon (24), for example, have shown, using natural river sediments, that there was no relationship between the biodegradation rate of several dialkyl quaternary ammonia compounds and their affinity for sediment surfaces. The models of Steen (44) and Ogram et al. (45) would predict, in general, that a relationship should exist.

A method for assessing the effect of partitioning on biodegradation with natural sediments, assuming that sorbed chemical is not degraded and that the concentration of catalytic units does not change over the course of an experiment, would be to examine "initial" rates of degradation in presence of sediment, since partitioning will affect starting concentrations. For the purpose of discussion, dissolved and sorbed chemical are assumed to be at equilibrium at the beginning of an experiment. Initial rates, in this case, are operationally defined as a significant decrease in the concentration of parent compound over a period of a few hours. Degradation rates will be directly related to starting dissolved concentration of the test chemical if the rates are measured over a time period in which very little of the total substrate has been converted to product. Theoretically, in experiments dealing with partitioning of the chemical to sediment, initial rates will decrease as the fraction sorbed increases.

Many chemicals, however, will desorb from sediment at a rate considerably faster than biodegradation rates and there will be a span of time when degradation kinetics are essentially in steady state because test chemical concentrations will be maintained by desorption (over short periods) at approximately initial levels. Operationally this will appear as a lag in the disappearance of parent compound relative to tests without sediment. The use of high sediment concentration and/or chemicals with high partitioning will mean that most of the chemical, at equilibrium, will be in the sorbed state and starting concentrations in the water will be initially maintained without significantly affecting total substrate concentration. This is illustrated, hypothetically, in Figure 10.5a. Absence of this lag in the disappearance of parent compound under these conditions would mean that either desorption is not much faster than the biodegradation rate or that biodegradation of the sorbed chemical was occurring.

In addition, the relationship between initial degradation rates and sediment concentration, assuming active degrader populations on the sediment, will not be linear, unless K_p, the partition coefficient, is zero (i.e., degradation rates are directly related to the biomass associated with the sediment). This is shown in the hypothetical plot in Figure 10.5b (sorption equilibrium is assumed as a prerequisite). Chemicals with small partition coefficients will give plots that appear linear at low sediment concentrations (i.e., most of the chemical is in the dissolved state) but that deviate from linearity at higher sediment concentrations. As partition coefficient increases from one chemical to the next, the biomass effect will be dwarfed because almost all of the chemical will be initially sorbed. If biodegradation of the sorbed chemical occurs, the effect of partitioning will be less pronounced and may even result in a completely linear relationship between initial degradation rates and sediment concentration. The control situation in this case would have to be a theoretical calculation based on equilibrium partitioning to sterile sediments.

Thus a variety of additional experiments are needed to fully characterize the effects of sorption on biodegradation rates. At this time it is not clear whether the current assumptions of bioavailability will apply to natural

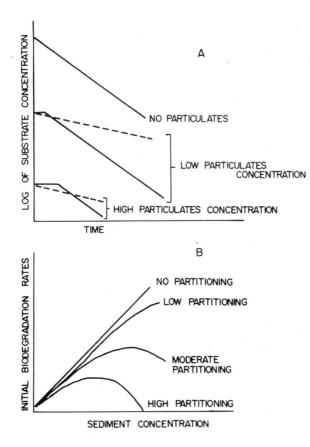

Fig. 10.5
Hypothetical curves representing the
effect of sorption and sediment concentration on biodegradation rates.

Fig. 10.5a
Effect of particulates concentration
assuming biodegradation of dissolved
chemical only and no microbial biomass
on particulate material. Solid lines;
desorption rates very rapid relative to
biodegradation rates. Dotted lines;
desorption rates approaching
biodegradation rates.

Fig. 10.5b
Effect of compounds with different
partition coefficients assuming biodegradation of dissolved chemical only
and microbial biomass on sediment is
capable of degrading dissolved chemical.

microbial communities on sediment surfaces and to a variety of compounds with differing chemical structures and sorption characteristics.

Assessing Biodegradation in Sediments

Biodegradation potentials, pathways and kinetics are generally determined in shake flasks using water-sediment suspensions. Sediments in nature are, of course, both in a suspended state and packed in a relatively static bed. In the latter state, a very definite ordering and structuring of physical and chemical properties occurs as a function of sediment depth. A set of parameters (such as pH, redox potential, temperature, nutrient status) for one layer of sediment may differ dramatically from an adjacent layer. Microbial activities in sediments are largely controlled by these parameters; numerous studies have characterized the quantitative and qualitative aspects of this control (see Nedwell and Brown (47)). The same control should likewise be exerted on the biodegradation of xenobiotic chemicals since pathways and mechanisms are largely borrowed for microbial transformation of naturally occurring organic materials. Information obtained from laboratory flask studies, however, typically relates an activity, or a degradation rate, only to a specific isolated parameter at a single point in space and time; thus it frequently fails to account for the interactive effect of these parameters and the temporal and spatial variability of these parameters in natural sediment beds. Application or extrapolation of this flask information, therefore, is complicated, but it must be attempted.

To date, little research has addressed this application or extrapolation problem. Exposure concentration estimations, particularly where a pollutant chemical is more rapidly biodegraded in sediments, must encompass methods which will allow extrapolation from simple test information to complex environmental situations. I believe that microcosm studies offer a very useful assessment method for this type of extrapolation. Many research investigations have convincingly shown that microcosms can reasonably simulate parts of nature (48) and, as such, may be useful in obtaining information on how fate processes operate in complex situations similar to those seen in the field. If we can delineate, through the use of laboratory flask information, the factors that control the fate of a chemical in a microcosm, then we are prepared to make predictions about fate in the field. As an example of this logic, we have compared the fate of fenthion (Baytex), an organophosphate insecticide, in laboratory shake flasks and microcosms containing estuarine sediments. Laboratory shake flask tests in our laboratory have shown that the presence of sediments greatly stimulates (half-life = 4 d) the biodegradation of fenthion in seawater (half-life = 22 d).

To assess the importance of this sediment effect in natural systems we simulated the sediment bed of a local saltmarsh in a microcosm. This was accomplished by taking intact sediment cores from the field and placing them directly in microcosm vessels. This method preserved the sediment-water interface and the physical integrity of the sediment bed. The vessels were kept at constant temperature in the laboratory and exposed to light regimes similar to that in the field. The water column was mixed with a glass rod attached to a constant rpm motor but so as not to resuspend sediment. A set of intact cores were sterilized with formalin. Fenthion at 200 µg/l was then spiked into the water column of each vessel and its disappearance rate monitored. After a specific incubation period, overlying water was removed and the sediment cores were frozen and fractionated into 0.5-cm layers. The concentrations of dissolved and sorbed fenthion in each layer were determined. A typical example of the concentration and distribution of fenthion in sterile and nonsterile microcosms is shown in Figure 10.6. Fenthion was lost from the water column more rapidly in active systems. Relative to the sterile microcosms, the pesticide was detected at much lower concentrations in the upper layers of sediment in the active microcosms but was found to a much greater sediment depth in active microcosms.

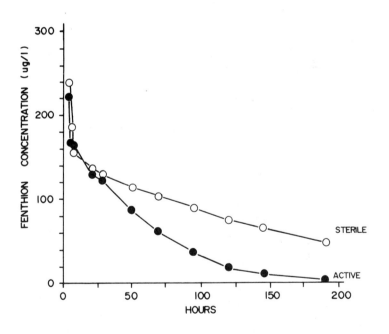

Fig. 10.6
Fate of fenthion in sediment-water microcosms.

Fig. 10.6a
Disappearance of fenthion in the water column
of sterile and nonsterile microcosms.

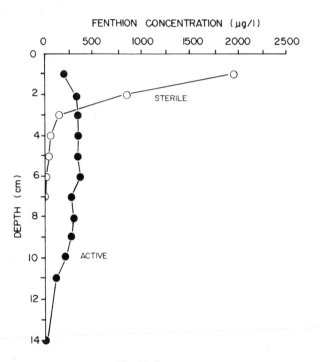

Fig. 10.6b
Distribution of fenthion in sterile and nonsterile
sediment after 190-h incubation.

The factors which affect the chemical's fate in these microcosms must be carefully considered. Kinetic information on each factor must be known if the effects due to biodegradation are to be sorted out. In the sterile systems, losses of fenthion from the water column could be the result of photolysis, hydrolysis, diffusion into sediment and sorption to sediment. Rates of abiotic degradation (neutral hydrolysis; pH independent in water and sediment) and photolysis were therefore separately determined in standard shake flask tests. Adsorption/desorption rates of fenthion to sediment were assumed to be instantaneous and an equilibrium partition coefficient was obtained in a batch sorption test. The partition coefficient was also assumed to be directly proportional to organic matter content in the sediment. Consequently, since the variation in organic content with depth in the sediment was known, the degree of partitioning in each sediment layer could be estimated. The amount of fenthion which rapidly adsorbs to particulate material at the sediment-water interface (i.e., independent of diffusion into sediment) was calculated from the total decrease of fenthion in the water column which occurred during the first 3 h of incubation (see Figure 6a). Lastly, relative diffusion rates for each sediment layer were estimated from the distribution of tritiated water in the sediment of the microcosms following (96 h) a single application of tritiated water to the water column. This procedure allowed the pore water mixing from turbulence, created by stirring of the water column, in the top sediment layers to be estimated. A preliminary summary of this data is given in Table 10.5.

Table 10.5   Preliminary Kinetic Information Used to Describe the Fate and Distribution of Fenthion in Sediment-Water Microcosms

Parameter[a]	Microcosms	
	Sterile	Nonsterile
Diffusion rates (cm^2/s $\times$ 10^{-5})	40[b]	140[b]
	2[c]	40[c]
	1[d]	8[d]
		1.4[e]
Partition coefficient (μg/kg (μg/L)$^{-1}$)	700	700
Initial fenthion concentration in sediment (μg/L)	707	1023
Photolysis rate constant (h^{-1})	0.0007	0.0013
Abiotic degradation rate constant,[f] water (h^{-1})	0.0001	0.0001
Abiotic degradation rate constant,[g] sediment (L/g h)	0.00006	0.00006
Biodegradation rate constant, water[h] (h^{-1})	0	0.0013
Biodegradation rate constant, sediment (L/g h)	0	0.00004

[a] Parameters derived from shake-flask tests using water and water-sediment slurries.
[b] Between water and the first sediment layer (0.5 cm).
[c] Between first sediment layer and the second sediment layer.
[d] Between seventh sediment layer and the eighth sediment layer.
[e] Between all remaining sediment layers.
[f] Disappearance rate of fenthion in seawater (pH 7.5) treated with 2% formalin.
[g] Disappearance rates of fenthion in seawater (pH 7.5) with different concentrations of sediment and treated with 2% formalin.
[h] First-order rate constant used because biomass was equal in flask and microcosm water.

All of this information was then used with a simple mass balance mathematical model to calculate the fate and distribution of fenthion in the sterile microcosm. Agreement between observed and calculated values was quite good if diffusion rates in the lower sediment layers were reduced to accommodate the difference in molecular size between fenthion and tritium and if rates in the top sediment layer were increased four-fold. We currently cannot explain the need for this elevated diffusion rate to accurately fit the observed data. This diffusion information was applied to the nonsterile microcosms.

In nonsterile microcosms, two additional processes affect the fate of the pesticide: biodegradation and bioturbation. Bioturbation was recognized because of considerably higher concentrations of tritium found in the sediment (following addition of tritiated water to the water column) as compared to the sterile microcosms. Diffusion rates in the model were adjusted in an attempt to accommodate the additional pore water mixing probably caused by bioturbation. This adjustment, however, was not sufficient to account for the high concentration of fenthion observed in the lower sediment layers of the nonsterile microcosm. Consequently, by assuming that bioturbation redistributed sediment particles containing sorbed fenthion, the model was run with several different diffusion rates in the sediment layers above 8 cm (bioturbation zone) until model calculations agreed approximately with the observed fenthion concentrations. Calculated and observed concentrations, under conditions in which biodegradation in the sediments was not considered, are shown in Figure 10.7 (curve A).

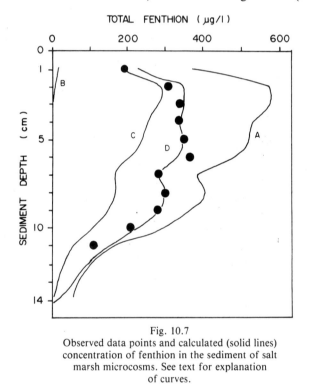

Fig. 10.7
Observed data points and calculated (solid lines)
concentration of fenthion in the sediment of salt
marsh microcosms. See text for explanation
of curves.

The effects of biodegradation were then evaluated. Second-order biodegradation rate constants (i.e., with respect to substrate concentration and sediment concentration) for water and sediment were determined from shake flask studies. Degradation in sediment was shown to be directly related to the concentration of organic detritus. Detritus was used as a surrogate measure of degrader biomass concentration in sediment; sand was assumed to be devoid of surface microorganisms which could degrade fenthion. Since the sand/detritus ratio varied with each sediment layer, biodegradation in the model was adjusted accordingly. Biodegradation of dissolved and sorbed chemical was considered in the model. The calculated distribution of fenthion in sediment using the sediment biodegradation rate constant derived from the shake flask studies is shown in Figure 10.7 (curve B). Obviously, the degradation rate is too high relative to what is actually occurring in the sediment bed. Arbitrarily reducing the rate constant 40-fold (i.e., suggesting that the kinetic information obtained from the flask was an artifact relative to more natural conditions) improves the model fit to the data (curve C). The best fit, however, is obtained when the degradation rate is adjusted to zero below the 8 cm depth in the sediment and graduated above 8 cm so that the rate is fastest in top layer and then decreases with depth (curve D). Specific depth-related environmental conditions in the intact sediment bed, such as redox potential, dissolved oxygen concentration and/or inorganic nutrient concentrations, could be responsible for the apparent range of degradation capabilities.

Assessing the biodegradation rates of xenobiotic chemicals in the natural sediments is, therefore, an understandably complicated process. The example with fenthion points out the level of detail required for a proper fate assessment. Although the microcosms appear to be an excellent method for carrying out these assessments, our work indicates the importance of a thorough knowledge of all relevant fate processes in these systems. Using a mathematical model to integrate these fate processes is an excellent mechanism for revealing where information is insufficient and for suggesting where new experiments are needed. Several other modeling studies, using soil systems (49), ponds (50, 51) and streams (52), have likewise shown the importance of the holistic approach and the extent of our ability to assess biodegradation process in intact sediment/soil systems. Bioturbation appears to greatly affect chemical distribution in sediment and, consequently, biodegradation. This aspect has been only minimally addressed in degradation studies in aquatic environments. Finally, studying sediments in shake flasks may give exaggerated estimates of biodegradation rates and the flask tests did not indicate the possible relative differences in rates of degradation with depth in the sediment. Additional work to quantitatively relate biodegradation kinetics in intact and suspended sediments tests is needed, since our knowledge in this area currently limits, significantly, our ability to assess biodegradation in natural aquatic systems.

The step of extrapolating microcosm data to the field may be a somewhat easier exercise than originally expected, since many of the subtle details of environmental fate in sediments can be elucidated with the use of microcosms and flask studies. Hydrodynamic factors typical of the field can be addressed with appropriate mathematical models and the representativeness of microcosms can be assessed through field calibration procedures (48). Thus, although field verification studies are definitely needed, interpretation of field study results will be greatly facilitated by the availability of the microcosm data.

Summary

Assessing the biodegradation of pollutant chemicals by microbial communities in natural sediments is a complex and exciting area of environmental fate research. The presence of a chemical in sediment, at least based on the available biodegradation literature information, is both good and bad depending on the chemical and perhaps on the design of the relevant experiments. In some cases, sorption to sediment surfaces appears to slow and perhaps stop degradation altogether, while in other cases the activity associated with microbial communities on sediment surfaces appears to be the only biological mechanism available, in some aquatic systems, for the transformation and decomposition of certain chemicals. Most likely, biodegradation in sediments is a unique balance between bioavailability, metabolic activity and certain environmental factors. Attempts to qualitatively and quantitatively understand this balance through laboratory investigations will require a unique combination of pure culture, flask, microcosm and field studies, but it is research that must be undertaken if we hope to properly estimate the environmental hazard of sediment associated pollutants. Future research, therefore, should concentrate on:

(a) Characterization of the biodegradation capabilities of microbial communities associated with sediments, particularly within the context that the degradation of pollutants will have to be interphased with the metabolic machinery that bacteria and fungi utilize to transform natural organic materials.

(b) Development of kinetic expressions for the biodegradation of chemicals to further account for sorption, microbial biomass (particularly the number and activity of specific degraders), and the regulation of metabolic pathways by environmental factors.

(c) Integrating, possibly through mathematical models and microcosm studies, biological information with the physiochemical characteristics of natural sediment beds particularly where factors such as tidal mixing, turbulent diffusion, bioturbation, oxidation-reduction potentials and sediment compaction and resuspension are important.

Obtaining the relevant information will not be easy, but considering that in many cases biodegradation may be the only process by which the concentration of pollutant chemicals can be reduced in sediments, the experimental pursuit is highly justified.

Acknowledgments—I thank Ms. Ellen O'Neill and Ms. Carol Monti for their technical assistance with the fenthion and chlorophenol experiments, respectively. The helpful discussions with Rick Cripe, Len Mueller and especially John Connolly are greatly appreciated. Don Ahearn, Paul Lefcourt, Rob Shrimp and John Rodgers are thanked for their review of this paper. The expert typing and editing of this manuscript by Valerie Caston is greatly appreciated and I wish her Happy Valentines Day! Portions of the research information presented in this paper were provided as part of cooperative agreement CR809370 between the Environmental Protection Agency, Gulf Breeze and Georgia State University.

REFERENCES

1. **Bull, A. T.** 1980. Biodegradation: Some attitudes and strategies of microorganisms and microbiologists. In D. C. Ellwood, et al., eds., *Contemporary Microbial Ecology.* Academic Press Inc., London, pp. 107-136.

2. **Sikka, H. C. and C. P. Rice.** 1973. Persistence of endothall in aquatic environment as determined by gas-liquid chromatography. *J. Agr. Food Chem.* **21**:842-846.

3. **Schaurte, W., J. P. Lay, W. Klein and F. Korte.** 1982. Long-term fate of organochlorine xenobiotics in aquatic ecosystems. *Ecotoxicol. Environ. Safety* **6**:560-569.

4. **Caplan, J. A., A. R. Isensee and J. O. Nelson.** 1984. Fate and effects of (^{14}C) fenvalerate in a tidal saltmarsh sediment ecosystem model. *J. Agr. Food Chem.* **32**:166-171.

5. **Chapman, R. A., C. M. Tu, C. R. Harris and C. Cole.** 1981. Persistence of five pyrethroid insecticides in sterile and natural, mineral and organic soil. *Bull. Environ. Contam. Toxicol.* **26**:513-519.

6. **Rawn, G. P., G. R. Webster and D. C. Muir.** 1982. Fate and permethrin in model outdoor ponds. *J. Environ. Sci. Hlth.* **B17**:463-486.

7. **Ward, C. T. and F. Matsumura.** 1978. Fate of 2, 3, 7, 8-tetrachlorodibenzo-*p*-dioxin (TCDD) in a model aquatic environment. *Arch. Environ. Contam. Toxicol.* **7**:349-352.

8. **Graetz, D. A., G. Chester, T. C. Daniel, L. W. Newland and G. B. Lee.** 1970. Parathion degradation in lake sediments. *J. Water Pollut. Control Fed.* **2**:R76-R94.

9. **Eichelberger, J. W. and J. J. Lichtenberg.** 1971. Persistence of pesticides in river water. *Environ. Sci. Technology* **5**:541-544.

10. **DeLaune, R. D., R. P. Grambrell and K. S. Reddy.** 1983. Fate of pentachlorophenol in estuarine sediment. *Environ. Pollut.* **6**:297.

11. **Pignatello, J. J., M. M. Martinson, J. G. Steiert, R. E. Carlson and R. L. Crawford.** 1983. Biodegradation and photolysis of pentachlorophenol in artificial freshwater streams. *Appl. Environ. Microbiol.* **46**:1024-1031.

12. **Lee, G. F., R. A. Hughes and G. D. Veith.** 1977. Evidence for partial degradation of toxaphene in the aquatic environment. *Water, Air, Soil Pollut.* **8**:479-484.

13. **Clark, J. M. and F. Matsumura.** 1979. Metabolism of toxaphene by aquatic sediments and a camphor-degrading pseudomonad. *Arch. Environ. Contam. Toxicol.* **8**:285-298.

14. **Smith, S. and G. H. Willis.** 1978. Disappearance of residual toxaphene in a Mississippi delta soil. *Soil Sci.* **126**:87-93.

15. **Williams, R. and T. Bidleman.** 1978. Toxaphene degradation in estuarine sediments. *J. Agr. Food Chem.* **26**:280-283.

16. **Kong, H. and G. S. Sayler.** 1983. Degradation and total mineralization of monohalogenated biphenyls in natural sediment and mixed bacterial culture. *Appl. Environ. Microbiol.* **46**:666.

17. **Miyzazki, S., H. C. Sikka and R. S. Lynch.** 1975. Metabolism of dichlobenil by microorganisms in the aquatic environment. *J. Agric. Food Chem.* **23**:365-368.

18. **Chocair, J. A. and L. J. Albright.** 1981. Heterotrophic activities of bacterioplankton and bacteriobenthos. *Can. J. Microbiol.* **27**:259-266.

19. **Griffiths, R. P., S. S. Hayasaka, T. M. McNamara and R. Y. Morita.** 1978. Relative microbial activity and bacterial concentrations in water and sediment samples taken in the Beaufort Sea. *Can. J. Microbiol.* **24**:1217-1226.

20. **Karl, D. M.** 1982. Microbial transformations of organic matter at oceanic interfaces: A review and prospectus. *EOS Revista Espanola de Entomolgia.* **63**:1073-1079.

21. **Lee, R. F. and C. Ryan.** 1979. Microbial degradation of organochlorine compounds in estuarine water and sediments. In A. W. Bourquin and P. H. Pritchard, eds., *Microbial Degradation of Pollutants in Marine Environments.* U.S. Environmental Protection Agency, EPA-600/9-79-012, Washington, D. C., pp. 443-450.

22. **Baker, M. D., C. I. Mayfield and W. E. Innis.** 1980. Degradation of chlorophenols in soil, sediment, and water at low temperature. *Water Res.* **14**:1765-1771.

23. **Larson, R. J.** 1984. Kinetic and ecological approaches for predicting biodegradation rates of xenobiotic organic chemicals in natural ecosystems. In M. J. Klug and C. A. Reddy, eds., *Current Perspectives in Microbial Ecology*. Am. Soc. for Microbiol., Washington, DC. pp. 677-686.

24. **Larson, R. J. and R. D. Vashon.** 1983. Adsorption and biodegradation of cationic surfactants in laboratory and environmental systems. *Dev. Ind. Microbiol.* **24**:425-434.

25. **Herbes, S. E.** 1981. Rate of microbial transformation of polycyclic aromatic hydrocarbons in water and sediments in the vicinity of a coal-coking waste water discharge. *Applic. Environ. Microbiol.* **41**:20-28.

26. **Walker, W. W., C. R. Cripe, P. H. Pritchard and A. W. Bourquin.** 1984. Biotic and abiotic degradation of dibutylphthalate in estuarine and freshwater environments. *Chemosphere* **13**:1283-1294.

27. **Fortmann, L. and A. Rosenberg.** 1984. Are solid surfaces of ecological significance to aquatic bacteria? *Chemosphere* **13**:53-65.

28. **Jones, T. W., W. M. Kemp, J. C. Stevenson and J. C. Means.** 1982. Degradation of atrazine in estuarine water/sediment sytems and soils. *J. Environ. Qual.* **11**:632-638.

29. **Simsiman, G. V. and G. Chesters.** 1975. Persistence of endothall in the aquatic environment. *Water, Air, Soil Pollut.* **4**:399-413.

30. **Van Veld, P. A. and J. C. Spain.** 1982. Degradation of selected xenobiotic compounds in three types of aquatic test sytems. *Chemosphere* **12**:86-97.

31. **Spain, J. C., P. Van Veld, C. A. Monti, P. H. Pritchard and C. R. Cripe.** 1984. Comparison of p-nitrophenol degradation in field and laboratory test systems. *Appl. Environ. Microbiol.* **48**:944-950.

32. **Novitsky, J. A.** 1983. Heterotrophic activity throughout a vertical profile of seawater and sediment in Halifax Harbor, Canada. *Appl. Environ. Microbiol.* **45**:1753-1760.

33. **Novitsky, J. A.** 1983. Microbial activity at the sediment-water interface in Halifax Harbor, Canada. *Appl. Environ. Microbiol.* **45**:1761-1766.

34. **Goulder, R.** 1977. Attached and free bacteria in an estuary with abundant suspended solids. *J. Appl. Bacteriol.* **43**:399-405.

35. **Li, A. Y. and F. A. DiGiano.** 1983. Availability of sorbed substrate on microbial degradation on granular activated carbon. *J. Water Pollut. Control Fed.* **55**:392-399.

36. **Gordon, A. S., S. M. Gerchakov and F. J. Millero.** 1983. Effects of inorganic particles on metabolism by a perphytic marine bacterium. *Appl. Environ. Microbiol.* **45**:411-417.

37. **Muir, D. C. G., N. P. Grift, A. P. Blouw and W. L. Lackhart.** 1980. Persistence of fluridone in small ponds. *J. Environ. Quality* **9**:151-156.

38. **Weber, J. B. and H. D. Coble.** 1968. Microbial decomposition of diquat adsorbed on montmorillonite and kaolinite clays. *J. Agr. Food Chem.* **16**:475-478.

39. **Simsiman, G. V. and G. Chesters.** 1976. Persistence of diquat in the aquatic environment. *Water Res.* **10**:105-112.

40. **Subba-Rao, R. V. and M. Alexander.** 1982. Effect of sorption on mineralization of low concentrations of aromatic compounds in lake water samples. *Appl. Environ. Microbiol.* **44**:659-668.

41. **Holm, T. R., M. A. Anderson, R. R. Stanforth and D. G. Iverson.** 1980. The influence of adsorption on the rates of microbial degradation of arsenic species in sediments. *Limnol. Oceanogr.* **25**:23-30.

42. **Perez, K. T., E. W. Davey, N. F. Lackie, G. E. Morrison, P. G. Murphy, A. E. Soper and D. L. Winslow.** 1985. Environmental assessment of a phthalate ester, di(2-ethyl hexyl) phthalate (DEHP), derived from a marine microcosm. In W. E. Bishop, R. D. Cardwell and B. B. Heidolph, eds., *Aquatic Toxicology and Hazard Assessment: Sixth Symposium*. ASTM STP 802. American Society for Testing and Materials, Philadelphia, pp. 180-191.

43. **Johnson, B. T. and W. Lulves.** 1975. Biodegradation of di-n-butylphthalate and di-2-ethylhexylphthalate in freshwater hydrosoil. *J. Fish. Res. Bd. Can.* **32**:333-339.

44. **Steen, W. C., D. F. Paris and G. L. Baughman.** 1980. Effects of sediment sorption on microbial degradation of toxic substances. In R. A. Barker, ed., *Contaminants and Sediments*. Ann Arbor Science Publishers, Ann Arbor, MI, pp. 477-482.

45. **Ogram, A. W., R. E. Jessup, L. T. Ou and P. S. C. Rao.** 1985. Effects of sorption on biological degradation rates of (2,4-dichlorophenoxy) acetic acid in soils. *Appl. Environ. Microbiol.* **49**:582-587.

46. **Gordon, A. S. and F. J. Millero.** 1985. Adsorption mediated decrease in biodegradation of organic compounds. *Microb. Ecol.* **11**:289-298.

47. **Fletcher, M. and K. C. Marshall.** 1982. Are solid surfaces of ecological significance to aquatic bacteria? *Adv. Microbiol. Ecol.* **6**:199-236.

48. **Pritchard, P. H. and A. W. Bourquin.** 1983. The use of microcosms for evaluation of interactions between pollutants and microorganisms. In C. C. Marchall, ed., *Advances in Microbial Ecology*, Volume 7. Plenum Press, New York, NY. pp. 133-215.

49. **Walker, A.** 1978. Simulation of the persistence of eight soil applied herbicides. *Weed Res.* **18**:305-310.

50. **Marshall, W. K. and J. R. Roberts.** 1971. Simulation modelling of the distribution of pesticides in ponds. *Scientific Criteria for Environmental Quality* (Canada) **2**:253-278.

51. **Rodgers, J. H., K. L. Dickson, F. Y. Saleh and C. A. Staples.** 1983. Use of microcosms to study transport, transformation and fate of organics in aquatic systems. *Environ. Toxicol. Chem.* **2**:155-167.

52. **Games, L. M.** 1982. Field validation of exposure analysis modeling system (EXAMS) in a flowing stream. In K. L. Dickson, A. W. Maki and J. Cairns Jr., eds., *Modeling the Fate of Chemicals in the Aquatic Environment*. Ann Arbor Science Publishers, Ann Arbor, MI, pp. 325-346.

Chapter 11

Synopsis of Discussion Session 2: Environmental Fate and Compartmentalization

D. DiToro (*Chairman*), F. Harrison, E. Jenne, S. Karickhoff and W. Lick

INTRODUCTION

In order to relate the mass discharge rate of a contaminant to its concentration in the water column and sediment, one must be able to predict the transport and chemical reactions of the contaminant along its path in the aquatic system. Pollutant fate in natural waters is highly dependent on sorptive behavior. Sorption can be involved directly, or indirectly, in pollutant degradation. The chemical reactivity of a pollutant in a sorbed state may differ significantly from that in solution, both in extent and chemical pathway. Moreover, natural sorbents may indirectly mediate solution-phase processes by altering the solution-phase pollutant concentration or by controlling pollutant release into the aqueous phase and thereby be the rate-limiting solution-phase reaction. In addition, natural sorbents "maintain" in solution a "buffered" suite of inorganic and organic species that can significantly affect pollutant reactivity in the aqueous phase.

An important key to describing pollutant fate in aquatic systems is an understanding of these sorptive processes. Many of the other reactions significant in the aquatic system have been discussed elsewhere (1, 2) and will not be discussed further here. However, a significant amount of work on the transport of particles has been done since the 1982 Pellston Conference—especially transport, entrainment and deposition, and aggregation of fine-grained sediment. This work has been reviewed in the present Conference (3).

TRANSPORT

In order to predict the fate and effects of pollutants discharged into an aquatic system, one must first be able to predict the transport and dispersal of the contaminant in order that the concentrations of the contaminant are known spatially and temporally. This is a first-order problem. At the same time, it is quantitatively not well understood because it involves many different processes with differing lengths and time scales. A model capable of accurately predicting the transport of particles in an aquatic system must include adequate information on at least the processes discussed below.

Wave Action and Currents

Wave action and currents are significant processes when considering sediment transport since (1) they are directly responsible for the transport of sediments; (2) they cause turbulence, which disperses sediments both vertically and horizontally; and (3) they cause a shear stress, resulting primarily from turbulence at the sediment-water interface, which is the primary cause of the resuspension of the sediments. Wave action is generally dominant in shallow water, whereas currents are more significant in deeper water.

Semi-empirical methods for predicting wave parameters are available although they may need further development for shallow water. More realistic wave models are presently being developed. Extensive work on numerical models of currents in oceans and lakes has been done. The most complete are three-dimensional and time-dependent. For many situations, two-dimensional (vertically-integrated), time-dependent models may be sufficient. The major need is for application and verification of these models.

Entrainment and Deposition

The processes that govern entrainment and deposition of sediments and the parameters on which they depend must be thoroughly understood. For entrainment, the most important of these parameters are (1) the turbulent stress at the sediment/water interface; (2) water content of the deposited sediment (time after deposition); (3) the composition of the deposited sediments including particle-size distribution, mineralogy, and organic content; (4) vertical distribution of sediment properties, i.e., manner of deposition of sediments; and (5) the activity of benthic organisms and bacteria.

The processes governing entrainment and deposition are becoming relatively well understood. A more accurate and realistic mathematical model of the sediment/water interaction that includes the dominant processes needs to be constructed and verified.

Aggregation and Disaggregation of Particles

A knowledge of particle sizes, settling speeds, and the effects of aggregation on these parameters is important in determining settling rates, deposition rates at the sediment/water interface, and the release of contaminants from aggregated particles to the surrounding waters. The aggregation of particles is a dynamic process with the state of aggregation at any particular location and time depending on the rates of aggregation and disaggregation, which in turn depend on state variables such as sediment concentration and shear. The processes that govern the rates of aggregation are becoming better understood, but the effect of shear stress and mass concentration on disaggregation of particles is not well understood and needs further investigation.

A completely general and accurate transport model would include all of the above processes. A calculation, although possible, would be incredibly complex and tedious. Because of the complexity of the transport problem, no general simplifying approximations are possible. Each problem must be considered separately and simplifications made accordingly. A few examples are given here.

In the very simplest approximation, perhaps it is sufficient to know whether the area of concern is an erosional or depositional site. This would depend on the range of particle sizes in the discharge, but even so, this could be determined readily either by observation, or, if this is not possible, by a combination of a wave action model and laboratory results on the entrainment of sediments.

A more complicated case might be discharge of a relatively small amount of pollutant near shore in shallow waters. A simple model of transport might be that the discharged sediment is always in suspension when the shear stress (caused by wave action) is greater than a critical shear stress and otherwise is stationary on the bottom. Aggregation-disaggregation is not significant. A simple model or observation of currents would be adequate.

The deep ocean dumping of contaminated material is a more complicated problem. The processes of aggregation and disaggregation will be important during the initial settling and subsequent resuspension of the material. A knowledge of entrainment rates is needed since currents will entrain the finer material. Wave action is probably not significant because of the depth. Currents will be significant in causing resuspension. They will tend to be more uniform than in shallow waters of lakes, but nevertheless, will have to be determined by some combination of observation and modeling.

The prediction of the transport of contaminants in a large shallow lake such as Lake Erie is one of the more difficult problems. In this case, all of the processes mentioned above (currents, wave action, entrainment and deposition, and aggregation-disaggregation) are significant and highly variable. To obtain a satisfactory definition of transport, one must incorporate all of the above processes to some extent in an overall transport model and then run this model for various winds, temperature, and seasonal conditions of significance throughout the year, a massive task if done properly. These and other cases are summarized in Table 11.1.

Table 11.1 Processes Which Generally Need to be Included in the Modeling of the Transport of Contaminants

	Wave action	Currents	Entrainment	Aggregation
Deep ocean dumping		X	X	X
Large shallow lake	X	X	X	X
River discharges		X	X	
Sludge outfall into deep water		X	X	X
Estuaries	X	X	X	X

EXPOSURE CONCENTRATION IN SEDIMENT

Background

It is our philosophy that a tiered approach to the screening of sediments for potential toxicity is necessary in order that those sediments that possess little contamination are cleared with the expenditure of a minimum of effort so that more resources can be expended on evaluating the potential toxicity of sediments containing sufficiently elevated levels of metals and/or organic chemicals to pose potential toxicity problems. We also believe that the screening should be based on a sound scientific basis, and the successive tiers of testing should progress logically towards state of the science capability for toxicity assessment. We find an approach based on thermodynamic principles to permit simultaneously a cost-effective tiered approach and to provide a sound basis for decision making.

The purpose of this section is to establish the framework within which the biota exposure concentrations can be calculated. A useful framework for predicting benthic organism body burden from bulk sediment concentration exists for neutral hydrophobic organic chemicals [4, 5]. The key ideas are that (1) thermodynamic potential gradients provide the driving force for organism uptake; (2) sediments and biota are composites of phases, each of which has an affinity for chemicals; (3) certain compartments (phases) in both biota and sediments dominate the sites of accumulation; (4) thermodynamic potentials (fugacity) appear to be the proper quantification for accumulation and for the expression of toxicity. Although these principles are common to organic chemicals and metals, the details of their application are sufficiently different for neutral organics, charged organics, and metals that we divide the discussion accordingly.

The fact that certain chemicals partition strongly to sediments is not an indication that they therefore exhibit negligible exposure concentration to biota. As shown below, their preferential association with sediment organic carbon is, for neutral organic chemicals, a direct indication that they have equal (to within a factor of 2) affinity to organism lipid. Therefore, these are exactly the chemicals that should receive the most attention.

Organic Chemicals

The potential for biological uptake of an organic pollutant derives from the thermodynamic activity (or fugacity) of the pollutant in the "exposure" environment. In this regard, if one has a fixed quantity of chemical in a fixed volume of water (V), the addition of a sediment (mass M) will result in a lowering of the chemical activity by the factor $(1 + \frac{M}{V} K_p)^{-1}$, with a corresponding lowering of biological uptake potential (K_p is the sediment/water distribution coefficient). At a state of sorption equilibrium, the chemical activities in the sorbed and solution phases are equal and, therefore, afford the same potential (in a thermodynamic sense) for biological uptake. The actual source will depend upon the relative kinetics of processes involved in various uptake modes, the physical properties of the chemical/sediment system (i.e., solids concentration, water solubility of the pollutant, etc.), and the life habit of the organism. Essentially, no kinetic modeling of biological uptake from sediment systems is available. A priori estimation of pollutant uptake from sediment via kinetic models is not likely realizable in the near future.

In lieu of a workable kinetic model, a plausible hypothesis for estimating worst case bioaccumulation of pollutants from sediments prescribes that the organisms will accumulate chemical until the pollutant activity in the organism equals that in the external "exposure" environment. This maximum concentration is independent of uptake pathway(s). Any pathway can be chosen for computational facility. For neutral hydrophobic organic chemicals, a very simple algorithm will provide estimates of maximal pollutant concentrations in the organism (C_{org}^{max}) in terms of the sorbed pollutant concentration on the sediment (C_{sed}). For these chemicals, the actual source of the chemical can be assumed to be the organic carbon component of the sediment (pollutant concentration per unit sediment organic carbon, C_{sed}/f_{oc}) and the uptake sink assumed to be organism lipid (pollutant concentration per unit organism lipid, C_{org}^{max}/f_l), where f_{oc} (f_l) denote mass fraction of organic carbon (lipid) in the sediment (organism). At equilibrium,

$$\gamma_l \frac{C_{org}^{max}}{f_l} = \gamma_{oc} \frac{C_{sed}}{f_{oc}} \qquad (1)$$

where γ_l (γ_{oc}) denote activity coefficients for the lipid (organic carbon) phases. Thus,

$$C_{org}^{max} = P_{l,oc} \frac{f_l}{f_{oc}} C_{sed} \qquad (2)$$

where $P_{l,oc} = \dfrac{\gamma_{oc}}{\gamma_l}$ denote the chemical preference for lipid vs. organic carbon. Since for hydrophobic chemicals, one would expect covariance of γ_{oc} and γ_l, the preference factor $P_{l,oc}$ should be chemical independent. Early testing from the Narragansett laboratory (5) suggests $P_{l,oc} \approx 2$. This approach provides for estimation of biological uptake maxima with no knowledge of K_p, BCF, or mode of uptake. The approach has not been thoroughly tested but merits a concerted testing effort (i.e., multiple sediments, organisms, chemicals, exposure conditions).

Unfortunately, this simple algorithm is not applicable to polar or charged chemicals due to the failure of both the organic carbon normalization hypothesis for the sediments and lipid normalization for the organism. The equilibrium approach remains valid—that is,

$$C_{org}^{max} = \frac{\text{BCF}}{K_p} C_{sed} \qquad (3)$$

but no reliable algorithm exists for a priori estimating (or extrapolating) either BCF or K_p.

Sediments may impact the kinetics of chemical uptake in numerous ways. In some cases, desorption may rate limit (i.e., retard) chemical uptake, whereas in other cases (in particular for chemicals of very limited water solubility), the flux of sediment-bound chemical to the active uptake organs (i.e., gill and gut surfaces) may enhance considerably the uptake rate. Kinetic studies should focus on elucidating the role of sediments in uptake.

Kinetics of organic chemical desorption can also be important. For neutral organic chemicals, it appears that eventually all the sorbed chemical can be desorbed. For highly hydrophobic chemicals, the kinetics of desorption may be very slow, so that when these chemicals are desorbed, they appear to exhibit a labile rapid desorption and a nonlabile or resistant-phase desorption (4). Desorption of the labile component appears to be very rapid (< 1 hr). The amount of chemical in the labile phase is a function of adsorption time, particle concentration and K_p. Although the details of the mechanisms and the proper mathematical formulations are still in question, desorption kinetics for hydrophobic chemicals ($K_{ow} > 10^4$) are certainly important.

Organism body burden is, for certain chemicals, the critical biological endpoint since their regulation by federal and state criteria and standards is governed by residue limits. PCBs are notable in this regard. For many other chemicals, however, the lethal and sublethal responses to acute and chronic exposures are important. The relationship between organism body burden and toxic effects is discussed below and in the summary of Section 3.

In order to quantify this relationship, it appears that thermodynamic frameworks are useful for both organic chemicals and metals. However, in lieu of the existence of such a theory relating body burden to toxicity, a relationship between the fugacity of the chemical, which at equilibrium is the same in both the sorbed and solution phases, and the toxicity to sediment-associated organism, which is developed experimentally, may suffice. It is important that toxicity data be developed within this framework.

TIERED EVALUATION OF EXPOSURE

Neutral Organic Chemicals

Tier I

(a)   Measure bulk sediment organic chemical concentration and sediment organic carbon. Predict the maximum body burden of any organism of concern directly from estimated and/or measured lipid content of the organism.

(b)   Estimate K_{oc} for the chemical and predict equilibrium interstitial water concentration. Compare to aquatic toxicity data for organisms.

Tier II

(a)   Measure body burdens of exposed infaunal organisms and verify Tier I predictions.

(b)   Measure interstitial water concentration and dissolved organic carbon (DOC). Calculate dissolved chemical concentration via K_{oc} (assume it to apply to DOC as well as sediment organic carbon). Compare to aquatic toxicity data.

Highly Polar and Charged Organic Chemicals

Tier I

(a)   If the chemical has a known desorption partition coefficient (K_p) and a known bioconcentration factor (BCF), use these to estimate body burden.

(b)   Using K_p, predict the interstitial water concentration and compare to aquatic toxicity concentration.

(*Note*: Unfortunately, for these classes of chemical, no good empirical methods appear to exist for estimating either K_p or BCF from sediment and chemical properties. It is known that cation exchange capacity, as well as sediment organic carbon, are important for K_p determination. For new chemical evaluations, this would be an extremely useful area of research.).

Tier II

(a) Measure body burdens of exposed infaunal organisms and verify Tier I predictions.

(b) Measure interstitial water concentration and dissolved organic carbon (DOC). Calculate dissolved chemical concentration via K_{oc} (assume it to apply to DOC as well as sediment organic carbon). Compare to aquatic toxicity data.

Metals

The application of the framework described above for calculating the exposure concentration for metals is, in principle, the same. However, the procedures are more difficult. It is well known that the total metal concentration of a sediment may not be entirely available. Metals in the crystal lattice that are released by complete chemical digestion are certainly not available via normal geochemical cycling in sediments. Thus, the issue of the fraction of desorbable metals is of first-order importance. Further, for metals in interstitial waters, a large fraction may be complexed to inorganic ligands and dissolved organic carbon, reducing their toxicity to organisms.

Perhaps the most vexing problem is assessing the potential availability of metals in anaerobic sediment to infaunal organisms that have created and live in aerobic microzones. This is discussed subsequently. We restrict the following discussion to the aerobic sediment layer.

Metal partitioning between solid/liquid phases and speciation in the liquid phase are critical features of metal behavior. We are operating on the hypothesis that organisms exposed to sediments are affected principally by the activity of a chemical species, e.g., the divalent metal cation activity or a few chemical species in the aqueous phase. The evidence supporting this hypothesis comes from recent studies of the toxicity of metals to algae and crustacea and from aerobic sediment exposure measurements. Conversely, it is well known that bulk heavy metal concentrations are not directly related to sediment toxicity (6, 7). A simple example would be the presence of metal precipitate of low solubility that contributes total metal to the sediment but is not directly contributing toxicity to organisms. Thus, two sediments with the same bulk total metal concentration can exhibit very different toxicity if the chemistry and, therefore, the activity of the divalent metal cation in the interstitial water is different.

Therefore, it appears that a fruitful approach is to relate desorbable, exchangeable metal to interstitial water metal activity and then to evaluate the toxicity of the metal activity to experimentally measured toxic endpoints.

Note that toxicity data must be correlated with free metal activity and not total dissolved metal (and certainly not as total bulk sediment metal concentration).

Given a sample of sediment, how does one calculate or estimate or measure the interstitial water metal activity of concern? Direct measurement of total dissolved metal in the pore waters of aerobic sediments is possible. If the pH and concentrations of the other complexing ligands, including dissolved organic matter, and their equilibrium constants are known, a simple calculation is possible. This requires careful sampling and a rather complete chemical analysis. For the dissolved organic carbon, an equilibrium constant must be estimated as well. This difficulty precludes a straightfoward evaluation, although for Tier I screening of aerobic sediment, it may suffice to assume that the total dissolved metal equals the total "available" metal.

As a result of discussions within our group during the week of the conference, what appears to be a workable Tier II framework has been developed that can relate exchangeable sorbed metal to interstitial water-metal activity directly. The assumptions are that (1) the total quantity of desorbable metal resides predominately in three solid phases: reactive iron and manganese oxides, and organic carbon; (2) the total quantity of desorbable metal in these phases can be measured via straightforward extraction procedures; (3) the quantity of each reactive phase can be measured by a straightforward extraction procedure; and (4) the intrinsic mass action equilibrium constants for metal-oxides and metal-organic matter sorption are known and are relatively independent of sediment origin and other interstitial water chemical properties. This assumption is the analog of the independence of organic chemical sorption, K_{oc}, to the properties of sediment organic carbon.

These four assumptions lead directly to an estimate of metal activity in the interstitial water without any further information or speciation calculations. The details of this proposed method are given below.

Let FeO_x, MnO_x, and Org-c be the concentration of the solid phase reactive iron and manganese oxides and organic carbon. Define the mass action sorption coefficients for divalent metal cation, Me^{++}, as:

$$K_{Fe} = \frac{(Me = FeO_x)}{(Me^{++})(FeO_x)} \tag{4}$$

$$K_{Mn} = \frac{(Me = MnO_x)}{(Me^{++})(MnO_x)} \tag{5}$$

$$K_{oc} = \frac{(Me = Org\ C)}{(Me^+)(Org\ C)} \qquad (6)$$

Let L_1, L_2, etc., be the concentrations of complexing ligands in the interstitial water (e.g., OH^-, Cl^-, DOC). The mass balance equation for the chemical system comprised of sediment-sorbed metal and interstitial water is:

$$Me_T = (1 - \phi)\ (Me = FeO_x + Me = MnO_x + Me = Org\ C)$$
$$+ \phi\ (Me^{++} + MeL_1 + MeL_2 + \ldots) \qquad (7)$$

where Me_T is the concentration of particulate bound-desorbable and interstitial water metal per unit bulk volume of sediment, and ϕ is the porosity of the sediment (volume fraction of interstitial water).

The key observation is that the absolute quantity of metal in the interstitial water is negligible relative to the sorbed metal. Measured ratios of total sorbed to total dissolved metal typically exceed 10^2 so that the mass balance equation can be simplified to yield:

$$Me_T = (1 - \phi)\ (Me = FeO_x + Me = MnO_x + Me = Org - C)\ \cdot \qquad (8)$$

Using the equilibrium expressions eqns (4)-(6) yields:

$$Me_T = Me^{++}\ (1 - \phi)\ (K_{Fe}\ (FeO_x) + K_{Mn}(MnO_x)) + K_{oc}(Org - C) \qquad (9)$$

so that:

$$Me^{++} = \frac{(Me_T)}{K_{Fe}(FeO_x) + K_{Mn}(MnO_x) + K_{oc}(Org - C)} \qquad (10)$$

where $(Me_T) = Me_T/(1-\phi)$, the total metal concentration per unit dry sediment.

It is important to realize that this equation gives the interstitial water metal activity directly in terms of readily measurable quantities: the total desorbable metal concentration, Me_T; and the sorbing phase concentrations, FeO_x, MnO_x, and Org-C all per unit sediment dry weight. The metal sorption mass action coefficients, K_{Fe}, K_{Mn}, and K_{oc}, are of course also required. The practical utility of the method depends upon the availability of these coefficients for the metals of concern and their relative independence of the details of the remaining sediment chemical and physical properties. Only further experimental work with real sediments will establish the utility of this approach. However, it is certainly a useful simplification and directly addresses the problem of speciation of metals in interstitial waters by showing that aqueous-phase speciation is correctly mirrored by the sorbed metal concentration, which itself is not significantly altered since the mass balance equation is dominated by the sorbed species.

It is interesting to note that eqn (10) is very similar in spirit to the expressions proposed for neutral hydrophobic chemicals. The bulk (desorbable, in this case) chemical concentration is normalized by the appropriate phase volumes. The activity coefficients are analogous to the sorption partition coefficients, and the result is the "active" chemical concentration, in this case the metal activity.

Whether this simple approach to speciation is practical—that is, whether it in fact can predict the metal activity in aerobic interstitial water—remains to be shown. However, it may provide a useful Tier II approach to the problem. The final link in the chain is to express toxicity in terms of metal ion activity as discussed above.

Tier III methods would involve either direct measurement of metal activity via specific ion electrodes or other electrochemical methods or full speciation computer models that account for all the macro and metal chemistry including sorption (e.g., MINTEQ, MINEQL, etc.).

ANAEROBIC SEDIMENTS

The difficulty with assessing the exposure concentration to bottom-dwelling and burrowing organisms is that they are exposed to either the aerobic surface layer or to aerobic microzones surrounding their burrows. However, direct measurements of interstitial water dissolved concentration may reflect the bulk anaerobic chemistry and not that of the aerobic microzones. Sorption relationships are dramatically altered because of the solubilization of the oxides of iron and manganese, although these presumably exist in the microzones. We can only speculate at this point what even a useful Tier I analysis would be.

Because little or no data are available on redox and pH gradients that exist from inhabited to non-inhabited areas of the sediment, there is large uncertainty as to whether the concentration in the sediment compartments available to the biota are typical of aerobic or anaerobic conditions. Data available on sediment indicate that the total metal concentrations of intact sediments are greater than those of a 24-h pH 4 nitric acid extraction, which are greater than those in the interstitial waters of anerobic sediments, which in turn are greater than those in the interstitial waters of aerobic sediments.

Although we do not know if metals are limited by sorption or solubility, since metal sulfides have low solubilities compared to other metal compounds, the solubility of metal sulfides will place an upper limit on the dissolved concentration of metals in estuarine and marine interstitial waters. This calculation may require use of the speciation and solubility portions of a computerized geochemical model. A Tier I calculation could disregard complexation by dissolved organics, which assumes that the metal present in bulk sediments will desorb or dissolve to provide a metal level limited either by solubility or by sorption. An upper limit value for dissolved sulfide could be used from the water body in question, if such data are available; otherwise, values from related water bodies could be used.

An alternative Tier I test of sediments would be to assume that a 24-h pH 4 nitric acid extraction represents the available chemical. Because the microzone is neither fully aerobic nor fully anaerobic, the maximum exposure concentration should be used for Tier I screening.

REFERENCES

1. **Bergman, H., R. Kimerle and A. Maki.** 1987. *Hazard Assessment for Complex Effluents.* SETAC Special Publication (in press).

2. **Pritchard, P. H.** 1987. Assessing the biodegradation of sediment associated chemicals. This volume, pp. 109-135.

3. **Lick, W.** 1987. The transport of sediment in aquatic systems. This volume, pp. 61-74.

4. **Karickhoff, S. W. and K. R. Morris.** 1987. Pollutant sorption: Relationship to bioavailability. This volume, pp. 75-82.

5. **Lake, J. L., N. Rubinstein and S. Pavigano.** 1987. Predicting bioaccumulation: Development of a simple partitioning model for use as a screening tool for regulating ocean disposal of wastes. This volume, pp. 151-166.

6. **Lee, G. F. and R. A. Jones.** 1987. Water quality significance of contaminants associated with sediments: An overview. This volume, pp. 1-34.

7. **Birge, W. J., P. C. Francis, J. A. Black and A. G. Westerman.** 1987. Toxicity of sediment-associated chemicals to freshwater organisms: Biomonitoring procedures. This volume, pp. 199-218.

Session 3

Biological Effects and Ecotoxicology of Sediment-Associated Chemicals

Chapter 12

Predicting Bioaccumulation: Development of a Simple Partitioning Model for Use as a Screening Tool for Regulating Ocean Disposal of Wastes

James L. Lake, Norman Rubinstein and Sharon Pavignano

INTRODUCTION

Simple bioconcentration models (1-5) and relatively complex food chain models (6-9) have been used to predict the accumulation of resistant organic compounds in aquatic organisms. Other models based on fugacities have been used to estimate distributions of compounds in idealized environments including organisms (10-12). Selection of a model depends on the availability of input data and the objectives of the study. The accuracy of the model prediction, however, depends upon (1) the accuracy of the input variables; (2) the validity of the model for predicting the modeled process; and (3) limitations in the range of applicability of the model including, for environmental studies, the importance of the process that is modeled in determining the total bioaccumulation in the organism.

Simple steady-state correlative models of the form

$$log\ BCF\ =\ a\ log\ P\ +\ b \tag{1}$$

where BCF = bioconcentration factor = (A) organism/(A) dissolved phase, P = n-octanol/water partition coefficient or aqueous solubility, and (A) = organic compound concentration, have been used in laboratory studies to predict the bioconcentration (uptake from water only) of organic compounds by fishes (1, 3, 13, 14), mussels (4, 5, 15) and other aquatic or marine organisms (16). For these applications, inaccuracies in model outputs result from measurement errors in aqueous solubilities (17) or n-octanol/water partition coefficient (*P*) (18) and from errors inherent in models developed from correlations with empirical data (19).

Although models of this type have been used to predict environmental concentrations of organic compounds in fishes (summarized by Bysshe (19)), they may be inadequate because of the above mentioned errors or because of the limited range of applicability of the model. Since this model was developed from steady-state concentrations and laboratory systems lacking suspended particulate material, the model cannot deal with the varying concentrations and forms of pollutants found in many environments. In addition, the model is too simple to account for changes in lipophilicity at high log *P* values, metabolism of compounds, and steric hindrances to accumulation (20). Further, the modeled process, bioconcentration, may not influence the total concentration in the organism significantly (i.e., approximately 90% of the accumulation of PCBs by trout in the Great Lakes was traced to uptake through food (6, 9)).

To predict the concentrations of organic contaminants in organisms and the importance of different exposure routes, relatively complex kinetic food chain models have been developed (6-9, 21). In general, these models predict the concentration of compounds in fishes by solving the following differential equation:

$$dc_f^i/dt = K_w^i \, c_w^i \, \alpha_w^i + k_{fd}^i \, c_{fd}^i \, \alpha_{fd}^i - k_2^i \, c_f^i \tag{2}$$

where

C_f^i = concentration of *i* in fish,

C_w^i = concentration of *i* in water,

C_{fd}^i = concentration of *i* in food,

k_w^i = rate constant for uptake from water,

k_{fd}^i = rate constant for uptake from food,

k_2^i = rate constant for depuration,

α_{fd}^i = uptake efficiency term for component *i* from food,

α_w^i = uptake efficiency term for component *i* from water.

The model's complexity increases if terms for growth rates, variable contaminant concentrations, length of food chains, and different storage compartments are incorporated. Even though these food chain models may be useful for estimating concentrations in organisms and for describing routes of accumulation (assuming they include and properly model the correct environmental processes), success depends upon the amount and accuracy of input data available. If the appropriate data are not available, then simplifying assumptions must be made to produce model estimations.

Models using fugacities (the tendency for a compound to escape from a phase) have been applied to predict the distributions of contaminants in idealized environments [10]. These models determine distributions of compounds in defined volumes of environmental compartments (suspended solids, water, biota, air, sediment and soil) by using fugacity capacities (analogous to heat capacities) and the concept that fugacities in all compartments are equal at equilibrium. These concepts have been applied in equilibrium steady-state models with and without advection [11] and in more complex steady-state nonequilibrium and non-steady-state nonequilibrium models [12].

One of the concerns in ocean disposal of dredged material and other mixed wastes containing persistent organic pollutants is the potential for bioaccumulation of contaminants and the consequent impact on the environment. Current regulations require that results of 10-d laboratory bioaccumulation studies be considered in the permit evaluation process [22]. Research, however, indicates that 10-d tests may be too short to reach steady-state organism concentrations [23]. In addition, laboratory testing is expensive and requires specialized facilities. More cost-effective methods are needed to serve as routine screening tools for ocean disposal evaluation. We are suggesting a model to predict the maximum concentration of organic contaminants that may be accumulated by indigenous biota at a disposal site. Our goal is to make these predictions using easily measured parameters thereby allowing regulators to test waste materials on a case-by-case basis. The model we propose for this task is a simple partitioning model that uses exposures of organisms to dissolved and particle-bound organic contaminants. This paper presents the theory behind the proposed model and discusses the technical difficulties in applying these predictions to real-world exposures.

THEORY

To develop the theoretical framework for the proposed model, we will consider a sessile marine organism (i.e., a filter-feeding bivalve) that is exposed to dissolved and particle-bound organic contaminants from a mixed waste (i.e., dredged material, sewage sludge). Assumptions are that no kinetic or steric barriers to the free exchange of contaminants are present; there is no flow; and the system has reached equilibrium with respect to component i, and therefore the fugacity (f) in all compartments is equal. Some of the distribution functions used here are described elsewhere [24].

$$f^i\text{atmosphere} = f^i\text{SPM} = f^i\text{dissolved(w)} = f^i\text{organism} \tag{3}$$

In this equation, the superscript denotes the component and the subscript denotes the phase. SPM is suspended particulate material. The fugacity of component i in the dissolved phase can be expressed as

$$f_w^i = \phi_w^i \, X_w^i \tag{4}$$

where ϕ_w^i = the fugacity coefficient of component i in the phase: in this case, this is a product of the reference-state fugacity (f_0^i) and the aqueous activity coefficient (γ_w^i).

X_w^i = mole fraction of component i in the aqueous phase.

· The organic carbon on the suspended particulate material (SPM) is the principal adsorption site for neutral organic compounds [25]. Similarly, the lipids or fats of the organism appear to be the principal site for adsorption of neutral organic compounds in organisms [26]. Parts of eqn (3) above can then be simplified:

$$f^i\text{SPM} = f_{oc}^i \tag{5}$$

and

$$f^i\text{organism} = f_l^i \tag{6}$$

where oc = organic carbon,
   l = lipids.

Bioconcentration refers to the process of organisms obtaining compounds from the dissolved phase (i.e., through gills). The lipid-normalized bioconcentration factor (BCF_L), using mole fractions, can be defined as:

$$BCF_L = Z \, C_L^i / C_w^i = X_l^i / X_w^i \tag{7}$$
$$L = \text{g lipid}$$
$$w = \text{g water}$$

where C is concentration, and the constant Z converts conventional concentrations to mole fractions in each phase. Rearranging and substituting from eqn (4) and cancelling fugacities (which are equal at equilibrium), the BCF_L can be expressed as a ratio of the fugacity coefficients.

$$BCF_L^i = \phi_w^i / \phi_l^i \tag{8}$$

Similarly, using eqn (5) and rearranging and substituting from eqn (4) organic carbon corrected K_ps (also called K_{oc}s) become

$$K_{oc} = Z \, C_{oc}^i / C_w^i = X_{oc}^i / X_w^i = \phi_w^i / \phi_{oc}^i \qquad (9)$$

$$OC = g \text{ organic carbon}$$

where Z converts conventional concentrations to mole fractions in each phase.

Since

$$\phi_{\text{phase}}^i = \gamma_{\text{phase}}^i f_o^i \qquad (10)$$

where f_o^i = reference state fugacity of component i, eqns (8) and (9) reduce to ratios of

$$\gamma_w^i / \gamma_{op}^i$$

where op = organic phase.

γ_{op} varies by relatively small amounts over a considerable range of log P (n-octanol/water partition coefficient) values. For example, for the organic phase octanol γ_{op} varies from 7.36 for phenanthrene to 5.50 for perylene [17]. γ_w, however, varies by approximately 2 orders of magnitude over the same range of log P values (i.e., $\sim 1.4 \times 10^6$ for phenanthrene to $\sim 100. \times 10^6$ for perylene). Therefore, the differences in γ_w over the range of compounds are primarily responsible for the increases observed in both K_ps [27] and BCFs [1] with increases in log P.

Bioaccumulation may be defined to include the processes of uptake of contaminants from water and food. For an organism accumulating compounds from both SPM and dissolved phases a normalized bioaccumulation factor (BAF$_L$) may be defined as:

$$\text{BAF}_L = C_L^i / (C_w^i + C_{oc}^i \varrho) \qquad (11)$$

where ϱ = concentration of SPM (expressed as grams OC/unit volume of suspension).

Converting to mole fractions, rearranging and substituting from eqn (4), and cancelling f_o^i (the reference state fugacity), this equation may be expressed as

$$\frac{l}{\text{BAF}} = \frac{\gamma_l^i}{\gamma_w^i} + \frac{\gamma_l^i \varrho}{\gamma_{oc}^i} \qquad (12)$$

Equation (12) yields interesting results because the quantity of a compound adsorbed in sediment-water systems is dependent upon the compound's solubility and the concentration of SPM [28]. For compounds with higher aqueous solubility or where SPM concentrations are low, a relatively greater abundance of the total compound will be in the dissolved phase. For these instances, the first term in eqn (12) will dominate, and the BAF distribution will increase with increases in γ_w^i (since γ_l^i remains relatively constant over this range of compounds). The result is the A part of the BAF_L (BAF_{lipid}) curve in Figure 12.1.

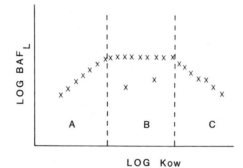

Fig. 12.1
Hypothetical steady-state
bioaccumulation curve for a
filter-feeding bivalve accumulating
organic compounds from the dissolved and
particle-bound phases. Log BAF_L
refers to the logarithm of the lipid
normalized bioaccumulation
factor. Log K_{ow} refers to the log
of the n-octanol/water
partition coefficient.

For compounds with low aqueous solubility or where SPM concentrations are high, a relatively greater abundance of the total compounds will be bound to the suspended particulates, and the second term in eqn (12) will be most important. Since γ_l^i and γ_{oc}^i are both for organic phases, they are assumed to be approximately constant, resulting in the level B part of the BAF_L curve (Fig. 12.1). The first term in eqn (12) continues to decrease with increasing log K_{ow} and becomes insignificant relative to the second term.

This view considers bioaccumulation as a redistribution of contaminants between sources (organic carbon of waste materials) and sinks (dissolved phase and lipids of organisms). For conditions in which the aqueous phase is not important as a sink, the BAF_L will depend on the concentration in the source (organic carbon) and sink (lipids of organisms). These conditions will result in maximum BAF_Ls (part B of curve in Fig. 12.1). Part C of the curve (Fig. 12.1) and the two points below the line in part B represent environmental accumulations for which model assumptions have been violated. These points represent compounds that have metabolic and/or steric factors that cause them to be metabolized or otherwise not effectively accumulated [20]. If this view of bioaccumulation is correct, then a *maximum* contaminant concentration in organisms may be predicted (corresponding to the level portion of the BAF_L curve) by measuring organic carbon and contaminant concentrations in the disposal material and the lipid content of the target organism.

$$C^{i}/g \text{ organism (dry wt)} \overline{\quad\quad\quad} (C^{i}/g \text{ SPM(dry wt)} /OC)L \quad\quad (13)$$

L = g lipid in organism/g organism (dry wt),

OC = g organic carbon/g SPM (dry wt).

This prediction would overestimate contaminant concentrations in the organism for parts A and C and the points below the level part of B (Fig. 12.1). In order to examine the applicability of this model, samples from field and laboratory studies were analyzed to determine exposure concentrations and the concentrations in organisms following exposure.

Methods

Mussels were exposed in dosing systems designed to maintain constant concentrations of suspended particulates and organisms were sampled when steady-state concentrations appeared to be attained [15, 29]. Organisms from the environment were obtained with clam rakes (*Nereis*) or by sieving bottom sediment (*Nephtys*, *Yoldia* and *Mulinia*). Sediments at these locations were sampled with jars or with a Smith-MacIntyre grab sampler. Lipid values were obtained by drying a portion of the extracts used for determining contaminant concentrations. Organic carbon analyses were performed on a carbon-hydrogen-nitrogen (CHN) analyzer. Contaminant concentrations were determined using capillary column electron capture detection gas chromatography. Identifications were checked using gas chromatography-mass spectrometry (Lake et al. (29)).

DISCUSSION

Moving from the idealized world of model development to evaluating the utility of the model in experimental systems (either laboratory or field) is difficult, because a number of assumptions made in model development are not easily satisfied. The assumptions of steady-state equilibrium and no-flow conditions required for model development are not readily attainable in laboratory systems because of the biological requirements of the organism (i.e., dissolved oxygen, food, removal of metabolites) and exposure system design problems (i.e., maintenance of suspended particulates, enclosing and

and measuring concentrations of compounds in the gaseous phase). Moreover, while some environmental exposure concentrations may be at steady state, they are not at equilibrium. Contaminants like PCBs that have low aqueous solubilities and high vapor pressures will partition to the atmosphere. This means that though steady-state concentrations may be obtained in flowing systems, the various phases may not be in equilibrium, and the fugacities in all phases may not be equal.

For generalization of model application, assumptions are made that lipids from different organisms and oc from different wastes are similar within each group, and that lipids of organisms and the oc of the waste are similar distributional phases. Research examining the solubility of organic compounds in various vegetable oils and fats demonstrates that the solubility values of the individual compounds are quite similar (30). These similarities in solubility support the assumption that lipids from different organisms may be viewed as similar distributional phases. For sediments and soils, research has shown that organic carbon is the most important distributional phase (18) and that partition coefficients normalized to organic carbon were highly invariant regardless of soil and sediment type (25). These observations suggest that organic carbon normalization as done in our proposed model may be justified. To the contrary, demonstrating that oc of SPM and lipids of organisms are equivalent organic phases is difficult, and certainly some differences in compound distributions will be expected between these two organic phases. We anticipate that these differences will be small, compared with the differences between the aqueous- and organic-phase distributions.

Other simplifying assumptions used in development of this model are that chemicals are freely exchanged between the various distributional volumes and that compounds behave conservatively. In reality, compound size (20) and structure (31) may influence accumulation, and portions of organic compounds present on suspended particulates may have kinetic or structural barriers to availability (32). In addition, many marine organisms contain enzyme systems capable of metabolizing organic compounds (33, 34).

One possible approach to dealing with these real-world difficulties is to include advective terms in the model or to use more complex nonequilibrium steady-state or nonequilibrium non-steady-state models (12). Alternatively, the simple model predictions can be compared with empirical exposure measurements to determine if the model is too simple, or nature (both laboratory and field) too complex to allow utilization of model predictions to estimate maximum bioaccumulation concentrations. Since our objective is to develop a model to estimate maximum bioaccumulation concentrations in an organism using easily measured parameters, the more complex model with its extensive data requirements is undesirable.

To assess bioaccumulation, BAFs have been calculated for worms living in, and mussels exposed to, suspensions of naturally contaminated sediments containing PCBs and other organic and inorganic contaminants. Since the exposures were maintained at constant levels and were relatively long term, steady-state conditions were presumed, and the BAFs for individual gas chromatographic PCB peaks were graphed versus Kovats retentions (Fig. 12.2). Kovats retention values express the gas chromatographic (gc) retention times of compounds relative to the gc retention time of n-alkanes. The number of carbon atoms in the n-alkane are used to determine the value of the index, i.e., decane n-(10 = 1,000); undecane n-(11 = 1,100) (35). The graphs in Figure 12.2 are representative of numerous other samples of mussels and infaunal polychaetes and sediments. Most of the differences in the magnitudes of BAF_Ls (between worms and mussels) are due to differences in solids density in suspended particulate (mussels) versus bedded (worms) exposures. Log K_{ow} (log n-octanol/water partition coefficient) would be a preferable independent variable, but log K_{ow}s were not available for all identified peaks, and since some peaks contained more than one compound, all peaks were not identified to the structural isomer level. In Figure 12.2, increases in Kovats retention would also show a corresponding increase in log K_{ow}. A rough estimate of the range of log K_{ow} in Figure 12.2 would be from ~4 to ~8.

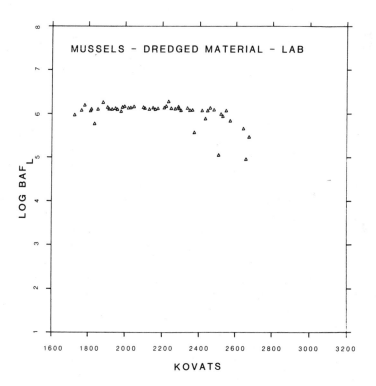

Fig. 12.2A

B

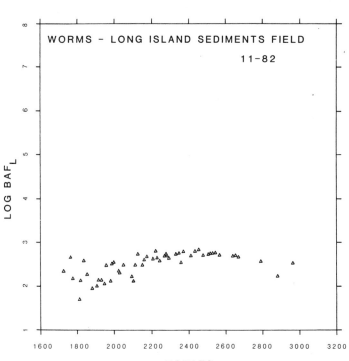

Fig. 12.2B
Lipid normalized bioaccumulation factors for A, mussels
(*Mytilus edulis*) and B, worms (*Nephtys incisa*). Kovats refers to
Kovats retention indices. An approximate range of
log K_{ow} (log octanol/water partition coefficient) for the worm
data (the more extensive data set) would be from ~4 to ~8.
Most of the difference in magnitude of BAF (between worms
and mussels) is due to differences in solids density in suspended
particulate (mussels) versus bedded (worms) exposures.

The concentration of particles is high in bedded sediments (worm study),
and the concentration of particles (SPM) was maintained at a relatively high
level in the mussel-dosing study. Therefore, in both examples, high percentages of the available compounds were bound to particulates. This is the
reason that the A part of the curve in the idealized example (Fig. 12.1) is not
evident above the scatter as a result of measurement errors (Fig. 12.2). Some
inhibition to accumulation of higher molecular weight PCBs is evident for
the mussels but is not as apparent for the worms (Fig. 12.2). This may
indicate differences in the ability of these organisms to accumulate the higher
molecular weight PCB compounds. Most significant, however, is the finding
that the log BAFs reach a constant level and do not continue to increase with

increases in Kovats retention (or log K_{ow}). Since similar curves are found in other samples of infaunal organisms and sediments, it appears that the theory describing uptake is supported.

Rearrangement of eqn (13) shows that the prediction of the maximum contaminant concentration in organisms relies on partitioning of contaminants between the organic carbon of the sediments and the lipids of the organisms.

$$\frac{C^{i}/g \text{ organisms (dry wt)}}{g \text{ lipid/g organisms (dry wt)}} = \frac{C^{i}/g \text{ SPM (dry wt)}}{g \text{ OC/g SPM (dry wt)}} \tag{14}$$

While the organic carbon and the lipids are approximately equivalent distributional phases, some differences in compound distributions are expected between them. We have calculated partitioning factors (PFs) defined as:

$$PF = \frac{C^{i}/g \text{ sed (dry wt)}/g \text{ oc/g sed (dry wt)}}{C^{i}/g \text{ organism (dry wt)}/g \text{ lipid/g organism (dry wt)}} \tag{15}$$

from some field and laboratory exposures and have compared them with each other and with a PF value derived from other research. If (1) organic carbon and lipids were equivalent distributional phases, (2) organic carbon and lipids from all sediment and organisms were the same, (3) the systems were in equilibrium with respect to contaminant i, (4) contaminant i was not metabolized, and (5) no sampling or analytical difficulties existed, then the PF should be equal to 1.0. McFarland (36) calculated a partitioning factor by utilizing equations correlating K_{oc} (25) and $BCF_{(lipid)}$ (from fish) (37) with K_{ow}.

$$\log K_{oc} = 0.989 \log K_{ow} - 0.346 \tag{16}$$

$$\log BCF_{(lipid)} = 0.980 \log K_{ow} - 0.063 \tag{17}$$

Assuming the first terms to be approximately equal and cancelling them, the following equation is derived:

$$PF = \frac{K_{oc}}{BCF_{(lipid)}} \cong 0.5 \tag{18}$$

The PF values for chlorinated compounds in Tables 12.1 and 12.2 represent the results of field and laboratory exposures where exposure concentrations

Table 12.1

Partitioning Factors (PF) for PCBs (as Aroclor 1254)

Exposure	Sample (date)	Material	PF($\pm$ SD) (Range)	a	References
Lab	*Mytilus edulis*	Heavily contaminated Providence River Sediment	0.40	1/1	(15)
Lab	*Mytilus edulis*	Heavily contaminated Black Rock Harbor Sediment	2.0	1/2	(29)
Lab	*Mytilus edulis*	Background contamination Lower Narragansett Bay (unfiltered laboratory seawater)	0.31	1/1	(15)
Field	*Nereis virens* (3/83)	Low level contamination Maine sediments	0.10	3/3	(29)
Field	*Nephtys incisa* (8/82-3/84)	Background samples from Reference stations and predump samples from Dredged disposal site Central Long Island Sound (9 stations total)	0.28($\pm$0.13) (0.10 $-$ 0.50)[b]		(29)
Field	*Mulinia lateralis* (8/17/82)	Reference station Central Long Island Sound	0.50	1/1	Present study
Field	*Nephtys incisa* (7/84)	Low level contamination Lower Narragansett Bay 2 samples	0.25(0.24 $-$ 0.26)	1/1	Present study
Field	*Yoldia limatula* (7/84)	Low level contamination Lower Narragansett Bay 2 samples	0.30 (0.29 $-$ 0.30)	1/1	Present study

[a] Due to small sample size of some organism samples, the number of replicates of sediments or suspended particulates (numerator) and the number of replicates of organism (denominator) used to calculate the PFs are indicated.

[b] Most PFs were calculated using three sediment samples and one organism sample at each location.

Table 12.2   Partitioning Factors (PF) for Some Chlorinated Compounds

Exposure	Sample (date)	Material	Chlordane	PF ($\pm$SD) (Range)	a	
Field	*Nephtys incisa* (7/84)	Low level contamination Lower Narrangansett Bay 2 samples	α - chlordane	0.24 (0.21 – 0.26)	1/1	Present study
		. . . .	γ - chlordane	0.17 (0.15 – 0.19)	1/1	Present study
		. . . .	o,p-DDD	0.21 (0.18 – 0.23)	1/1	Present study
		. . . .	p,p-DDD	0.24 (0.23 – 0.25	1/1	Present study
		. . . .	Tetrachlorodiphenyl ether	0.30 (0.27 – 0.33)	1/1	Present study
Field	*Yoldia limulata* (7/84)	Low level contamination Lower Narrangansett Bay 2 samples	α - chlordane	0.25 (0.23 – 0.26)	1/1	Present study
		. . . .	γ - chlordane	0.22 (0.20-0.23)	1/1	Present study
		. . . .	o,p-DDD	0.24 (0.23 – 0.24)	1/1	Present study
		. . . .	p,p-DDD	0.25 (0.25 – 0.25)	1/1	Present study
		. . . .	Tetrachlorodiphenyl ether	0.22 (0.21 – 0.22)	1/1	Present study

[a] Due to small sample size of some organism samples, the number of replicates of sediments or suspended particulates (numerator) and the number of replicates of organism (denominator) used to calculate the PFs are indicated.

were constant and could be established (i.e., concentrations in sediment were used for infauna; concentrations in SPM in dosing systems were used for mussels). Considering the differences in organisms, exposures, and the factors mentioned above, the agreement of the PFs with each other and with the one in eqn (18) is quite good. This appears to indicate that modeling bioaccumulation as a redistribution of contaminants between oc of sediments and lipids of organisms is justified for at least some nonpolar chlorinated organics, organisms, and exposures.

Further tests of the model will include laboratory exposures with different organisms (covering a range of lipid types and feeding modes) and solid-phase wastes (covering a range of types and concentrations of organic contaminants and organic carbon values). These studies will attempt to maintain constant exposure concentrations, maximize the availability of the contaminant for bioaccumulation, and continue until organism concentrations are at steady state. Additional field samples of organisms and sediment will be obtained where field exposure conditions are perceived to be constant. Also, the results of appropriate (steady-state) laboratory and field bioaccumulation studies from the literature will be examined and used where sufficient data are available (i.e., lipid values, TOC, concentrations in organisms, concentrations of contaminants in waste). This comparative approach will allow a body of empirical evidence to be developed to evaluate the utility of model predictions.

The criterion for success of the proposed model is how accurately it predicts maximum concentrations of organic compounds bioaccumulated from a given waste. A simple partitioning model is presented for making these estimations. Since the ultimate use of this model is as a first-level screen to assess the advisability of ocean disposal of a given waste, errors in estimation should ensure maximum environmental protection. If the theory is correct, the predicted maximum concentration accumulated will be greater than or equal to the actual measured organism concentrations. This provides a conservative estimate for regulatory purposes of environmental protection.

Our approach to testing the theory is to compare model predictions with empirical data from laboratory and field exposures and from the literature. This approach has yielded data for some chlorinated organics that show agreement of preference factors from environmental and laboratory exposures with each other and with one calculated from the literature. From these data, a preliminary assessment can be made that the predictive model appears to work for at least some compounds, organisms, and exposures. The transition from model development and empirical testing to application in regulation can be made only when a sufficient number of comparisons between predicted and measured organism concentrations have been documented to allow the risk associated with model predictions to be evaluated prudently.

Acknowledgments—The authors wish to acknowledge the participants in the workshop on bioaccumulation held at the U.S. Army Waterways Experiment Station, Vicksburg, Mississippi, July 1983, and to thank S. W. Karickhoff for his valued input.

REFERENCES

1. **Veith, G. D., D. L. Defoe and B. V. Bergstedt.** 1979. Measuring and estimating the bioconcentration factor of chemicals in fish. *J. Fish. Res. Board Can.* **36**:1040-1048.

2. **Neely, W. B., D. R. Branson and G. E. Blau.** 1974. Partition coefficient to measure bioconcentration potential of organic chemicals in fish. *Environ. Sci. Technol.* **8**:1113-1115.

3. **Chiou, C. T., V. H. Freed, D. W. Schmedding and F. L. Kohnert.** 1977. Partition coefficient and bioaccumulation of selected organic chemicals. *Environ. Sci. Technol.* **11**:475-478.

4. **Ernst, W.** 1977. Determination of the bioconcentration potential of marine organisms—A steady state approach to bioconcentration data for seven chlorinated pesticides in mussels (*Mytilus edulis*) and their relationship to solubility data. *Chemosphere* **6**:731-740.

5. **Geyer, H., P. Sheehan, D. Kotzias, D. Freitag and F. Korte.** 1982. Predictions of ecotoxicological behavior of chemicals: Relationship between physio-chemical properties and bioaccumulation of organic chemicals in the mussel *Mytilus edulis*. *Chemosphere* **11**:1121-1134.

6. **Weininger, D.** 1978. Accumulation of PCBs in lake trout in Lake Michigan. Ph.D. thesis, University of Wisconsin, Madison.

7. **Norstrom, R. J., A. E. McKinnon and A. S. W. de Freitas.** 1976. A bioenergetics-based model for pollutant accumulation by fish, simulation of PCB and methyl mercury residue levels in Ottawa River yellow perch (*Perca flavescens*). *J. Fish. Res. Board Can.* **33**:248-267.

8. **Thomann, R. V.** 1981. Equilibrium model of fate of microcontaminants in diverse aquatic food chains. *Can. J. Fish. Aquat. Sci.* **38**:280-296.

9. **Thomann, R. V. and J. P. Connolly.** 1984. Model of PCB in the Lake Michigan lake trout food chain. *Environ. Sci. Technol.* **18**:65-71.

10. **Mackay, D.** 1979. Finding fugacity feasible. *Environ. Sci. Technol.* **13**:1218-1223.

11. **Mackay, D. and S. Paterson.** 1981. Calculating fugacity. *Environ. Sci. Technol.* **15**:1006-1014.

12. **Mackay, D. and S. Paterson.** 1982. Fugacity revisited. *Environ. Sci. Technol.* **16**:654A-660A.

13. **Neely, W. B., D. R. Branson and G. E. Blau.** 1974. Partition coefficient to measure bioconcentration potential of organic chemicals in fish. *Environ. Sci. Technol.* **8**:1113-1115.

14. **Kenaga, E. E. and C. A. I. Goring.** 1980. Relationship between water solubility, soil sorption, octanol-water partitioning, and bioconcentration of chemicals in biota. In J. G. Eaton, P. R. Parrish and A. C. Hendricks, eds., *Aquatic Toxicology: Third Symposium*. ASTM STP 707. American Society for Testing and Materials, Philadelphia, PA. pp. 78-115.

15. **Pruel, R. J., J. L. Lake, W. R. Davis and J. G. Quinn.** Availability of polychlorinated biphenyls and polycyclic aromatic hydrocarbons in the blue mussel, *Mytilus edulis,* from artificially resuspended estuarine sediment. In J. M. Capuzzo and D. R. Kester, eds., *Ocean Processes in the Marine Pollution*, Vol. 1. Krieger Publishing, Melbourne, FL (in press).

16. **Southworth, G. W., J. J. Beauchamp and P. K. Schmeider.** 1978. Bioaccumulation potential of polycyclic aromatic hydrocarbons in *Daphnia pulex*. *Water Res.* **12**:973-977.

17. **Mackay, D., A. Bobra, W. Y. Shiu and S. H. Yalkowsky.** 1980. Relationships between aqueous solubility and octanol-water partition coefficients. *Chemosphere* **9**:701-711.

18. **Karickhoff, S. W., D. S. Brown and T. A. Scott.** 1979. Sorption of hydrophobic pollutants on natural sediments. *Water Res.* **13**:241-248.

19. **Bysshe, S. E.** 1982. Bioaccumulation factor in aquatic organisms. In W. J. Lyman, W. F. Reehl and D. H. Rosenblatt, eds., *Handbook of Chemical Property Estimation Methods.* McGraw-Hill, New York, pp. 5-1 to 5-30.

20. **Tulp, M. T. M. and O. Hutzinger.** 1978. Some thoughts on aqueous solubilities and partition coefficients of PCB and the mathematical correlation between bioaccumulation and physio-chemical properties. *Chemosphere* 7:849-860.

21. **Bruggeman, W. A., L. B. J. M. Marton, D. Kooiman and O. Hutzinger.** 1981. Accumulation and elimination kinetics of di-, tri- and tetrachlorobiphenyls by goldfish after dietary and aqueous exposure. *Chemosphere* 10:811-832.

22. **U.S. Environmental Protection Agency/U.S. Army Corps of Engineers (EPA/USACE).** 1977. The ecological evaluation of proposed discharge of dredged material into ocean waters; Implementation Manual for Section 103 of PL 92-532. Environmental Effects Lab, USACE, Waterways Experiment Station, Vicksburg, MS.

23. **Rubinstein, N. I., E. Lores and N. R. Gregory.** 1983. Accumulation of PCBs, mercury and cadmium by *Nereis virens, Mercenaria mercenaria* and *Palaemonates pugio* from contaminated harbor sediments. *Aquatic Toxicol.* 3:249-260.

24. **Karickhoff, S. W. and K. R. Morris.** 1986. Pollutant Sorption: Relationship to Bioavailability. This volume pp. 75-82.

25. **Karickhoff, S. W.** 1981. Semi-empirical estimation of sorptions of hydrophobic pollutants on natural sediments and soils. *Chemosphere* 10:833-846.

26. **Schneider, R.** 1982. Polychlorinated biphenyls (PCBs) in cod tissues from the Western Baltic: Significance of equilibrium partitioning and lipid composition in the bioaccumulation of lipophilic pollutants in gill-breathing animals. *Verlag Paul Parey* 2:69-79, Hamburg, Germany.

27. **Banerjee, S., S. H. Yalkowsy and S. C. Valvani.** 1980. Water solubility and octanol/water partition coefficients of organics. Limitations of the solubility-partition coefficient correlation. *Environ. Sci. Technol.* 14:1227-1229.

28. **Dexter, R. N.** 1976. An application of equilibrium adsorption theory to the chemical dynamics of stable organic molecules in the marine environment. Ph.D. thesis. University of Washington, Seattle.

29. **Lake, J., G. Hoffman and S. Schimmel.** 1985. Bioaccumulation of contaminants from Black Rock Harbor dredged material by mussels and polychaetes. Technical Report D-85-2, prepared by the Environmental Protection Agency, Narragansett, RI, for the U.S. Army Engineer Waterways Experiment Station, Vicksburg, MI.

30. **Dobbs, A. J. and N. Williams.** 1983. Fat solubility—a property of environmental relevance? *Chemosphere* 12:97-104.

31. **Shaw, G. R. and D. W. Connell.** 1984. Physiochemical properties controlling polychlorinated biphenyl (PCB) concentrations in aquatic organisms. *Environ. Sci. Technol.* 18:18-23.

32. **DiToro, D. M. and L. M. Horzempa.** 1982. Reversible and resistant components of PCB adsorption-desorption isotherms. *Environ. Sci. Technol.* 16:594-602.

33. **Malins, D. C.** 1977. Biotransformation of petroleum hydrocarbons in marine organisms indigenous to the Arctic and Subarctic. In D. A. Wolfe, ed., *Fate and Effects of Petroleum Hydrocarbons in Marine Organisms and Ecosystems.* Pergamon Press, New York, pp. 47-51.

34. **Stegeman, J. J.** 1981. Polynuclear aromatic hydrocarbons and their metabolism in the marine environment. In H. V. Gelboin and P. O. P. Ts'o, eds. *Polycyclic Hydrocarbons and Cancer*, Vol. 3. Academic Press, New York, pp. 1-60.

35. **Kovats, E.** 1958. Gas chromatographic characterization of organic compounds. 1. Retention indexes of aliphatic halides, alcohols, aldehydes and ketones. *Helv. Chem. Acta* 1915-32.

36. **McFarland, V. A.** 1984. Activity-based evaluation of potential bioaccumulation for sediments. In R. L. Montgomery and J. W. Leach, eds., *Dredging and Dredged Material Disposal*, Vol. 1. American Society of Civil Engineers, New York, pp. 461-467.

37. **Konemann, H. and K. van Leewen.** 1980. Toxicokinetics in fish: Accumulation and elimination of six chlorobenzenes by guppies. *Chemosphere* 9:3-19.

Chapter 13

The Application of a Hazard Assessment Strategy to Sediment Testing: Issues and Case Study

John H. Gentile and K. John Scott

INTRODUCTION

The Marine Protection, Research, and Sanctuaries Act (Public Law 92-532) was passed by Congress in 1972. This law states that it is the policy of the United States to regulate the dumping of all types of materials into ocean waters and to prevent or strictly limit the dumping of any material that would adversely affect human health, welfare, the marine environment or ecological systems. The implementation of this law through the issuance of permits is jointly shared by the U.S. Environmental Protection Agency (EPA) and the U.S. Army Corps of Engineers (USACE).

The EPA has responsibility for establishing and applying criteria for reviewing and evaluating ocean dumping permits. Briefly, in establishing such criteria, consideration is given but not limited to the following: demonstrating a need for ocean dumping; effects on human health and welfare; effects on fisheries resources; ecosystem effects with respect to transport and fate of materials, changes in marine ecosystem diversity, productivity, and population dynamics; the amounts, concentrations, and persistence of contaminants; appropriate alternative methods of disposal and the environmental and economic impacts of such alternatives (Section 102).

The issuance of permits for ocean disposal of dredged material is the responsibility of the USACE and is described in Section 103 of PL-92-532. There are three phases to this process. Initially, the USACE must notify EPA of its intention to issue a permit; this is followed by the acquisition of the technical information necessary to evaluate compliance with the criteria established by EPA; and finally, a joint review and concurrence of EPA with the initial determination conducted by the USACE. Should there be a lack of concurrence, the EPA decision will prevail.

The permits issued under Title I-Ocean Dumping of PL-92-532 shall designate and include the type and amount of material to be authorized for dumping, the location where dumping shall take place, and the length of time for which the permit is valid. In addition, the EPA and USACE may choose to include any special provisions deemed necessary to ensure compliance with the established criteria.

In 1977 the EPA published final regulations and criteria for ocean dumping. These regulations reflected the experience of the Agency in the implementation of the Act, and specified in more detail the considerations that go into determining whether a permit will be issued. The regulations also improved the site designation process, and incorporated recent scientific advances into the establishment of criteria. These changes in no way amend the original intent of the act to eliminate the ocean dumping of unacceptable materials. Part 227, Subpart B of the regulations addresses the environmental impact of ocean disposal of dredged materials. It defines the types of environmental information that will be required by EPA to make a determination of the suitability of the material for ocean disposal that is consistent with the criteria established by EPA.

DREDGED MATERIAL IMPLEMENTATION MANUAL— ISSUES

In order to implement the regulations in a manner consistent with the criteria, the EPA/USACE prepared technical guidance in the form of an implementation manual (1). The manual specified which test procedures were to be followed in collecting the necessary information to be used in making a disposal decision that is consistent with the regulatory criteria. Among the procedures detailed in the manual were those for chemically characterizing the proposed dredged material, determining the acute toxicity of liquid, suspended particulate, and solid phases, for estimating the potential for contaminant bioaccumulation, and for describing the initial mixing during disposal.

These procedures represented a balance between the technical state of the art and what was routinely implemented at the time of preparation; they were never intended to be inflexible standards. The recommended test methods were chosen to provide technical information that was consistent with the criteria specified in the regulations. Because the technical state of the art is a dynamic process, it was recognized from the outset that the implementation manual would require both augmentation and revision on a periodic basis.

Several years of operational experience with the sediment test methods and the interpretive guidance provided in the manual has led to the identification of the following technical issues:

- The toxicity test methods are limited to measuring mortality over a short (4 to 10 d) exposure period from which long-term assessments of impact are to be inferred. The species commonly selected for these tests are often insensitive. If mortality is to be a useful endpoint for regulating sediment disposal, only the most sensitive species should be used.

- The testing of a liquid phase is only relevant to assessing the short-term water-column effects that might occur during the disposal operation and is of questionable utility. In addition, the preparation of the liquid phase (elutriate test) does not have credible scientific support for use as a measure of contaminant bioavailability.

- The solid phase test design does not permit the determination of an exposure-response relationship (e.g., LC_{50}, EC_{50}, etc.) which is essential for comparing the toxicity of different sediments and for conducting risk assessments.

- Generally, the selection of species for all the testing including sediment toxicity and bioaccumulation is ecologically inappropriate.

- The currently utilized experimental design is not appropriate for assessing effects outside the disposal site.

- There is no strategy or rationale for synthesizing or interpreting the toxicological or bioaccumulation information. Test methods have two primary functions, that of detection and assessment (prediction). In the case of detection methods, the purpose is to determine if a change in some biological response can be quantified as a function of exposure to a particular material or compound. This is simply the ability to detect a change in a response as a function of exposure. It does not address the problem of the significance of that measure beyond the level of biological organization at which it was conducted. Assessment methods attempt to address the "so what" question and are generally integrative in nature. Responses such as growth, survival, and reproduction are measured that have direct bearing on the long-term integrity of populations, and thus are viewed as having more predictive value.

- There is the question of interpreting the ecological significance of tissue residues resulting from the bioaccumulation of contaminants. This issue has two components: first is the ability to develop predictive models for contaminant bioavailability and subsequent tissue residues from sediment contaminants; and second is the need to determine the biological and ultimately the ecological significance of these residues. This residue-effects link remains undefined and, as such, allows only for the quantification of a contaminant in an organism as a tissue residue but tells us nothing concerning its biological significance unless it is directly related to some measure of biological response.

● Finally, there is the issue of defining the limits of applicability of laboratory data as used to assess impacts occurring in the field (field verification).

The above issues and limitations have led the USACE and EPA to reexamine the state of the art in the ecotoxicology of sediment testing for the evaluation of dredged-material disposal in the marine environment. The intent of this evaluation is to increase and improve the test methods and endpoints being used and to determine the limits of applicability of laboratory tests to the environment through field verification. The evaluation process utilizes a hazard assessment strategy to synthesize data from laboratory measures of biological effects with estimates of environmental contaminant concentrations in order to make predictions of risk to populations and communities. The purpose of this paper is to describe a decision rationale for ocean disposal that is based upon a predictive hazard assessment strategy and to present a site-specific case study designed to evaluate the applicability of the strategy for the disposal of dredged materials.

STRATEGY OF OCEAN DISPOSAL

If the ocean disposal of contaminated sediments is to become a waste management option, then decision makers must be provided with a rationale that is both expeditious and scientifically sound. In addition to providing guidance for the acquisition and interpretation of information, the rationale must include a predictive assessment procedure so that environmental consequences of a disposal decision can be made a priori. The disposal decision must have a prescribed level of confidence when compared to alternative disposal options. In order for the rationale to be useful, it must functionally reflect the primary components that comprise the regulatory process.

The functional description of such a rationale must be capable of coupling pollutant source inputs with ecosystem impacts as illustrated in Figure 13.1.

Fig. 13.1
Functional description of the regulatory process

Ecosystem impacts are typically deduced from tests on individual species from which forecasts are made to the population and community levels of biological organization. This can be accomplished by interfacing the source inputs (e.g., pollutant type and loading rate) to the biological effects through measurements of environmental exposure. Pollutant source input rates are then used as source terms for fate and transport models that are used to predict pollutant concentration isopleths on defined spatial and temporal scales. This form of output is directly compatible with biological effects measurements in the laboratory that are typically described as functions of pollutant concentration and exposure duration. A hazard assessment occurs when the probability of environmental impact is estimated from a comparison of the environmental contaminant exposure with contaminant concentrations producing biological effects (Fig. 13.2).

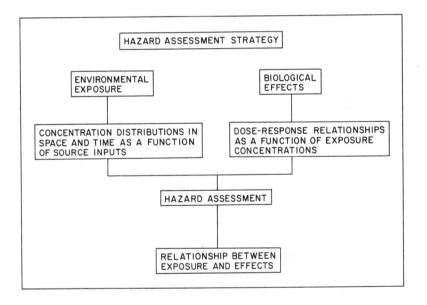

Fig. 13.2
Hazard assessment strategy: exposure and effects definitions.

Bierman et al. (2) proposed a research strategy for ocean disposal that supports a decision rationale based upon a predictive hazard assessment approach. This rationale contains several components, each involving the acquisition and synthesis of information that leads to a regulatory decision (Fig. 13.3). The objective of this rationale is to produce a systematic, scientifically sound basis for estimating the potential impact of waste contaminants that are disposed in marine and estuarine environments.

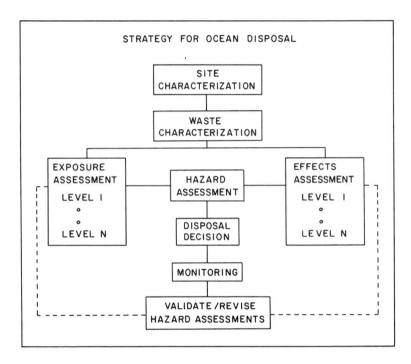

Fig. 13.3
Decision rationale for ocean disposal.

One of the principal features of a hazard assessment strategy is that the effects and exposure components are tiered (Fig. 13.3). The ordering of the tiers implies increasing degrees of complexity, resolution, and predictive confidence (3). An additional assumption is that there are clearly defined criteria for decisions at each level of the hierarchy that triggers the need for increased data acquisition at the higher, more complex tiers. A prediction of hazard is made by comparing the predicted environmental exposure concentration of the material (exposure assessment) and the concentration producing biological effects in the laboratory studies (effects assessment) at each tier. When properly synthesized, these data provide a confidence-bounded estimate of the probability (risk) of unacceptably altering the aquatic environment as the result of the disposal of the waste. If either the confidence of the prediction or the probability of risk is unacceptable, then additional levels of testing can be conducted. This strategy places the biological tests in a hierarchy of complexity that reflects a continuum from detection to assessment methods with increasing predictive confidence; furthermore, the strategy directly relates the contaminant concentrations causing effects to the predicted environmental concentrations determined at the same hierarchical level. This approach permits the development of a series of testable hypotheses that are amenable to both laboratory and field verification.

HAZARD ASSESSMENT APPLICATION TO DREDGED MATERIAL—A CASE STUDY

The EPA and USACE are jointly conducting a comprehensive Field Verification Program (FVP) to evaluate the risks associated with the ocean disposal of dredged material. This study can be viewed as a site- and waste-specific application of the ocean-disposal rationale (Fig. 13.3). The objective of the FVP is to evaluate and field verify exposure and effects methodologies that are used in predicting the consequences of dredged-material disposal in an aquatic environment.

This case study offers the opportunity to evaluate the application of the hazard assessment strategy to predicting the impact of an ocean-disposed material and to develop and test hypotheses relevant to the issues highlighted as limitations in the present regulatory approach. This paper will focus on the hierarchical nature of the effects assessment (sediment toxicity testing) component of the strategy. In so doing, the discussion will illustrate how toxicity and bioaccumulation data can be coupled with environmental exposure information to make a prediction of hazard.

The effects component of the FVP is constructed in a hierarchical manner proceeding from the individual species level of biological organization to the population level and ultimately to effects on communities. Within the individual species level of organization, a variety of endpoints are being examined that in themselves represent a continuum from short-term responses at the biochemical level of organization to integrative whole organism responses on survival, growth, and reproduction from which population inferences can be made through the use of demographic techniques.

The aquatic disposal site for the FVP is an historical site known as the Central Long Island Sound (CLIS) Disposal Site, which is located approximately 15 km off the coast of Connecticut (Fig. 13.4). This area was previously known as the New Haven Dump Site and was first site designated in Long Island Sound after passage of the Ocean Dumping Act (1972). The disposal site is rectangular in shape, measures 3.7 km in the east-west direction and 1.8 km north-south, with a depth of 25 m. Tidal currents typically dominate the east-west water motion resulting in surface currents of 20 to 40 cm/s, and near bottom currents of 5 cm/s, which have a net northerly drift. Sedimentology of the disposal and adjacent reference site is primarily silt-clay with a mean grain size of 0.013 mm. Salinities range from 24 to 30 mg/kg. Temperatures cover the annual seasonal range with thermal stratification occurring from April to September. The suspended solids

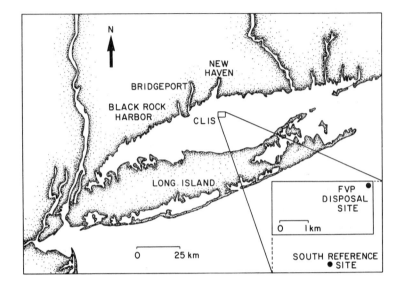

Fig. 13.4
Locations of the Black Rock Harbor dredged-material site,
Bridgeport, CT, and the field verification disposal site in
Central Long Island Sound (CLIS) dumpsite.

concentration measured at 1 m above the bottom typically averages 5 mg/L, with storm associated values reaching 25 mg/L.

The dredged site for the field verification study is Black Rock Harbor (BRH), Connecticut where maintenance dredging was conducted to provide a channel 46 m wide and 5.2 m deep at mean low water. Of approximately 150,000 m^3 of material removed, 57,000 m^3 were disposed of in the north easterly corner of the CLIS disposal site. The BRH dredged material contains substantial concentrations of both organic and inorganic contaminants. The PCB concentration is 6,800 ng/g, and the PAH concentrations range from 3,400 ng/g for the naphthalenes to 13,400 ng/g for the benzopyrenes and 22,800 ng/g for the benzanthracenes. The total organic carbon was 5.6%, and the total petroleum (F$_1$-fraction) hydrocarbons were 13,200 ng/g. The principal inorganic contaminants present in BRH dredged material are copper (2,300 μg/g), chromium (1,430 μg/g), zinc (1,200 μg/g), lead (380 μg/g), nickel (140 μg/g), cadmium (23 μg/g), and mercury (1.7 μg/g).

The approach taken in the effects assessment component of the FVP is illustrated in Figure 13.5.

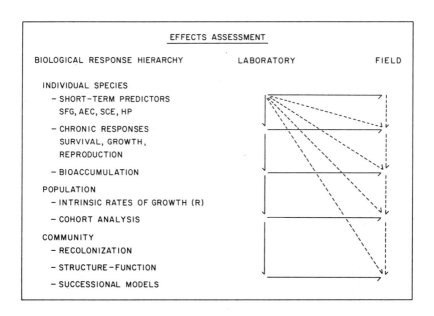

Fig. 13.5
Schematic of the FVP design showing the hierarchy of biological
responses, the laboratory-field interrelations, and the coupling of
exposure and effects for hazard assessment.

The hierarchy of test methods is being conducted on selected indigenous or introduced epibenthic and infaunal species (Table 13.1) under both laboratory and field exposure conditions. The horizontal arrows illustrate the concept of lab to field verification of responses while the vertical arrows illustrate the concept of predictability among responses and between the levels of biological organization. In addition, certain species are being used in only laboratory studies to test the hypothesis that one can predict long-term effects from short-term responses.

The field verification of laboratory responses will be addressed by simulating environmental exposures for sediments and suspended solids in the laboratory and then comparing these response data with those collected from the same species along an exposure gradient at the disposal site. Although this may at first glance seem rather straightforward, there are a variety of complex technical questions involved in any laboratory simulation, ranging from the definition of the contaminant exposure as a continuous or modulating function to the degree with which one simulates oxidation or weathering of test sediments.

Table 13.1
Indigenous, Introduced and Surrogate Study Species for
the Field Verification Program

ANNELIDS

Nepthys incisa
Nereis virens
Neanthes arenaceodentata

MOLLUSCS

Mytilus edulis
Yoldia limatula
Mulinea lateralis

ARTHROPODS

Ampelisca abdita
Mysidopsis bahia
Neomysis americanus
Mysidopsis bigelowi
Homarus americanus

FISHES

Menidia menidia
Cyprinodon variegatus
Ammodytes americanus
Paralichthys dentatus
Pseudopleuronectes americanus

To assess the biological consequences of tissue residues resulting from the bioaccumulation of contaminants, these studies will compare the tissue residues for selected organic and inorganic contaminants with the hierarchy of effects measurements from both laboratory- and field-exposed organisms. This basically is a correlative analysis approach that is consistent with the fact that the waste material contains multiple contaminants from which it would be impossible to unequivocally determine a cause-effect relationship on a contaminant specific basis.

PRELIMINARY HAZARD ASSESSMENT FOR THE FVP CASE STUDY

To illustrate how this approach can be utilized to make an initial prediction of hazard, exposure data has been synthesized from the post-disposal field studies on the suspended solids concentrations at 1 m above the bottom. These data will form the basis of our predicted environmental exposures for the waste material under three sets of conditions: normal background suspended solids concentrations, average storm conditions, and a storm of hurricane intensity and duration. The effects component consists solely of laboratory studies of solid and suspended solids concentrations that elicited a specific response (Table 13.2). These laboratory studies did not simulate

Table 13.2 FVP—Endpoints and Organisms Tested

Response criterion	Taxon	Exposure	
Acute toxicity	Amphipod, *Ampelisca abdita*	SL,SU	4 days
	Mysid, *Mysidopsis bahia*	SL,SU	4 days
Burrowing behavior	Polychaete, *Nephtys incisa*	SL	10 days
	Bivalve, *Yoldia limatula*	SL	10 days
Genetic (SCE)	Polychaete, *Nephtys incisa*		
	Polychaete, *Neanthes arenaceodentata*		
Biochemical (AEC)	Polychaete, *Nephtys incisa*	SL	10 days
	Mussel, *Mytilus edulis*	SU	26-28 days
Bioenergetics (SFG)	Mussel, *Mytilus edulis*	SU	28 days
(growth efficiency)	Polychaete, *Nephtys incisa*	SL	10 days
Clearance rate	Mussel, *Mytilus edulis*	SU	28 days
Growth	Amphipod, *Ampelisca abdita*	SU	6-8 weeks
Reproduction	Mysid, *Mysidopsis bahia*	SL,SU	4 weeks
Population growth	Amphipod, *Ampelisca abdita*	SU	6-8 weeks
	Mysid, *Mysidopsis bahia*	SL,SU	4 weeks
Bioaccumulation	Mussel, *Mytilus edulis*	SU	28 days
	Polychaete, *Nereis virens*	SL	28 days
Pathology	Polychaete, *Nephtys incisa*	SL	10 days
	Polychaete, *Neanthes arenaceodentata*	SL	10 days
	Amphipod, *Ampelisca abdita*	SU	4 days
	Mussel, *Mytilus edulis*	SU	28 days

environmental exposure conditions, so they do not represent a field verification. This initial comparison of the concentrations that have been found to cause biological effects with predicted environmental exposure concentrations is what typically might be generated in a regulatory situation and upon which initial disposal decisions might be made.

Data on effects assessment have been synthesized for the mysid shrimp, *Mysidopsis bahia,* and the infaunal amphipod, *Ampelisca abdita*, and presented in conjunction for three field exposure conditions. The effects isopleths constructed for these species include responses ranging from 96h LC_{50} values to chronic mortality, reduced reproduction, and intrinsic rates of population growth. Predicted field exposures, as a function of BRH suspended sediment concentration (ordinate) and exposure duration (abscissa) are illustrated by the horizontal bars. The effects isopleth for *M. bahia* indicate that suspended-solids exposures required to produce acute toxicity would only be approached under a worst-case conditions that would result from a storm of hurricane intensity and duration. Normal background suspended-solids concentrations of 5 mg/l are approximately fifty times below those required to produce acute effects ($LC_{50} = 250$ mg/l) in *M. bahia* of experiencing chronic effects from exposures to background concentrations of BRH suspended sediments may be expressed in terms of the amount of overlap between the exposure and effects concentrations. It is obvious that the greater the distance between the exposure and effects concentrations the less the probability of risk. Risk is quantified through the use of a probability function which is defined as the degree of overlap resulting from the intersection of the variances about the exposure and effects concentrations.

In contrast, *A. abdita* could experience acute mortalities in the environment under worst-case exposure conditions, experience sublethal effects under typical storm conditions, and have direct effects on population growth at normal background suspended-solids concentrations if they consisted totally of resuspended BRH sediments. Thus, the probability of risk based on this species would be very high and lead to the conclusion of no margin of safety with direct impacts.

The following conclusions can be deduced from this preliminary example. First, the absolute effect concentrations are species dependent, and second, the relationship between short- and long-term responses is also species dependent. For example, the acute/chronic ratio for *M. bahia* is approximately 8.0, yet a similar comparison for *A. abdita* results in a value of 18.0. A similar analysis of data for other species will give an estimate of species variability in terms of the absolute sensitivity to the dredged material and between short- and long-term responses. This is important since the comprehensive assessment of risk will have to account for the uncertainties associated with differences in species sensitivity.

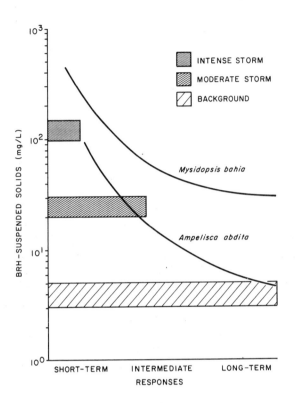

Fig. 13.6
Initial comparison of effects isopleths for
M. bahia and *A. abdita* with three
predicted environmental exposures of BRH
suspended sediments.

FIELD VERIFICATION

The objective of the field verification component of this study is to obtain a measure of the limits of applicability between laboratory predictions and factual field measurements of the same responses. The null hypothesis is that there are no significant differences in the response measured in the laboratory and the field given that the exposure conditions are analogous. Three endpoints, scope for growth (SFG), adenylate energy charge (AEC), and bioaccumulation measured in *M. edulis*, were chosen to illustrate the comparison of laboratory and field responses (Table 13.3). Similar data are also available for other species and biological responses.

Table 13.3   Field and Laboratory Response Data for the Caged Mussel, *Mytilus edulis*

(The endpoints are scope for growth (SFG) in J/h, adenylate energy charge (AEC) in μmoles/g wet weight, and bioaccumulation of total PCB in ng/g dry weight.)

Response	Exposure	Duration	Concentration mg/l	Control	Exposed	Relative difference	Significance[a]
Scope for growth J/h	Laboratory #1	28 days	12	2.53	-2.32	4.85	a
	Laboratory #2	28 days	12	10.22	0.51	9.71	a
	Field T + 8	32 days	10-15	0.51	-5.41	-5.92	a
	T + 12	28 days	5-10	1.27	-0.65	1.92	NS
Adenylate energy charge μmole/g wet (wt)	Laboratory #1	28 days	12	0.83	0.75	0.08	a
	Laboratory #2	28 days	12	0.83	0.80	0.03	a
	Field T + 8	32 days	10-15	0.78	0.82	-0.04	ND
	T + 12	28 days	5-10	0.87	0.80	0.07	ND
Bioaccumulation	Laboratory	28 days	10	84	3,000	2,916	a
Total PCB ng/g	Field T + 8	32 days	10-15	477	700	233	NS
	T + 16	27 days	5-10	553	873	320	NS

a = Differences are significant at $P > 0.05$.
NS = Not significant.
ND = Statistical significance not determined.

Field exposure conditions were measured at 1 m above the bottom at the disposal site during disposal and at selected time intervals following disposal through the use of in situ continuous monitoring arrays and the chemical analysis of discrete water samples. This information was used to design subsequent laboratory studies that attempted to simulate field exposure conditions. The data, summarized in Table 13.3, indicates a close correspondence between the laboratory and field results for the three endpoints. Statistically significant decreases were recorded for scope for growth in mussels exposed in the lab and field to a total suspended solids concentration of 12-15 mg/l consisting of an average of 10% BRH sediment. When the field exposure concentration (T + 12 weeks) decreased, no effects were noted. Adenylate energy charge, on the other hand, did not show a significant change as a result of exposure to BRH contaminated sediments either in the laboratory or the field.

PCB tissue residues showed a significant increase during and immediately after disposal (T + 2 weeks) which agreed with data derived from the laboratory simulation studies but declined rapidly in the field when exposures decreased. This and other data from this study indicate that biological responses measured in the laboratory and field can be expected to agree, given analogous contaminant exposure regimes. These results are encouraging since in conducting risk analyses, laboratory data are used to predict field responses with the specific assumption that the results of laboratory and field data are comparable.

SUMMARY

The hazard assessment strategy provides the researcher with a conceptual framework for identifying, collecting, synthesizing and interpreting data which then can be used to formulate informed environmental decisions. A tiered hazard assessment is presented for determining and field verifying the impacts of open water disposal of dredged material. Results from this program indicate that it is possible to conduct a field verification on a hierarchy of biological methods and that there is excellent correspondence between laboratory and field responses. This concurrence is dependent upon a variety of factors the most important of which is the prediction and measurement of environmental exposure conditions which are essential for the proper design and conduct of laboratory studies. Our ability to measure biological responses in the laboratory and the field is a function of the type of response being measured and the level of biological organization being studied. Biological test methods for evaluating sediment impacts has improved dramatically during the past 5 years (Swartz, this vol., p. 183). Many of these methods can be used in both laboratory and field situations, and can be arranged in a hierarchy for testing purposes (Anderson et al., this vol., p. 267).

The synthesis of laboratory effects data with either predicted or measured field exposure data into an estimate of hazard or risk was done for two species. The success of this analysis is dependent upon our ability to predict or measure the environmental exposure concentrations in various environmental compartments. However, to extrapolate the risk from individual species to populations and communities is a much more difficult task but one which must receive attention. In keeping with the legislative mandate of the Marine Protection, Research, and Sanctuaries Act, the environmental manager is responsible for maintaining and protecting indigenous populations, not individuals.

REFERENCES

1. **U.S. Environmental Protection Agency/U.S. Army Corps of Engineers (EPA/USACE).** 1977. The ecological evaluation of proposed discharge of dredged material into ocean waters; Implementation Manual for Section 103 of PL 92-532. Environmental Effects Lab, USACE, Waterways Experiment Station, Vicksburg, MS.

2. **Bierman, V. J., Jr., J. H. Gentile, J. F. Paul, D. C. Miller and W. A. Brungs.** 1984. Research strategy for ocean disposal: Conceptual framework and case study. Society of Environmental Toxicology and Chemistry, Special Publication No. 1 (in press).

3. **Cairns, J., Jr., K. L. Dickson and A. W. Maki** (eds.). 1978. *Estimating the Hazard of Chemical Substances to Aquatic Life.* ASTM STP-657. American Society for Testing and Materials, Philadelphia, PA.

Chapter 14

Toxicological Methods for Determining the Effects of Contaminated Sediment on Marine Organisms

Richard C. Swartz

INTRODUCTION

Most chemicals and waste materials discharged into the marine environment contain fractions that eventually accumulate in sediment. Uncertainty about the ecological significance of sediment contamination to benthic, epibenthic, and pelagic communities has prompted development of methods for testing sediment toxicity. The Environmental Protection Agency (EPA)/Army Corps of Engineers (USACE) Implementation Manual [1], mandated by the 1977 Ocean Dumping Regulations, provided the first guidance for sediment bioassays. During the next 7 years, a great variety of ecotoxicological methods were used to evaluate sediment contamination. This paper reviews these methods and evaluates their application in regulatory, monitoring, and research programs.

OVERVIEW

Representative methods for determining the toxicity of sediment to marine organisms are summarized in Table 14.1. Most of these techniques were not developed specifically for use in permit programs of regulatory agencies. Response criteria range from biochemical through community levels of biological organization. With some exceptions, the response criteria fall into four groups: larval development and survival, behavior (especially sediment

Table 14.1   Bioassay Methods for Determining the Effects of Contaminated Sediment on Marine Organisms

Response criterion	Species	Collection site of tested sediment; or toxic material added to control sediment	Sediment phase[a]	Reference
Detoxifying enzyme induction	Polychaete, *Capitella capitata*	benz(a)anthracene crude oil	So, 2000 g, FT, 10 wk	(2)
Bioluminescence inhibition	Bacterium, *Photobacterium phosphoreum*	Puget Sound	Li, St, 15 min	(56)
Chromosome damage	Fish, *Salmo gairdnerii*	Puget Sound	Li, St, 3 d	(3, 57)
Egg/larval development	Fish, *Hypomesus pretiosus*	Puget Sound	Su, Li, 80 g, 35 d +	(4)
Larval development	Bivalve, *Crassostrea gigas*	Puget Sound	Su, 15 g, St, 2 d	(5)
Larval physiology,	Shrimp, *Palaemonetes pugio*; polychaete, *Polydora sp.*	Charleston Harbor	Su, Li, St, 1 min- 30 d	(6)
Larval mortality	Fishes, *Brevoortia tyrannus, Lagodon rhomboides, Paralichthys dentatus, P. albigutta, P. lethostigma, Leiostomus xanthurus, Micropogon undulatus*	Charleston Harbor	Li, St, 4 d	(7)
Respiration	Oligochaeta, *Monopylephorus cuticulatus*	Puget Sound	So, Li, St, 8 h	(3)
Osmoregulation	Fish, *Oncorhynchus kisutch*	Grays Harbor	Su, FT, 9 d	(8)
Sediment avoidance	Echinoderms, *Brisaster latifrons Chiridota laevis*; squat-lobster, *Munida quadrispina*; crabs, *Cancer magister, Hemigrapsus oregonesis, Chionoecetes bairdi*; shrimps, *Crangon alaskensis, Pandalus danae*; bivalve, *Macoma inconspicua*; amphipod, *Corophium slamonis*	NW Canadian coast	So, St, 1-3 d	(9)

Table 14.1 (contd.)

Response criterion	Species	Collection site of tested sediment; or toxic material added to control sediment	Sediment phase[a]	Reference
Sediment avoidance	Amphipod, *Rhepoxynius* spp	sewage, Zn, Cd, Palos Verdes Shelf	So, Ft, 3 d	[10]
Sediment avoidance	Crab, *Carcinus maenas*; shrimp, *Crangon crangon*; fishes, *Solea solea, Pomatoschistus minutus*	parathion, malathion	So, 3.3 L, FT, 10 min	[11]
Burrowing behavior	Fish, *Ammodytes hexapterus*	crude oil	So, Ft, 4 d	[12]
Burrowing behavior	Bivalve, *Mercenaria mercenaria*	crude oil	So, 40 L, FT, 4 d	[13]
Burrowing behavior	Bivalve, *Protothaca staminea*	Cu	So, 3.3-7.1 L, FT. 2 d +	[14]
Burrowing behavior	Bivalves, *Cerastoderma edule, Abra alba, Macoma baltica*; polychaetes, *Nereis diversicolor, Scoloplos armiger*	parathion, malathion	So, 3.3 L, FT, 20 h	[11]
Feeding activity	Polychaete, *Arenicola cristata*	kepone	So, 62 L, St, 6 d	[15]
Predator avoidance	Bivalve, *Protothaca staminea*	crude oil	So, 135 L, Fe, 13-29 d	[16]
Reproduction, mortality	Mysid shrimp, *Mysidopsis bahia*	East Bay	Su, FT, 4-28 d	[17]
Pathology, mortality	Fish, *Leiostomus xanthurus*	Elizabeth River	So, FT, 8 d	[58]
Acute mortality	Amphipod, *Rhepoxynius abronius* bivalves, *Protothaca staminea, Macoma inquinata*; Polychaete, *Glycinde picta*; Cumacea	Puget Sound Raritan River Houston Ship Channel	So, 1.8 L, FT, 10 d	[18]
Acute mortality	Amphipod, *Rhepoxynius abronius*	Commencement Bay, WA Palos Verdes Shelf sewage sludge, CD	So, 0.18 L, St, 10 d	[19, 20, 21, 22, 59, 60]
Acute mortality	Fishes, *Fundulus heteroclitus, Leiostomus xanthurus*; bivalve, *Mya arenaria*	Baltimore Harbor	Su, St, 2 d	[23]

Table 14.1 (Contd.)

Response criterion	Species	Collection site of tested sediment; or toxic material added to control sediment	Sediment phase[a]	Reference
Acute mortality	Fishes, Menidia menidia, Cyprinodon variegatus, Fundulus similis	New York Harbor	So, Su, 3.5 L, FT, 52 d	[24]
Acute mortality	Shrimp, Crangon septemspinosa; polychaete, Nereis virens	endrin, endosulfan, dieldrin, chlordane DDT, PCB, HCB	So, 500 g, St, 4-12 d	[25, 26]
Acute mortality	Shrimp, Palaemonetes pugio	Hampton Roads Duwamish River, WA Los Angeles Harbor	So, Su, Li, St, 4-10 d	[27, 28]
Acute mortality	Copepods, Acartia tonsa, Tigriopus californicus; shrimps, Palaemonetes pugio, P. vulgaris, Mysidopsis bahia; amphipod, Parahaustorius sp; isopod; Sphaeroma quadridentatum	Duwamish River, New York Harbor, Long Island Sound	So, Su, LI, St, 4-15 d	[29]
Acute mortality, pathology	Amphipod, Ampelisca abdita	Cu, As	metals in seawater, 4-6 d	[30]
Mortality	Shrimp, Crangon spp: crab, Cancer magister; mussel, Mytilus californianus; tunicate, Ascidia ceratodes; lobster, Homarus americanus	San Francisco Bay	Su, FT, 3 weeks	[31]
Population growth	Nematodes, Chromadorina germanica, Diplolaimella punicea	New York Bight	So, 0.06 L, St, 2 wk	[32]
Full life cycle Community recolonization	Polychaete, Capitella capitata Macrobenthos	Puget Sound sewage sludge	So, Su, Li, St, 35 d + So, 6.7 L, Fe, 4 wk	[4, 61] [33]

Table 14.1 (Contd.)

Response criterion	Species	Collection site of tested sediment; or toxic material added to control sediment	Sediment phase[a]	Reference
Community recolonization	Macrobenthos	barite; PCB	So, 3 L, FT, 10 wk	[34, 35, 36]
Community recolonization	Macrobenthos	crude oil	So, 6.8 L, FE, 15 mo	[37]
Community/population density	Macrobenthos	sewage sludge	So, 4 m^2 plots, FE, 8 mo	[38]
Community structure	Macrobenthos	sewage sludge	So, 1.2 m^2 plots, FE, 12 mo	[39]
Microcosm structure/function	Macrobenthos	phthalate ester	So, FT, 30 d	[40]
Microcosm structure/function	Macrobenthos and meiobenthos	fuel oil, oil mousse; Narragansett Bay	So, 800 L +, FT, 7 mo +	[41, 42, 43, 44, 45]

[a] So = Solid; Su = Suspended; Li = Liquid Sediment Mass (L or g); Fe = Field Exposure; Seawater Supply: St = Static; FT = Flow-through; Duration

avoidance and burrowing), acute mortality, and community structure or function. Effects on biochemistry, genotoxicity, physiology, and population dynamics have been examined less frequently. A great variety of fish and benthic invertebrate species have been tested. Most assessments of sediment toxicity involve more than one species, although they are usually tested separately by adding a known number of specimens to the exposure chamber. When community response is examined, densities are not manipulated and animals are transferred from the field as intact assemblages or allowed to recruit as larvae from the water column in flow-through or field exposures.

Bioassays have been applied to material collected from contaminated field sites or to unpolluted sediment to which toxic materials have been added. Sediments from many contaminated field sites have now been examined, but bioassays with specific chemicals are not common. The intact ("solid phase") sediment is usually tested, and suspended or liquid sediment phases are sometimes examined, especially in relation to dredge material disposal. In contrast to the literature on the bioaccumulation of pollutants, relatively few reports concern relationships between sediment toxicity and the geo-chemistry of contaminants. The quantity of sediment and duration of exposure vary greatly: from $<< 1$ L to > 800 L and from 10 min to more than a year, respectively. Both static and flow-through seawater systems have been employed successfully. The simpler methods (i.e., acute mortality in static seawater systems with relatively small quantities of sediment) are commonly used in regulatory programs. Field validation of the ecological relevance of such systems is rare. The more logistically complicated methods (i.e., macrobenthic community response during long-term exposures) appear to provide a better simulation of natural benthic ecosystem.

RESEARCH NEEDS AND ISSUES

Test Species Selection

The primary criteria for test species selection should be the species' ecological importance and its relative sensitivity to sediment contamination. Laboratory bioassays and field surveys have shown that many species survive well in nature where sediment is acutely toxic to more sensitive species [19, 46, 47]. It is possible to select indigenous, ecologically significant, and easily collected taxa that will survive in almost any sediment. Some of these taxa (e.g., *Macoma, Nereis, Glycinde*) are included on the EPA/USACE [1] list of recommended species for solid-phase bioassays. Such species may be appropriate for some sublethal response criteria, but they should be avoided for acute toxicity bioassays because of their tolerance to sediment contaminants.

The best guide for bioassay species selection lies in the wealth of data available about the distribution of the macrobenthos near sites of contaminated sediment. On the basis of the extensive macrobenthic surveys that have been conducted in the Southern California Bight, Mearns and Word (48) identified nineteen species that were dominant in control or background areas and decreased in abundance with increasing proximity to sewage discharges. Twelve of the nineteen species were either phoxocephalid or ampeliscid amphipods. The distribution of amphipods in Puget Sound (19), the New York Bight (49), and many other coastal sites indicates that many amphipods are appropriately sensitive to sediment contaminants. This sensitivity has been confirmed in a variety of bioassays (10, 19, 30, 33, 46). The field distributions of many untested taxa also indicate promise as bioassay species, e.g., brittlestars (*Amphiodia*), polychaetes (*Sthenelanella*) and phoronids (*Phoronis*) (48).

Experimental comparison of the effects of contaminated sediment on different marine organisms is very relevant to test species selection. Shuba et al. (29) reported that the copepod *Acartia tonsa* was more sensitive than the copepod *Tigriopus californicus* to the liquid and suspended phases of Duwamish River, Washington, sediment. They also found that the isopod *Sphaeroma quadridentatum*, the mysid *Mysidopsis bahia*, the amphipod *Parahaustorius* sp., and larvae of the shrimp *Palaemonetes* spp. were more sensitive than adult *Palaemonetes* and the bivalves *Rangia cuneata* and *Mercenaria mercenaria* to the solid phase of several field sediments. Swartz et al. (18) found that the amphipod *Rhepoxynius abronius* and several cumaceans were more sensitive than two molluscs (*Macoma inquinata*, *Protothaca staminea*) or a polychaete (*Glycinde picta*) to sediment from the Houston Ship Channel and other field sites. Additional comparisons like these are desirable not just for the selection of test species, but also to elucidate factors controlling the distribution of different invertebrates along pollution gradients.

Two factors that are correlated with a species' sensitivity to sediment contaminants are its phylogenetic position and its relation to the substrate. For example, amphipods and other crustaceans are generally more sensitive than molluscs and polychaetes. Within phylogenetic groups, the infauna appear more sensitive than the epifaunal, demersal, or pelagic biota. Among the infauna, the most sensitive taxa are those that burrow beneath the sediment surface without extending tubes or body parts into overlying water, e.g., phoxocephalid amphipods. Such animals are greatly dependent on the quality of interstitial water and sediment particulates.

Shuba et al. (50) discussed several criteria for selection of test species and a rating system for comparing the suitability of several species. The criteria include whether the species: is indigenous to the disposal site or closely related to an indigenous species; is available through collecting or purchase; has a toxicological data base; can be maintained and cultured in the laboratory; is ecologically and economically important; has a wide geographic distribution; and is compatible with other test species. Many species that may be appropriate for sediment bioassays do not meet these criteria simply because the focus of pollution ecology has historically been on water quality rather than sediment quality. Culture methods and toxicological data bases have not been developed for many infaunal animals. As these data are developed, they should include information about the effects of natural sediment variables (organic content, particle size distribution, interstitial water salinity, etc.) on test species. Analysis of sediment bioassay data should be able to discriminate effects of chemical contamination from those attributable to habitat alteration. Use of indigenous species promotes site specificity and allows comparison of sediment toxicity with the field distribution of the test species. However, indigenous species that are tolerant of sediment contaminants should not be selected.

Selection of Response Criterion

Bioassays with mortality as an endpoint are usually not very sensitive to sediment contaminants, but these bioassays have the advantage of being more easily interpreted than sublethal tests (46). Such bioassays are included in the EPA/USACE Implementation Manual (1) and are widely applied in permit programs for ocean disposal. The key to the success of these acute, lethal bioassays lies in the selection of sensitive test species like phoxocephalid and ampeliscid amphipods.

Sublethal responses are generally more sensitive, but their application in permit programs is sometimes limited by the "So what?" syndrome. The ecological significance of induction of an enzyme system or changes in the rate of respiration is less obvious than mortality, genotoxicity, or reductions in population density and species richness. However, sublethal endpoints can indicate exposure to bioavailable contaminants in sediment, and the biological processes through which benthic organisms attempt to adapt to the exposure. Behavioral responses are particularly relevant. Oakden et al. (10) found that phoxocephalid amphipods avoided sediment contaminated with metals and sewage and concluded that behavioral avoidance might explain the absence of phoxocephalids near sewage outfalls. Pearson et al. (16) demonstrated that shallow burial and slower reburrowing of the clam *Protothaca staminea* in oiled sediment led to increased predation on the clam by Dungeness crabs (*Cancer magister*). They noted that this ecological effect would not have been predicted by standard toxicity tests.

Use of macrobenthic or meiobenthic assemblages has many advantages over use of one or a few test species. These designs are usually based on larval recruitment or collection of entire assemblages, and therefore no a priori species selection is required. Scaling of micro- and mesocosms with respect to size, duration, and ecological components generally provides a much better simulation of natural benthic ecosystems. These systems are highly desirable research tools, but their application in routine regulatory programs is limited by logistical constraints.

Research is needed to compare the efficacy of different bioassay methods. Chapman et al. (3, 4, 5, 46, 47) compared several methods as part of a comprehensive survey of sediment toxicity in Puget Sound. These methods include acute lethality of an oligochaete, amphipod, polychaete, and fish; effects on the respiration of an oligochaete; success of oyster and fish larval development; chromosomal damage to fish gonad cells (anaphase aberration); and alterations in the population dynamics of a polychaete. They found that individual tests gave different toxic responses to sediment from the same station. Acute lethality was fairly rare and restricted to a few sensitive species like phoxocephalid amphipods. They believed that different tests might be sensitive to different contaminants. The anaphase aberration test for genotoxicity was very sensitive to sediment extracts. Despite differences in relative sensitivity there was a good correspondence among the methods in geographic trends of sediment toxicity. Chapman and Long (46) concluded that several bioassays with different response criteria should be used to evaluate toxicity.

Interlaboratory comparisons of bioassay methods are needed to establish confidence in the reproducibility of results. Five laboratories in the Puget Sound area have recently conducted phoxocephalid amphipod bioassays on seven different sediments collected from four field sites or contaminated artificially with three concentrations of cadmium (60). There was good agreement among the laboratories in the relative and absolute toxicity of the test sediments.

Experimental Design

Test duration and size of the bioassay chamber vary greatly with the response criterion and test species (Table 1). Ten-day exposures seem appropriate for acute mortality tests as recommended by the EPA/USACE Implementation Manual (1, 22). Sublethal effects, especially behavioral changes, are evident usually in less than 10 d. Impacts on populations or communities depend on longer-term biological processes (recruitment, reproduction, growth, competition) that usually require several months or longer for completion. An exception is the bioassay developed recently by Tietjen and Lee (32) that measures impacts on the population dynamics of nematodes after 2 weeks' exposure to test sediment. Meiobenthic taxa like nematodes are rarely used in sediment bioassays or benthic pollution ecology in general. Because of their abundance, ecological significance, and short life histories, the meiobenthos offer experimental opportunities that should be exploited (32).

Differences between flow-through and static seawater systems may not be important for acute sublethal and lethal bioassays. Such tests generally do not attempt to simulate transport processes of bioturbation and physical mixing of sediment. Since molecular diffusion is a minor short-term mechanism of exchange of solutes between interstitial and overlying waters (51), contaminant concentrations remain essentially constant during acute bioassays. Static systems are not acceptable if an attempt is made to simulate the fate of contaminants.

Geochemical properties of sediments change during storage. Drying, freezing, and cold storage affect toxicity (52, 53, 54). Swartz et al. (22) recommended that sediment be stored at 4°C for no more than 5 d before the initiation of bioassays.

Effects of storage are only one manifestation of the lack of knowledge about geochemical changes in sediment in the laboratory. In particular, redox potential and partitioning of contaminants between interstitial water and particles are altered under laboratory conditions, but rarely monitored during bioassays. The ecological significance of bioassays depends on the generally untested assumption that contaminant bioavailability in the laboratory simulates conditions in natural benthic ecosystems. This uncertainty is especially serious for bioassays in which contaminants are experimentally added to sediment.

There are several applications of sediment bioassays in marine pollution ecology. As part of permitting programs, they are used to determine the acceptability of dredged materials and other wastes for ocean disposal (1). This involves a comparison of the toxicity of the waste material to a sediment that represents the disposal site or some other reference condition. It is possible to add complex wastes like sewage sludge to uncontaminated sediment and obtain an estimate of the relative toxicity of wastes from different sources (20). Unfortunately, measures of the toxicity of complex wastes cannot be easily related to specific chemicals. The question of causal factors can be directly examined by adding known quantities of specific chemicals to nontoxic sediment (25, 26, 59). Bioassay results are used to estimate LC_{50}s or related measures of safe concentrations. This provides at least a conceptual basis for sediment quality criteria. However, the success of this relatively new research area depends largely on the problem of relating toxicity to a measure of the bioavailable fraction of bulk sediment concentrations. Spatial distributions of sediment toxicity near disposal sites or urbanized embayments can be identified by bioassays of field samples collected at many stations in the area of interest (3, 4, 19, 21, 23, 27). Such surveys can efficiently locate "hot spots" where more comprehensive study is warranted. They can be applied in monitoring programs to detect unacceptable perturbations, e.g., beyond a dump site boundary; or in research projects where sediment toxicity can be compared with sediment contamination or the structure of benthic communities. Sediment bioassays are also used to address more fundamental questions about processes of degradation and recovery of contaminated benthic ecosystems (37, 45). This research usually involves field exposures or larger laboratory systems that attempt to simulate natural conditions accurately.

It is difficult to identify specific causes of toxicity associated with sediment collected from field sites. There are three general problems: (1) to discriminate effects of sediment chemical contaminants (metals, PCBs, etc.) from effects of natural sediment properties (organic carbon, grain size, etc.), (2) to discriminate effects of different chemical contaminants, and (3) to identify and quantify interactions between different contaminants and between contaminants and natural sediment properties. The first problem is important because organic enrichment and chemical contamination have different regulatory implications (20). The structure of benthic communities is greatly affected by the organic carbon content of sediment (62). In chemically contaminated benthic habitats such as the Palos Verdes Shelf, California, organic enrichment and anaerobic sediment conditions appear to have a greater anthropogenic influence on the benthos than chemical contaminants (21). If sediment bioassays are used to detect toxicity associated with chemical pollutants, it is necessary to demonstrate that the bioassay is not responding to other substrate alterations. The second problem arises because polluted marine habitats are often affected by many chemical contaminants. A method has recently been proposed to identify a toxicity "apparent effect threshold" (AET) for individual chemicals based on chemical and bioassay data for field-collected sediment samples (63). The AET is defined as the contaminant concentration above which significant sediment toxicity would always be expected. The AET method is most useful when sediment toxicity is associated primarily with single factors. The third problem, concerning synergistic or other interactions, has received little attention. Some method of quantifying these interactions must be developed before it will be possible to identify the joint toxic action of combinations of chemicals in sediment. Fortunately, some conceptual and mathematical models exist for describing interaction between chemicals dissolved in water (64, 65, 66). It may be possible to apply these models directly to interstitial water or to particle-bound contaminant concentrations after accounting for partitioning to the aqueous phase.

There are inherent limitations in the ability of acute laboratory bioassays to reflect the full range of ecological consequences of sediment contamination. Nonetheless, such bioassays are being used in hazard assessment protocols, often in the initial tier of biological evaluation. When bioassays determine acceptability for ocean disposal of waste material, their results should be applied in an environmentally protective manner. If there is any evidence for acute toxicity, ocean disposal should be denied or the waste material should be subjected to more comprehensive analyses. It may therefore be inappropriate to use normal statistical conventions to compare control and test material data. If a null hypothesis of no difference between control and test means is tested at the 95% significance level (alpha = 0.05), there can be no more than a 5% chance of rejecting a true null hypothesis (type I error) (55). However, the probability of not rejecting a false null hypothesis (type II error) can be very large (as great as 95%). Normal statistical criteria therefore can permit the ocean discharge of waste materials that are in reality more toxic than the bioassay control. The solution to this problem is counterintuitive to most scientists, i.e., the significance level must be decreased rather than increased. The probability of a type II error (and the probability of discharging toxic wastes) would be greatly reduced if the null hypothesis of no difference between control and tests means is rejected at alpha values exceeding 0.50, i.e, if the probability of a type I error is 50% or more. Alternative statistical methods that directly define the change of a type II error could also be used.

Field Validation

The ecological relevance of several bioassays has been tested by comparing the toxicity of sediment along a pollution gradient with biological communities or sediment contamination along the same gradient (10, 19, 21, 32, 47). Swartz et al. (19) found that phoxocephalid amphipods were absent from the industrialized waterways of Commencement Bay, Washington, where sediment was acutely toxic in laboratory bioassays to the phoxocephalid, *Rhepoxynius abronius*, but phoxocephalids were ubiquitous in the deeper portion of the bay where sediment was not toxic. Similarly, the survival of *R. abronius* in bioassays of sediment from the Palos Verdes Shelf, California, near the Los Angeles County sewage outfall, was significantly correlated with the spatial distribution of phoxocephalids, the structure of benthic communities, and sediment contamination (21). Oakden et al. (10) showed that phoxocephalids avoided contaminated sediment from the Palos Verdes Shelf. Tietjen and Lee (32) reported significant correlations between chemical contamination and the growth of nematode populations in sediment collected along a pollution gradient in the Hudson-Raritan estuarine area adjacent to the New York Bight.

The ecological relevance of marine mesocosms is dependent on the strategy of either capturing or precisely simulating a natural system. The mesocosms of the Marine Ecosystem Research Laboratory at the University of Rhode Island simulate both benthic and pelagic dynamics of Narragansett Bay (41-44). This simulation is greatly enhanced by the initial placement of intact sediment cores in the mesocosm and the addition of Narragansett Bay water to the mesocosm during experiments.

Research Priorities

There are innumerable methodological problems that should be resolved. These include comparative study of species sensitivity and different response criteria, development of better sediment storage methods, more realistic incorporation of sediment bioassays into hazard assessment protocols, and field validation of bioassay results. However, sediment bioassay methods are sufficiently advanced that research should now be directed toward more fundamental questions. Three research needs that have a high priority are (1) the relation between toxicity and the bioavailability of sediment contamination, (2) the importance of interactions between contaminants in determining sediment toxicity, and (3) the functional significance of sediment toxicity, particularly in relation to trophic dynamics and benthic-pelagic coupling. Bioassay methods that concern sublethal and lethal effects on individuals may present the best experimental approach to the first two of these research areas. Clearly, functional changes in benthic ecosystems should be addresses in larger-scale and longer-term experiments with micro- and mesocosms.

REFERENCES

1. **Environmental Protection Agency/U.S. Army Corps of Engineers.** 1977. Ecological evaluation of proposed discharge of dredged material into ocean waters; Implementation manual for section 103 of PL 92-532. Environmental Effects Laboratory, U.S. Army Engineer Waterways Experiment Station, Vicksburg, MS.

2. **Lee, R. F., S. C. Singer, K. R. Tenore, W. S. Gardner and R. M. Philpot.** 1979. Detoxification system in polychaete worms: Importance in the degradation of sediment hydrocarbons. In W. B. Vernberg, A. Calabrese, F. P. Thurberg and F. J. Vernberg, eds., Marine Pollution: Functional Responses. Academic Press, New York, pp. 23-37.

3. **Chapman, P. M., G. A. Vigers, M. A. Farrell, R. N. Dexter, E. A. Quinlan, R. M. Kocan and M. Landolt.** 1982. Survey of biological effects of toxicants upon Puget Sound biota. I. Broad-scale toxicity survey. NOAA Technical Memorandum OMPA-25, Boulder, CO.

4. **Chapman, P. M., D. R. Munday, J. Morgan, R. Fink, R. M. Kocan, M. L. Landolt and R. N. Dexter.** 1983. Survey of biological effects of toxicants upon Puget Sound biota. II. Tests of reproductive impairment. NOAA Technical Report NOS 102 OMS 1, Rockville, MD.

5. **Chapman, P. M. and J. D. Morgan.** 1983. Sediment bioassays with oyster larvae. *Bull. Environ. Contam. Toxicol.* 31:438-444.

6. **DeCoursey, P. J. and W. B. Vernberg.** 1975. The effect of dredging in a polluted estuary on the physiology of larval zooplankton. *Water Res.* 9:149-154.

7. **Hoss, D. E., L. C. Coston and W. E. Schaaf.** 1974. Effects of sea water extracts of sediments from Charleston Harbor, S. C., on larval estuarine fishes. *Estuar. Coastal Mar. Sci.* 2:323-328.

8. **Kehoe, D. M.** 1983. Effects of Grays Harbor estuary sediment on the osmoregulatory ability of coho salmon smolts (*Oncorhynchus kisutch*). *Bull. Environ. Contam. Toxicol.* 30:522-529.

9. **Chang, B. D. and C. D. Levings.** 1976. Laboratory experiment on the effects of ocean dumping on benthic invertebrates. I. Choice tests with solid wastes. Fisheries and Marine Service Technical Report No. 637, West Vancouver, B.C.

10. **Oakden, J. M., J. S. Oliver and A. R. Flegal.** 1984. Behavioral responses of a phoxocephalid amphipod to organic enrichment and trace metals in sediment. *Mar. Ecol. Progr. Series* 14:253-257.

11. **Mohlenberg, F. and T. Kiorboe.** 1983. Burrowing and avoidance behaviour in marine organisms exposed to pesticide-contaminated sediment. *Mar. Pollut. Bull.* 14:57-60.

12. **Pearson, W. H., D. L. Woodruff, P. C. Sugarman and B. O. Olla.** 1984. The burrowing behavior of sand lance, *Ammodytes hexapterus:* Effects of oil-contaminated sediment. *Mar. Environ. Res.* 11:17-32.

13. **Olla, B. L. and A. J. Bejda.** 1983. Effects of oiled sediment on the burrowing behaviour of the hard clam, *Mercenaria mercenaria. Mar. Environ. Res.* **9**:183-193.

14. **Phelps, H. L., J. T. Hardy, W. H. Pearson and C. W. Apts.** 1983. Clam burrowing behaviour: Inhibition by copper-enriched sediment. *Mar. Pollut. Bull.* **14**:452-455.

15. **Rubinstein, N.I.** 1979. A benthic bioassay using time-lapse photography to measure the effect of toxicants on the feeding behavior of lugworms (Polychaeta: Arenicolidae). In W. B. Vernberg, A. Calabrese, F. P. Thurberg and F. J. Vernberg, eds., *Marine Pollution: Functional Responses.* Academic Press, New York, pp. 341-351.

16. **Pearson, W.H., D. L. Woodruff, P. C. Sugarman and B. L. Olla.** 1981. Effects of oiled sediment on predation on the littleneck clam, *Protothaca staminea,* by the Dungeness crab, *Cancer magister. Estuar. Coastal Shelf Sci.* **13**:445-454.

17. **Nimmo, D. R., T. L. Hamaker, E. Matthews and W. T. Young.** 1982. The long-term effects of suspended particulates on survival and reproduction on the mysid shrimp, *Mysidopsis bahia,* in the laboratory. In G. F. Mayer, ed., *Ecological Stress and the New York Bight: Science and Management.* Estuarine Research Federation, Columbia, SC, pp. 413-422.

18. **Swartz, R. C., W. A. DeBen and F. A. Cole.** 1979. A bioassay for toxicity of sediment to marine macrobenthos. *J. Water Pollut. Control Fed.* **51**:944-950.

19. **Swartz, R. C., W. A. DeBen, K. A. Sercu and J. O. Lamberson.** 1982. Sediment toxicity and the distribution of amphipods in Commencement Bay, Washington, USA. *Mar. Pollut. Bull.* **13**:359-364.

20. **Swartz, R. C., D. W. Schults, G. R. Ditsworth and W. A. DeBen.** 1984. Toxicity of sewage sludge to *Rhepoxynius abronius,* a marine benthic amphipod. *Arch. Environ. Contam. Toxicol.* **13**:207-216.

21. **Swartz, R. C., D. W. Schults, G. R. Ditsworth, W. A. DeBen and F. A. Cole.** 1985. Sediment toxicity, contamination, and macrobenthic communities near a large sewage outfall. In T. P. Boyle, ed., *Validation and Predictability of Laboratory Methods for Assessing the Fate and Effects of Contaminants in Aquatic Ecosystems.* ASTM STP 865. American Society for Testing and Materials, Philadelphia, PA, pp. 152-175.

22. **Swartz, R. C., W. A. DeBen, J. K. P. Jones, J. O. Lamberson and F. A. Cole.** 1985. Phoxocephalid amphipod bioassay for marine sediment toxicity. In R. D. Cardwell, R. Purdy and R. C. Bahner, eds., *Aquatic Toxicology and Hazard Assessment: Seventh Symposium.* ASTM STP 854. American Society for Testing and Materials, Philadelphia, PA, pp. 284-307.

23. **Tsai, C., J. Welch, K. Chang, J. Shaefer and L. E. Cronin.** 1979. Bioassay of Baltimore Harbor sediments. *Estuaries* **2**:141-153.

24. **Rubinstein, N. I., W. T. Gilliam and N. R. Gregory.** 1983. Evaluation of three fish species as bioassay organisms for dredged material testing. EPA-600/X-83-062. U.S. Environmental Protection Agency, Gulf Breeze, FL.

25. **McLeese, D. W. and C. D. Metcalfe.** 1980. Toxicities of eight organochlorine compounds in sediment and seawater *Crangon septemspinosa. Bull. Environ. Contam. Toxicol.* **25**:921-928.

26. **McLeese, D. W., L. E. Burridge and J. Van Dinter.** 1982. Toxicities of five organochlorine compounds in water and sediment in *Nereis virens. Bull. Environ. Contam. Toxicol.* **28**:216-210.

27. **Alden, R. W., III and R. J. Young, Jr.** 1982. Open ocean disposal of materials dredged from a highly industrialized estuary: An evaluation of potential lethal effects. *Arch. Environ. Contam. Toxicol.* **11**:567-576.

28. **Lee, G. F. and G. M. Mariani.** 1976. Evaluation of the significance of waterway sediment-associated contaminants on water quality at the dredged material disposal site. In F. L. Mayer and J. L. Hamelink, eds., *Aquatic Toxicology and Hazard Evaluation.* ASTM STP 634. American Society for Testing and Materials, Philadelphia, PA, pp. 196-213.

29. **Shuba, P. J., H. E. Tatem and J. H. Carroll.** 1978. Biological assessment methods to predict the impact of open-water disposal of dredged material. Technical Report D-78-50. U.S. Army Engineer Waterways Experiment Station, Vicksburg, MS.

30. **Scott, K. J., P. P. Yevich and W. S. Boothman.** Toxicological methods using the benthic amphipod *Ampelisca abdita* Mills. (Submitted to *Aquat. Toxicol.*)

31. **Peddicord, R. K.** 1980. Direct effects of suspended sediments on aquatic organisms. In R. A. Baker, ed., *Contaminants and Sediments,* Vol. 1. Ann Arbor Science Publishers, Ann Arbor, MI, pp. 501-536.

32. **Tietjen, J. H. and J. J. Lee.** 1984. The use of free-living nematodes as a bioassay for estuarine sediments. *Mar. Environ. Res.* 11:233-251.

33. **MacKenzie, C. L., Jr., D. Radosh and R. N. Reid.** 1983. Reduced numbers of early stages of Mollusca and Amphipoda on experimental sediments containing sewage sludge. *Coastal Ocean Pollution Assessment News* 3:7-8.

34. **Hansen, D. J. and M. E. Tagatz.** 1980. A laboratory test for assessing impacts of substances on developing communities of benthic estuarine organisms. In J. G. Eaton, P. R. Parrish and A. C. Hendricks, eds., *Aquatic Toxicology.* ASTM STP 707. American Society for Testing and Materials, Philadelphia, PA, pp. 40-57.

35. **Hansen, D. J.** 1974. Aroclor 1254: Effect on composition of developing estuarine animal communities in the laboratory. *Mar. Sci.* 18:19-33.

36. **Tagatz, M. E. and M. Tobia.** 1978. Effects of barite (BaSO₄) on development of estuarine communities. *Estuar. Coastal Mar. Sci.* 7:401-407.

37. **Vanderhorst, J. R., J. W. Blaylock, P. Wilkinson, M. Wilkinson and G. Fellingham.** 1980. Effects of experimental oiling on recovery of Strait of Juan De Fuca intertidal habitats. EPA-600/7-81-008. U.S. Environmental Protection Agency, Washington, DC.

38. **Young, D. K. and M. W. Young.** 1978. Regulation of species densities of seagrass-associated macrobenthos: Evidence from field experiment in the Indian River estuary, Florida. *Mar. Res.* 36:569-593.

39. **Eleftheriou, A., D. C. Moore, D. J. Basford and M. R. Robertson.** 1982. Underwater experiments on the effects of sewage sludge on a marine ecosystem. *Neth. Sea Res.* 16:465-473.

40. **Perez, K. T.** 1983. Environmental assessment of a phthalate ester, di(2-ethylhexyl) phthalate (DEHP), derived from a marine microcosm. In W. E. Bishop, R. D. Cardwell and B. B. Heidolph, eds., *Aquatic Toxicology and Hazard Assessment.* ASTM STP 801. American Society for Testing and Materials, Philadelphia, PA, pp. 180-191.

41. **Elmgren, R., G. A. Vargo, J. F. Grassle, J. P. Grassle, D. R. Heinle, G. Langlois and S. L. Vargo.** 1980. Trophic interactions in experimental marine ecosystems perturbed by oil. In J. P. Giesy, Jr., ed., *Microcosms in Ecological Research.* U.S. Department of Energy, Symposium Series 52, Washington, DC, pp. 779-800.

42. **Grassle, J. F., R. Elmgren and J. P. Grassle.** 1981. Response of benthic communities in MERL experimental ecosystems to low level chronic additions of No. 2 fuel oil. *Mar. Environ. Res.* 4:279-297.

43. **Oviatt, C., J. Frithsen, J. Gearing and P. Gearing.** 1982. Low chronic additions of No. 2 fuel oil: Chemical behavior, biological impact and recovery in a simulated estuarine environment. *Mar. Ecol. Prog. Ser.* 9:121-136.

44. **Oviatt, C. A., M. E. Q. Pilson, S. W. Nixon, J. B. Frithsen, D. T. Rudnick, J. R. Kelly, J. F. Grassle and J. P. Grassle.** 1984. Recovery of a polluted estuarine system: A mesocosm experiment. *Mar. Ecol. Prog. Ser.* 16:203-217.

45. **Kuiper, J., P. DeWilde and W. Wolff.** 1984. Effects of an oil spill in outdoor model tidal flat ecosystems. *Mar. Pollut. Bull.* 15:102-106.

46. **Chapman, P. M. and E. R. Long.** 1983. The use of bioassays as part of a comprehensive approach to marine pollution assessment. *Mar. Pollut. Bull.* 14:81-84.

47. **Chapman, P. M., R. N. Dexter, R. M. Kocan and E. R. Long.** 1985. An overview of biological effects testing in Puget Sound, Washington: Methods, results and implications, in R. D. Cardwell, R. Purdy and R. C. Bahner, eds., *Aquatic Toxicology and Hazard Assessment: Seventh Symposium.* ASTM STP 854. American Society for Testing and Materials, Philadelphia, PA, pp. 344-363.

48. **Mearns, A. J. and J. Q. Word.** 1982. Forecasting effects of sewage solids on marine benthic communities. In G. F. Mayer, ed., *Ecological Stress and the New York Bight: Science and Management.* Estuarine Research Federation, Columbia, SC, pp. 495-512.

49. **National Marine Fisheries Service.** 1972. The effects of waste disposal in the New York Bight. Final Report, Section 2: Benthic studies. NTIS Rept. No. AD739532. Sandy Hook Sports Fisheries Mar. Lab., Highland, NJ.

50. **Shuba, P. J., S. R. Petrocelli and R. E. Bentley.** 1981. Considerations in selecting bioassay organisms for determining the potential environmental impact of dredged material. Technical Report EL-81-8. U.S. Army Engineer Waterways Experiment Station, Vicksburg, MS.

51. **Christensen, J. P., A. H. Devol and W. M. Smethie, Jr.** 1984. Biological enhancement of solute exchange between sediments and bottom water on the Washington continental shelf. *Continental Shelf Research* 3:9-23.

52. **Lee, G. F. and R. H. Plumb.** 1974. Literature review on research study for the development of dredged material disposal criteria. Contract Report D-74-1. U.S. Army Engineer Waterways Experiment Station, Vicksburg, MS.

53. **Lee, G. F. and R. A. Jones.** 1982. Dredged material evaluations: Correlation between chemical and biological evaluation procedures. *J. Water Pollut. Control Fed.* 54:406.

54. **Dillon, T.** 1983. The effects of storage temperature and time on sediment toxicity. Poster presented at Fourth Annual SETAC Meeting, Arlington, VA. U.S. Army Engineer Waterways Experiment Station, Vicksburg, MS.

55. **Sokal, R. R. and F. J. Rohlf.** 1981. *Biometry*, 2nd ed. W. H. Freeman and Company, San Francisco, CA.

56. **Schiewe, M. H., E. G. Hawk, D. I. Actor and M. M. Krahn.** 1985. Use of a bacterial bioluminescence assay to test toxicity of contaminated marine sediments. *Can. J. Fish. Aquat. Sci.* 42:1244-1248.

57. **Landolt, M. L. and R. M. Kocan.** 1984. Lethal and sublethal effects of marine sediment extracts on fish cells and chromosomes. *Helgolander Meersunters.* 37:479-491.

58. **Hargis, W. J., Jr., M. H. Roberts, Jr. and D. E. Zwerner.** 1984. Effects of contaminated sediments and sediment-exposed effluent water on an estuarine fish: acute toxicity. *Mar. Environ. Res.* 14:337-354.

59. **Swartz, R. C., G. R. Ditsworth, D. W. Schults and J. O. Lamberson.** 1986. Sediment toxicity to a marine infaunal amphipod: cadmium and its interaction with sewage sludge. *Mar. Environ. Res.* 18 (in press).

60. **Mearns, A. J., R. C. Swartz, J. M. Cummins, P. A. Dinnel, P. Plesha and P. M. Chapman.** 1986. Inter-laboratory comparison of a sediment toxicity test using the marine amphipod, *Rhepoxynius abronius. Mar. Environ. Res.* 18 (in press).

61. **Chapman, P. M. and R.Fink.** 1984. Effects of Puget Sound sediments and their elutriates on the life cycle of *Capitella capitata. Bull. Environ. Contam. Toxicol.* 33:451-459.

62. **Pearson, T. H. and R. Rosenberg.** 1978. Macrobenthic succession in relation to organic enrichment and pollution of the marine environment. *Oceanogr. Mar. Biol. Ann. Rev.* 16:229-311.

63. **Tetra Tech, Inc.** 1985. Commencement Bay nearshore/tideflats remedial investigation. Washington State Dept. Ecology, Olympia, WA. TC-3752, Vol. 1, section 4.2.

64. **Sprague, J. B.** 1970. Measurement of pollutant toxicity to fish. II. Utilizing and applying bioassay results. *Water Res.* 4:3-32.

65. **Voyer, R. A. and J. F. Heltshe.** 1984. Factor interactions and aquatic toxicity testing. *Water Res.* 18:441-447.

66. **Christensen, E. R. and C. Y. Chen.** 1985. A general noninteractive multiple toxicity model including probit, logit, and Weibull transformations. *Biometrics* 41:711-725.

Chapter 15

Toxicity of Sediment-Associated Metals to Freshwater Organisms: Biomonitoring Procedures

Wesley J. Birge, Jeffrey A. Black, Albert G. Westerman and Paul C. Francis

INTRODUCTION

It has been well established that many metals and persistent organic compounds that enter natural water resources accumulate to high levels in bottom sediments. Cadmium, mercury, zinc, and various other heavy metals have been reported at concentrations of 100 mg/kg or more in bottom sediments of numerous U.S. waterways (1-5). Moreover, it is known that sediment-bound metals are released to interstitial water or back into the water column at rates sufficient to exert adverse effects on resident biota. For example, Jernelov (6) showed that fish exposed to bottom sediments spiked to 100 μg Hg/g accumulated tissue mercury to 3 μg/g in 15 d. As discussed below, other investigators also have shown that sediment-associated chemicals affect not only the benthos but also epibenthic species. Concerning the latter, it is important to note that bottom sediments serve as spawning substrates for many species that inhabit the water column, and thus it is necessary to determine whether the remobilization rates of sediment-associated chemicals are sufficient to affect reproductive behavior, egg hatchability, or development of early life stages (e.g., mortality, teratogenesis, bioconcentration). In addition to lethality and abnormal development, bioconcentration may be particularly important as fish embryos and larvae are known to accumulate high chemical residues over a short period of time. For example, trout and catfish embryos and larvae exposed to 0.1 to 0.78 µg Hg/L in a continuous-flow system concentrated tissue mercury at 45 to 100 times the exposure level per day of treatment (7).

The effects of toxicant body burden on survival may be more critical for early life stages that lack significant quantities of adipose tissue. Certain metallic and lipophilic organic compounds that accumulate in lipid reservoirs of healthy adult stages are more prone to exert effects on essential biochemical processes in growing and differentiating embryos and larvae (8, 9). Thus it is possible that early life stages represent one of the most sensitive target sites for sediment-released chemicals on epibenthic species, especially those that spawn directly upon or close to bottom sediments. The two-fold purpose of this paper is to present a synoptic review of biomonitoring procedures used for evaluating the toxic potential of contaminated freshwater sediments and to discuss in more detail embryo-larval sediment bioassay methods.

APPLICATION OF TOXICOLOGICAL BIOMONITORING TO FRESHWATER SEDIMENTS

Following the study by Jernelov (6), Gillespie and Scott (10) performed static sediment bioassays with mercury using 40-L aquaria. A natural sediment containing 0.24 µg/g mercury (dry weight) was spiked to a level of 50 µg/g mercury, using mercuric chloride and mercuric sulfide. In the various tests, only small fractions of the sediment-bound mercury were released, judging from the tissue concentrations found in male guppies (*Poecilia reticulata*) used as experimental organisms. Uptake was substantially greater when sediment was spiked with mercuric chloride rather than with mercuric sulfide.

In 1977, Prater and Anderson (11) used 96-h bioassays to assess the biological impact of heavily-, moderately- and non-polluted sediments from Lake Superior harbors. They concluded that sediment-associated chemicals pose a threat to aquatic ecosystems and that sediment bioassays can be successfully used in hazard assessment. In six of eight cases, they also obtained a good correlation between sediment bulk chemistry and results of their sediment toxicity tests. In the same year, Prater and Anderson (12) used 96-h sediment bioassays in studies of the Otter Creek in Ohio, and they concluded that such procedures constitute a valid and practical approach for evaluating the hazards of sediment-sequestered toxicants. Wentsel et al. (13-15) also developed sediment bioassay procedures using the midge (*Chironomus tentans*) to differentiate the biological impact of contaminated versus uncontaminated sediments. Sediments used in the bioassays were characterized on the basis of bulk chemistry, and good correlations were established between sediment metal concentrations and biological effects observed in the sediment tests. Test endpoints for midge larvae involved avoidance reactions, emergence, and length and weight.

Chu-fa et al. (16), working with sediment pollution in the northern Chesapeake Bay, performed 24- and 48-h acute sediment bioassays with the mummichog (*Fundulus heteroclitus*), the spot (*Leiostomus xanthurus*), and the soft-shell clam (*Mya arenaria*). Test results were used to differentiate the study area into highly, moderately and slightly toxic zones, based upon sediment contamination. Mortality observed in the sediment tests increased proportionately with concentrations of numerous metals and certain organics (e.g., PCB) contained in the sediment samples. This further supports the premise that bulk sediment chemistry is important in criterion development for regulating sediment contamination.

Nebeker et al. (17) recently put together an extensive review of the literature on sediment bioassay procedures, and they have taken a leading role in developing a suite of sediment bioassay tests for quantifying the effects of sediment contamination on *Daphnia* and certain benthic species. Their principal objective has been to delineate methods for determining the acute and chronic effects of contaminated sediments on freshwater organisms, including *Daphnia magna, Hyalella azteca, Gammarus lacustris, Chironomus tentans,* and *Hexagenia limbata.* The results of their investigations further support the rationale that sediment bioassays constitute a feasible and economical approach for quantifying the hazardous effects of contaminated sediments on aquatic biota. Attention has also been given to the effects of sediment-sequestered chemicals on early life stages of fish and amphibian species (18, 19). Embryos and larvae that develop in close contact with the sediment may receive exposure from the water column, the interstitial (pore) water, or by the ingestion of sediment particulates, and they may well be as susceptible and perhaps less well-adapted than benthic organisms to toxicant stress from sediment-associated chemicals.

In view of the above studies, it is apparent that sediment contamination is an important environmental problem. On the positive side, sediments sequester and substantially reduce the bioavailability of metals and many organic contaminants. On the other hand, sediments can accumulate a substantial storehouse of toxic chemicals, which may desorb slowly over a prolonged period of time (6). Furthermore, via hydrological transport, contaminated sediment can be distributed well beyond its point of origin. Benthic organisms obviously are directly affected by sediment-associated chemicals, either by the ingestion of particulate matter or by exposure to interstitial water, and the potential exists for both acute and chronic effects. As remobilization from the sediment into the overlying water generally occurs slowly for most chemicals, it does not appear that acute toxicity is a major problem for adult stages of biotic species that primarily inhabit the water column. However, with long-term exposure, sediment-released chemicals may bioconcentrate to appreciable levels in fish (6, 10), and such exposure may lead to deleterious chronic effects. With respect to fish and amphibians, developmental stages that develop in close association with the sediment are more susceptible to sediment contamination. Eggs, embryos, and larvae of these and probably other epibenthic species may suffer appreciable frequencies of mortality and teratogenesis or may bioconcentrate desorbed toxicants at moderate to high rates. Considering the complexity of adsorption-desorption kinetics and the fact that sediment contamination usually involves mixtures of chemicals, toxicological biomonitoring offers considerable promise for quantifying biological effects of sediment-associated chemicals, as well as for developing or evaluating regulatory criteria. Based on the current literature, it is also evident that there has been substantial progress in developing procedures for conducting sediment toxicity tests (17, 19, 20).

USE OF FRESHWATER FISH AND AMPHIBIAN EARLY LIFE STAGES FOR MONITORING SEDIMENT TOXICITY

Procedures

Attention here is devoted to a more detailed consideration of freshwater sediment studies that focus on uptake and release of metals and initial procedures for conducting sediment bioassays using fish and amphibian embryo-larval stages. Species included in these investigations were the largemouth bass (*Micropterus salmoides*), goldfish (*Carassius auratus*), rainbow trout (*Salmo gairdneri*), leopard frog (*Rana pipiens*), and the narrow-mouthed toad (*Gastrophyrne carolinensis*).

Sediments used for test purposes were collected from second and third-order streams in the Kentucky River drainage system (18, 19). The major source of sediment was from Steele's Run, a second-order limestone-bed stream which supported a healthy aquatic ecosystem and which was largely free of contamination. Steele's Run sediment was used in metal-enrichment experiments and as a control in toxicity tests with natural sediments having varying degrees of contamination. Naturally contaminated bottom sediments were collected from Shelby Branch, Town Branch, and Wolf Run at sites previously described (21). These streams had suffered significant degrees of metal influx at the time of study, resulting in marked reductions in aquatic biota in Town Branch and Wolf Run and moderate decreases in density and diversity of fish and macroinvertebrates in Shelby Branch (21, 22).

Sediment samples were restricted to the upper 5 cm and were collected 1 to 2 m from shore at water depths of 0.5 to 1 m. Studies were conducted with fresh sediment to prevent alteration in sediment composition likely to occur with prolonged storage or freezing. Properties of the Steele's Run sediment are summarized in Table 15.1. Texture analysis was conducted by Bouyoucos' hydrometer procedure as modified by Cox (23). Determination of pH was made on a 1:1 mixture of sediment and distilled water, according

Table 15.1

Physical-Chemical Characteristics of Control Sediment from Steele's Run

Parameter	Mean ± SD (n = 2)	Parameter	Mean ± SD (n = 2)
pH (range)	7.6 – 7.7	Total cadmium (mg/kg)	1.02 ± 0.05[a]
Specific gravity	2.55 ± 0.21	Total mercury /mg/kg)	0.052 ± 0.002[a]
% Organic matter	2.26 ± 0.47	Total zinc	108.2 ± 2.1[a]
% Volatile solids	5.82 ± 0.63	Total iron (%)	5.52 ± 0.27[a]
% Total nitrogen	0.18 ± 0.02	Extractable iron (mg/kg)	283.0 ± 61.4
		Textural classification	Sandy loam
Cation exchange capacity (Meq/100 g)	13.58 ± 0.13	Sand (%)	52.6 ± 3.4
Exchangeable phosphorus (mg/kg)	114.5 ± 6.4	Silt (%)	35.4 ± 4.7
Exchangeable potassium (mg/mg)	84.0 ± 2.8	Clay (%)	12.0 ± 1.3
Exchangeable calcium (mg/kg)	2,270 ± 28		

[a] $n = 3$.

Reprinted with permission from Francis, Birge and Black, Effects of cadmium-enriched sediment on fish and amphibian embryo-larval stages, *Ecotoxicology and Environmental Safety*, **8**, 378-387 (1984).

to the procedure reported by Jackson (24). Specific gravity values were obtained by the method of Carlisle and Caldwell (25), and percent volatile solids was analyzed by the procedure given in Standard Methods (26). Total metal concentrations were determined using acid digestion procedures and atomic absorption spectrophotometry (AAS), as discussed below. Extractable iron was determined by AAS analysis, following the HCl/H_2SO_4 extraction procedure of Perkins (27, 28). Percent organic matter, percent total nitrogen, cation exchange capacity, and exchangeable calcium, phosphorus, and potassium were determined by the Soils Testing Laboratory, University of Kentucky College of Agriculture, using standardized soil analytical procedures (29).

Static sediment toxicity tests were performed using Pyrex deep petri dishes (Fig. 15.1). The glass inlet tube (4 mm I.D.) was used to add water by gravity flow at the onset of the test and thereafter to provide gentle aeration sufficient to maintain dissolved oxygen at or near saturation. Sediment was layered to a depth of 2 cm and covered with 350 mL of toxicant-free reconstituted water (30). The test water had pH of 7.5-8.0 and was used at a hardness of 200 mg/L $CaCO_3$, except that a hardness of 100 mg/L was maintained in studies originally reported by Francis et al. ((20), Table 15.2, Fig. 15.2). Routine AAS monitoring of water used for test purposes failed to reveal detectable concentrations of cadmium, mercury, or zinc.

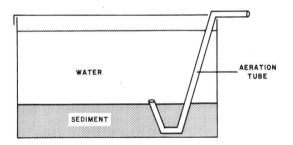

Fig. 15.1
Deep petri dish used in short-term embryo-larval
sediment toxicity tests. The glass inlet tube was
used to add reconstituted water by gravity flow
to avoid disrupting the sediment layer. The latter
was 2 cm in depth. After adding water for this
static test, the inlet was used to supply moderate
aeration. Reprinted with permission from
Francis, Birge and Black, Effects of cadmium-
enriched sediment on fish and amphibian
embryo-larval stages, *Ecotoxology and
Environmental Safety*, **8**, 378-387 (1984).

Table 15.2   Effects of Cadmium-enriched Sediment on Embryo-Larval Stages of Aquatic Vertebrates

Water Species	Sediment (mg/kg)		Total cadmium concentration	Tissue (µg/g)	Percent survival
	Nominal	Actual mean (±SD)	Water (µg/L) Mean ±SD		
Goldfish	0.0 (control)	0.97 ± 0.28	0.5 ± 0.5	0.12	90
	1.0	1.93 ± 0.9	2.2 ± 0.9	0.25	87
	10.0	10.48 ± 0.55	2.3 ± 1.9	0.36	90
	100.0	89.7 ± 7.9	3.2 ± 1.4	0.79	83
	1,000	1,008 ± 20	68.6 ± 1.7	4.73	88
Leopard frog	0.0 (control)	1.04 ± 0.06	1.0 ± 0.6	0.36	98
	1.0	2.8 ± 0.14	1.1 ± 0.8	0.44	98
	10.0	11.48 ± 0.21	2.1 ± 4.4	0.70	97
	100.0	96.8 ± 2.4	4.4 ± 1.8	3.44	97
	1,000	1,074 ± 14	76.5 ± 17.1	12.91	98
Largemouth bass	0.0 (control)	1.06 ± 0.02	0.5 ± 0.3	0.06	83
	1.0	2.21 ± 0.11	2.0 ± 0.9	0.24	76
	10.0	11.30 ± 0.09	2.7 ± 0.9	0.78	79
	100.0	97.6 ± 3.4	6.1 ± 2.4	14.79	78
	1,000	1,079 ± 27	43.9 ± 8.0	60.1	62*

[a] Survival frequencies were determined 4 d after hatching.
Data taken from Francis et al. (19) except that survival frequencies and tissue concentrations are not control adjusted. Survival frequencies that differ significantly ($p = 0.05$) from control values are designated with an asterisk.
Reprinted with permission from Francis, Birge and Black. Effects of cadmium-enriched sediment on fish and amphibian embryo-larval stages, *Ecotoxicology and Environmental Safety*, **8**, 378-387 (1984).

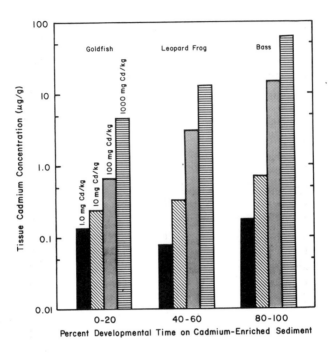

Fig. 15.2
Control-adjusted tissue cadmium concentrations for
hatched organisms exposed to metal-enriched sediment.
The length of time each test species was in direct
contact with the sediment is expressed as a percentage
of its exposure period. Reprinted with permission
from Francis, Birge and Black, Effects of cadmium-
enriched sediment on fish and amphibian embryo-larval
stages, *Ecotoxicology and Environmental Safety*,
8, 378-387 (1984).

    Minimum egg sample size was set at 100, and except for the rainbow trout,
exposure was initiated at or soon after fertilization and maintained for 4 d
after hatching. Hatching times for warmwater species were 3, 3.5, 3.5 and 4
d for the narrow-mouthed toad, goldfish, largemouth bass and leopard frog,
respectively, and this gave overall exposure periods of 7 to 8 d. Tests with
trout eggs were initiated at the early eyed stage and continued for 10 d
beyond hatching, giving an exposure period of 20 d. Studies were conducted
in walk-in environmental rooms, and temperature was regulated at 20° to
22.5°C for warmwater species and at 13° to 14°C for trout. Test water was
monitored at 1- to 2-d intervals for temperature, dissolved oxygen, water
hardness, and pH, using a mercury thermometer or YSI telethermometer
(model 42SC), YSI oxygen meter (model 51A), Orion divalent cation electrode

(model 92-32), and a Corning digital pH meter (model 110). Animal populations were inspected daily to gauge extent of development and to remove dead organisms. Depending on the length of the exposure period, water samples for metal analyses were collected at 2- to 4-d intervals, acidified with HNO_3, and analyzed using atomic absorption spectrophotometry. As the water was free of any appreciable turbidity, unfiltered samples were used for metal assays, and total metal concentrations were expressed as mean values with standard deviations for the full exposure period. Test responses included tissue metal concentrations, egg hatchability, and frequencies of gross terata appearing in newly hatched populations, as well as frequencies of larval survival or mortality expressed 4 d after hatching for warmwater species and 10 d after hatching for trout. In determining survival frequencies for trout, grossly teratic larvae were counted as mortalities. Tests were performed using two or three replicates. The ANOVA test was used to calculate probability values (p).

In metal-spiking tests performed with Steele's Run sediment, enrichment levels ranged from 0.1 to 1,000 mg metal/kg sediment and metals were added as reagent grade chloride salt solutions using the following procedure. After decanting excess water, 30 g of wet sediment were dried at 100°C for 24 h and weighed to determine percent moisture content. A sample of drained sediment equivalent to 250-g dry weight was placed in a 500-mL Erlenmeyer flask. Using a 25-mL aliquot of distilled, deionized water, the selected test metal was added at a sufficient concentration to obtain the desired sediment-enrichment level. Control cultures each received a 25-mL aliquot of metal-free water. Flasks were sealed with parafilm and either maintained on a shaker for 5 d (18) or shaken manually twice daily for 7 d (19) to facilitate homogeneous distribution of the added metal. After mixing, the sediment suspensions were filtered under vacuum using no. 1 Whatman paper, as described previously (19). Subsequent to filtration, the moisture content of the sediment was approximately 15% to 20%. After removal of excess water, the prepared sediment was placed in the bioassay exposure chambers and covered with reconstituted water as described above. Sediment exposure chambers were allowed to stand overnight or for at least 10 h prior to introducing test organisms.

In all cases, natural and enriched sediments were analyzed for cadmium, mercury, or zinc by AAS, using a Perkin-Elmer model 503 unit equipped with a mercury analysis system and a graphite furnace (model HGA 2100). Sediment-metal analyses were performed in triplicate at the beginning and end of each test. Initial and final values were averaged for results given in Table 15.2, but data presented in Tables 15.3 and 15.4 represent initial sediment metal concentrations. Prior to analysis, sediment samples were digested using a 1:1 mixture of sulfuric and nitric acid (26, 27). Fish larvae were analyzed after testing for whole-body metal concentrations by AAS. Prior to analysis, larvae were washed thoroughly with deionized water and

Table 15.3   Effects of Mercury-Enriched Sediment on Early Life Stages of the Rainbow Trout

Sediment (mg/kg)		Total mercury concentration			Percent survival[a]
		Water (μg/L)		Tissue (μg/g)	
Nominal	Actual mean ± SD	Mean ± SD	Range	Mean ± SD	
0.0 (control)	0.052 ± 0.002	0.11 ± 0.08	0.0 - 0.3	0.024 ± 0.004	94
0.1	0.180 ± 0.021	0.25 ± 0.18	0.0 - 0.5	0.036 ± 0.008	70
1.0	1.050 ± 0.043	0.15 ± 0.12	0.0 - 0.3	0.041 ± 0.001	45*
10.0	12.10 ± 1.47	1.83 ± 2.98	0.2 - 6.3	0.269 ± 0.129	23*
100.0	106.7 ± 9.9	6.40 ± 7.32	0.4 - 15.7	0.902 ± 0.059	0*

[a] Treatment was initiated at the early eyed-egg stage and continued through 10 d posthatching, giving an exposure period of 20 d. Survival frequencies that differ significantly ($p = 0.05$) from the control value are designated with asterisks.
Reprinted with permission from Birge, Black, Westerman and Hudson, The effects of mercury on reproduction of fish and amphibians, in *The Biogeochemistry of Mercury in the Environment*, (3), ed. J. O. Nriagu, Elsevier/North Holland Biomedical Press, 1979.

Table 15.4 Effects of Cadmium- and Zinc-enriched Sediments on Early Life Stages of the Rainbow Trout

Metal	Sediment enrichment concentration[a] (mg/kg)	Total metal concentration Mean ± SD		Percent survival[b]
		Sediment (mg/kg)	Water (μg/L)	
Cadmium	0.0 (control)	1.02 ± 0.05	2.1 ± 1.3	94
	0.1	1.45 ± 0.10	4.7 ± 5.4	80
	1.0	2.15 ± 0.10	6.8 ± 3.8	53*
	10.0	12.7 ± 0.3	7.5 ± 4.2	37*
	100.0	121.0 ± 1.9	7.2 ± 3.4	11*
Zinc	0.0 (control)	108.2 ± 2.1	19.4 ± 4.2	94
	1.0	115.9 ± 1.0	17.0 ± 10.3	88
	10.0	121.4 ± 2.1	21.2 ± 10.0	75*
	100.0	210.6 ± 1.9	32.3 ± 16.1	56*
	1,000.0	1,157.0 ± 5.0	122.8 ± 34.3	28*

[a] Calculated sediment enrichment levels, based on total metal, for cadmium or zinc used in separate experiments.
[b] Treatment was initiated at the early eyed-egg stage and continued through 4 d post-hatching, giving an exposure period of 20 d. Survival frequencies that differ significantly ($p = 0.05$) from control values are designated with asterisks.

pooled to give 0.3 to 1.0 g samples that were digested with a 1:1 mixture of concentrated nitric and perchloric acid (19).

Results

In pilot experiments (18), fertilized eggs of the narrow-mouthed toad and goldfish were exposed to sediments spiked with cadmium, mercury, or zinc at levels of 0.1 to 100 mg/kg sediment. Eggs were deposited directly on the surface of the sediment. A comparison of results revealed that toxicity for the three sediment-associated metals increased in the order of Zn, Cd and Hg. As compared with responses observed for control organisms, cadmium- and mercury-spiked sediments substantially reduced egg hatchability and larval survival. However, frequencies of mortality did not necessarily correlate with increasing sediment-metal concentrations. Except in studies with zinc, exposed organisms consistently exhibited abnormal development, and the number of teratic larvae generally increased with metal-enrichment levels. For example, frequencies of terata for the narrow-mouthed toad ranged from 1% to 11% and 11% to 23% in tests with cadmium and mercury, respectively.

Subsequent tests with cadmium were repeated in triplicate with each of three organisms, including the largemouth bass, leopard frog, and goldfish. Particular attention was given to the distribution of cadmium in sediment, water, and test organisms (Table 15.2). A similar study was conducted in duplicate with mercury-spiked sediments, starting with early eyed stages of the rainbow trout (Table 15.3). Aqueous concentrations of cadmium and mercury were quite low in all tests, and this is consistent with low metal desorption rates observed in other investigations (31-34). Average metal concentrations in water overlying sediments enriched to 1.0 to 100 mg/kg were 1.1 and 6.1 μg/L for cadmium and 0.15 to 6.40 μg/L for mercury. At a concentration of 1,000 mg Cd/kg sediment, aqueous total cadmium ranged from 43.9 to 76.5 μg/L. In each of the three experiments summarized in Table 15.2, there was a positive correlation between aqueous and sediment cadmium concentrations ($r = 0.99$). A high correlation coefficient ($r = 0.99$) was also obtained between aqueous and sediment mercury in studies with the trout (Table 15.3). Averaging data from the three tests with cadmium, only 0.12%, 0.03%, 0.007% and 0.009% of this metal was remobilized from sediments enriched at respective concentrations of 1.0, 10.0, 100 and 1,000 mg Cd/kg. The percentage of sediment-associated mercury remobilized into the water column ranged from 0.2% at 0.1 mg/kg to 0.008% at 100 mg/kg. The low rates at which these metals moved back to the overlying water may have been due in part to the high sorption capacity of the sediment. As predicted by the Langmuir isotherm, the sorption capacity of the Steele's Run sediment for cadmium was calculated to be 4,800 to 4,900 mg/kg. Therefore, even at a sediment concentration of

1,000mg Cd/kg, no more than about 20% of the sorption capacity would have been utilized. Consistent with the views of Jenne and Zachara (35), factors affecting the metal sequestering capacity of the Steele's Run sediment (Table 15.1) likely included its soil type (47.4% silt and clay), content of organic matter (2.26%), and the presence of iron and manganese oxides or similar inorganic compounds that form additional metal sorption sites.

Early life stages of goldfish, leopard frog and largemouth bass exposed to spiked sediments were observed to accumulate cadmium (Table 15.2). For example, tissue cadmium concentrations were 0.06, 0.24, 0.78, 14.79 and 60.1 μg/g for bass stages exposed for 7 d to sediments enriched at concentrations of 0.0, 1.0, 10.0, 100 and 1,000 mg Cd/kg, respectively. As shown in Table 15.3, mercury residues in trout alevins also increased with sediment spiking levels. Thus, for both mercury and cadmium, rates of tissue uptake correlated closely with aqueous and sediment metal concentrations ($r = 0.98$). It should be noted that metal uptake values were based on whole-body concentrations and that, prior to analysis, animals were rinsed to reduce surface-absorbed metal. Exposure to cadmium-spiked sediments usually did not result in appreciable mortality to leopard frog or goldfish larvae (Table 15.2). Though only moderate reductions in survival occurred with the bass (Table 15.2), positive correlations were observed between frequencies of mortality and cadmium concentrations in tissue ($r = 0.89$), water ($r = 0.93$) and sediment ($r = 0.92$). In the 20-d test with trout early life stages, a marked dose-response relationship was observed between mortality and sediment-mercury concentrations (Table 15.3). Compared to mean control survival of 94%, survival of trout alevins dropped to 70% when sediment was spiked with 0.1 mg Hg/kg, and this likely represented an actual response. When sediment mercury was increased to 1 mg/kg, there was a significant drop in survival to 45% ($p = 0.05$). In summarizing results obtained with the trout, frequencies of mortality correlated significantly with log-transformed mercury concentrations in tissue ($r = 0.93$), water ($r = 0.90$) and sediment ($r = 0.99$). As opposed to results obtained with adult trout (7), mortality of developmental stages varied directly with metal body burden.

Sediment toxicity tests with the trout also were conducted using cadmium and zinc (Table 15.4). As in all other experiments, sediment-metal concentrations decreased only slightly over the duration of the test. However, desorption rates for cadmium and zinc were sufficient to exert appreciable frequencies of mortality in exposed populations. Using log-transformed values for cadmium, there was not a good linear relationship between aqueous and sediment concentrations, primarily due to the seemingly low water cadmium value obtained at the highest enrichment level (Table 15.4). However, good correlations existed between zinc concentrations in water and sediment ($r = 0.99$), and between frequencies of mortality and sediment concentrations for both cadmium ($r = 0.94$) and zinc ($r = 0.95$).

In summary of results discussed above, aqueous and sediment metal concentrations usually correlated well and high correlations consistently occurred when sediment metal concentrations were compared with tissue metal values or with frequencies of mortality. In addition, in cadmium experiments with the bass and in mercury studies with the trout, mortality increased with metal body burden.

Results of the sediment-enrichment experiments presented in Table 15.2 indicate that the duration of time in which fish and amphibian developmental stages were in contact with the sediment affected the magnitude of biological responses. When bass eggs were placed into exposure chambers, they settled onto the sediment surface and the larvae remained there during the experiment. Eggs of the leopard frog were transferred to the exposure chambers in small clumps of algae and were suspended in the water column until hatching. At hatching, frog larvae moved to the sediment surface for the last 4 d of the exposure period. Goldfish eggs also were suspended above the sediment in filamentous algae and, subsequent to hatching, the larvae usually remained in the overlying water or rested at the sides of the test chambers. As illustrated in Figure 15.2, tissue cadmium concentrations, which increased in the order of goldfish, leopard frog and largemouth bass, were closely related to sediment contact time. We should also note that in these experiments only the bass suffered significant mortality. This likely was not a result of differences in cadmium tolerance, but rather to variations in the duration of contact with the metal-enriched sediment. This interpretation is supported by results of earlier embryo-larval toxicity tests performed without sediment, in which LC_{50} values for the 7- to 8-d exposure periods were 0.17 and 1.64 mg Cd/L with the goldfish (36) and the largemouth bass (37), respectively. It is also likely that the high frequencies of mortality and tissue-metal residues observed for the trout were due in part to the organisms' close contact with the sediment throughout the longer exposure period (i.e., 20-d) used for this species.

The relationship between sediment contact time and biological effects could have resulted from ingestion of contaminated sediment particles by early larvae and/or exposure to sediment interstitial water. The first possibility is supported by evidence that cadmium bound to ingested sediment particles can accumulate in animal tissues (38). The second assumption is based on the probability that organisms situated on metal-contaminated sediment experience at least some direct contact with interstitial water, which may contain elevated metal concentrations (33, 39). For instance, Peddicord (33) found that concentrations of several metals (e.g., Cd, As) were five to eighteen times higher in interstitial water than in sediment elutriates, and Prosi (40) reported that cadmium in interstitial water from polluted river sediments contributed substantially to high cadmium residues in sludge worms.

To evaluate further the use of fish and amphibian early life stages for sediment biomonitoring, tests with trout eyed eggs were conducted using natural sediments collected from 0.5 to 1.0 mile study areas on four local streams affected by different levels of metal contamination. These sediment tests were set up in duplicate using toxicant-free reconstituted water. As described above, the four streams included Steele's Run (control site), Shelby Branch, Wolf Run, and Town Branch (18, 21, 22). Of fifteen known resident species of fish none was detected in the impacted section of Town Branch; only two were found in Wolf Run, whereas all fifteen were present on Steele's Run. Of seventeen resident fish species reported for Shelby Branch, fourteen remained in the contaminated section of the stream. Based on these data, the percent retention of resident species was 0, 13, 82 and 100 for Town Branch, Wolf Run, Shelby Branch and Steele's Run, respectively. Taking streams in the same order, trout survival frequencies observed in the sediment toxicity tests were 1%, 30%, 81% and 94%, and a high correlation ($r = 0.96$) existed between these test results and fish collection data. Though it would have been desirable to have based such correlations on more quantitative faunistic surveys for fish and other species (e.g., macroinvertebrates), these results support the use of sediment toxicity tests for estimating ecological impact of contaminated sediments.

Cadmium, mercury, and zinc concentrations that produced effects (mortality, bioconcentration) on trout stages were generally in the same ranges for both enriched and natural sediments. Based on three replicate measurements, Town Branch metal concentrations expressed in mg/kg (mean $\pm$ S.D.) for mercury, cadmium, and zinc were 0.3 $\pm$ 0.003, 7.1 $\pm$ 0.2 and 299 $\pm$ 3 respectively. These values approached or exceeded concentrations for each of the three metals that produced significant frequencies of mortality in the enrichment studies reported for trout (Tables 15.3 and 15.4). Concentrations of metals determined for Steele's Run sediment are presented as control values in Tables 15.3 and 15.4. Wolf Run sediment contained cadmium and zinc at 1.4 $\pm$ 0.1 and 147 $\pm$ 1 mg/kg and, in respective order, these values approached and exceeded actual metal concentrations that produced significant frequencies of trout mortality in the enrichment experiments (Table 15.4). In Shelby Branch sediment, concentrations of mercury and zinc did not exceed values obtained from the Steele's Run control sediment. However, cadmium was present at 1.45 $\pm$ 0.05 mg/kg, and this equaled the actual cadmium sediment concentration reported at the lowest enrichment level (Table 15.4). Furthermore, trout survival frequencies for Shelby Branch sediment and Steele's Run sediment enriched to the same cadmium concentration were 81% and 80%, respectively. Control survival obtained with Steele's Run sediment was 94%. Obviously other contaminants which were not analyzed could have altered the biological responses observed in the comparative biomonitoring study with the four natural sediments. Nevertheless, in the trout studies, there was a reasonably consistent dose-response relationship for cadmium, mercury, and zinc for both enriched and naturally contaminated sediments.

There also was surprisingly good agreement between metal desorption rates for the enriched and the non-spiked natural sediments, and this resulted in approximately the same ranges of aqueous cadmium, mercury, and zinc for comparable sediment metal concentrations. In enrichment studies with mercury (Table 15.3), sediment concentrations of 0.05 to 0.18 mg Hg/kg produced aqueous mercury concentrations of 0.11 to 0.25 μg/L. By comparison, aqueous mercury concentrations varied from 0.11 to 0.22 μg/L for natural sediment mercury concentrations of 0.05 to 0.27 mg/kg. For cadmium concentrations that ranged from 1.4 to 12.7 mg/kg and 1.4 to 7.1 mg/kg for spiked and non-spiked sediments, the corresponding aqueous cadmium values varied from 4.7 to 7.2 μg/L and 3.1 to 5.3 μg/L. When Steele's Run sediment was spiked with zinc to concentrations 116, 121 and 211 mg/kg, the aqueous zinc values were 17.0 to 32.3 μg/L (Table 15.4), whereas in natural sediments which contained 80 to 299 μg Zn/kg, aqueous zinc ranged from 10.3 to 19.9 μg/L.

CONCLUSIONS

Considering results reported by other investigators (16, 17, 41, 42) and our own data (18, 19), there is substantial evidence that toxicological biomonitoring constitutes a practical and reliable means for identifying polluted sediments and evaluating degrees of contamination that pose environmental risk. In the toxicity tests with early life stages discussed above, the relationships among sediment metal concentrations, aqueous metal concentrations, and biological effects were quite similar for both spiked and naturally contaminated sediments. These findings indicate that bulk chemistry is indeed important in evaluating the hazard of sediment-associated metals and that enrichment tests, in which sediments can be spiked with an exponential series of toxicant concentrations, should prove useful in the development and/or verification of regulatory criteria. Judging from the information available, it would appear that sediment contamination frequently involves a mixture of chemicals and that desorption rates for the different toxicants may vary independently depending on their concentrations and physical-chemical properties, the composition of the sediment, the structure, density, and spatial distribution of the biomass, and numerous other environmental variables (18, 19, 31-34, 43). One distinct advantage of the direct biomonitoring approach is that it provides a measurement of the net toxicity for all chemicals present under any given set of site-specific conditions. As in evaluations of the biological effects of complex effluents (44, 45), accurate assessments of sediment contamination may be difficult to achieve using numerical criteria (46) for individual toxicants.

With respect to epibenthic species, it is probable that early life stages generally are more vulnerable to the effects of sediment-released chemicals than are mature organisms. Greatest concern exists for bottom spawning fish and amphibians whose eggs and larvae may develop in close association with the sediment. Such organisms may receive toxicant exposure from the water column, the sediment interstitial water, and from the ingestion of contaminated particulates, and perhaps such life cycle stages should be considered together with infaunal species in developing regulatory strategies for sediment-sequestered toxicants.

It is also evident that fish and amphibian early life stages are opportune test organisms for sediment biomonitoring. They can be used in simple and cost-effective test systems, and they exhibit clearly detectable endpoints, such as egg hatchability, mortality, teratogenesis (abnormal development), and toxicant bioconcentration. Developmental stages of fish and amphibians rapidly bioaccumulate metals and hydrophobic organics (8, 9, 19, 47) and may constitute an effective model system for evaluating the relationship between toxicant body burden and biological effects. As noted in the experiments with spiked sediments, there was a close correlation between tissue concentrations of mercury and mortality frequencies with trout embryo-larval stages. Toxicant uptake studies with early life stages may provide a means of identifying sediment-released toxicants which, given long-term exposure, may bioconcentrate in adult fish sufficiently to impair reproduction or produce other deleterious effects. The bioconcentration potential of sediment-released chemicals is not fully understood. Factors that effect desorption rates and temporal and spatial variations of toxicants in the water column complicate estimates of bioconcentration based on octanol-water partition coefficients (43).

Tentative threshold contaminant concentrations for cadmium, mercury, and zinc have been estimated to be 31, 0.8 and 760 mg/kg on a dry weight basis for sediments containing 4% total organic carbon (TOC). The Steele's Run sediment used in experiments reported in Tables 15.3 and 15.4 contained 2.26% organic matter and this converts to 1.31% TOC (48). Normalizing for organic carbon, as recommended (46) would result in dividing the above criteria by a factor of about three. On this basis, the adjusted criteria for cadmium and zinc would still be above sediment metal concentrations that produced biological effects in trout, but the value for mercury would provide a closer fit to data presented in Table 15.4. However, such comparisons must be undertaken with caution, as tests with trout early life stages are substantially less sensitive when initiated with eyed eggs rather than with freshly fertilized eggs (49).

The Sixth Pellston Workshop served to stress the importance of and raise unresolved questions about sediment-associated chemicals, both with respect to their effects on the chemical ecology of aquatic life and their relevance to hazard assessment. While considerable progress has been made, further investigations will be required to elucidate and properly model the sorption kinetics of important categories of aquatic toxicants and to develop optimum biomonitoring procedures.

REFERENCES

1. **Perhac, R. M. and C. J. Whelan.** 1972. A comparison of water, suspended solid, and bottom sediment analyses for geochemical prospecting in a northeast Tennessee zinc district. *J. Geochem. Explor.* **1**:47-53.

2. **Hancock, H. M.** 1970. Investigations on the influence of the effluent from a concentrated industrial complex on a large river. Kentucky Project Report No. 4-48-R.

3. **Bondietti, E. A., F. H. Sweeton, T. Tamura, R. M. Perhac, L. D. Hulett and T. J. Kneip.** 1974. Toxic metals and sediments. In W. Fulkerson, W. D. Shults and R. I. Van Hook (compilers), First Annual NSF Trace Contaminants Conference. ORNL-NSF-EATC-6, pp. 176-194.

4. **Bondietta, E. A., R. M. Perhac, R. H. Sweeton and T. Tamura.** 1973. Toxicity of metals in sediments. In *Ecology and Analysis of Trace Contaminants*, Oak Ridge National Laboratory, pp. 151-186.

5. **Reynolds, W. R. and D. A. Thompson.** 1975. Occurrence and distribution of clay minerals and trace metals in the bottom sediment of Biloxi Bay, Mississippi. Water Resources Research Institute, Mississippi State University, State College, MS.

6. **Jernelov, A.** 1970. Release of methyl mercury from sediment with layers containing inorganic mercury at different depths. *Limnol. Ocean.* **15**:958-960.

7. **Birge, W. J., J. A. Black, A. G. Westerman and J. E. Hudson.** 1979. The effects of mercury on reproduction of fish and amphibians. In J. O. Nriagu, ed., *Biogeochemistry of Mercury in the Environment.* Elsevier/North Holland Biomedical Press, Amsterdam, pp. 629-655.

8. **Birge, W. J., J. A. Black and B. A. Ramey.** 1981. The reproductive toxicology of aquatic contaminants. In J. Saxena and F. Fisher, eds., *Hazard Assessment of Chemicals: Current Developments*, Vol. 1. Academic Press, Inc., New York, NY, pp. 59-115.

9. **Birge, W. J., J. A. Black and A. G. Westerman.** 1979. Evaluation of aquatic pollutants using fish and amphibian eggs as bioassay organisms. In *Animals as Monitors of Environmental Pollutants.* National Academy of Sciences, Washington, DC, pp. 108-118.

10. **Gillespie, D. C. and D. P. Scott.** 1971. Mobilization of mercuric sulfide from sediment into fish under aerobic conditions. *J. Fish. Res. Board Canada* **28**:1807-1808.

11. **Prater, B. L. and M. A. Anderson.** 1977. A 96-h sediment bioassay of Duluth and Superior Harbor Basins (Minnesota) using *Hexagenia limbata, Asellus communis, Daphnia magna,* and *Pimephales promelas* as test organisms. *Bull. Environ. Contam. Toxiciol.* **18**:159-169.

12. **Prater, B. L. and M. A. Anderson.** 1977. A 96-hour bioassay of Otter Creek, Ohio. *J. Water Pollut. Control Fed.* **48**:2099-2106.

13. **Wentsel, R., A. McIntosh, W. P. McCafferty, G. Atchison and V. Anderson.** 1977. Avoidance response of midge larvae (*Chironomus tentans*) to sediments containing heavy metals. *Hydrobiologia* **55**:171-175.

14. **Wentsel, R., A. McIntosh and G. Atchison.** 1977. Sublethal effects of heavy metal contaminated sediment on midge larvae (*Chironomus tentans*). *Hydrobiologia* **56**:153-156.

15. **Wentsel, R., A. McIntosh and W. P. McCafferty.** 1978. Emergence of the midge *Chironomus tentans* when exposed to heavy metal contaminated sediment. *Hydrobiologia* **57**:195-196.

16. **Chu-fa, T., J. Welch, K. Y. Chang, J. Shaeffer and L. E. Cronin.** 1979. Bioassay of Baltimore Harbor sediment. *Estuaries* **2**:141-153.

17. **Nebeker, A. V., M. A. Cairns, J. H. Gakstatter, K. W. Malueg, G. S. Schuytema and D. F. Krawczyk.** Biological methods for determining toxicity of contaminated freshwater sediments to invertebrates. *Environ. Toxicol. Chem.* (*in press*).

18. **Birge, W. J., J. A. Black, A. G. Westerman, P. C. Francis and J. E. Hudson.** 1977. Embryopathic effects of waterborne and sediment-accumulated cadmium, mercury, and zinc on reproduction and survival of fish and amphibian populations in Kentucky. Research Report No. 100. U.S. Department of the Interior, Washington, DC.

19. **Francis, P. C., W. J. Birge and J. A. Black.** 1984. Effects of cadmium-enriched sediment on fish and amphibian embryo-larval stages. *Ecotox. Environ. Safety* **8**:378-387.

20. **Swartz, R. C.** Toxicological methods for determining the effects of contaminated sediment on marine organisms. This volume, pp. 183-198.

21. **Birge, W. J., A. G. Westerman and J. A. Black.** 1975. Sensitivity of vertebrate embryos to heavy metals as a criterion of water quality. Phase III. Use of fish and amphibian eggs and embryos as bioindicator organisms for evaluating water quality. Research Report No. 91. U.S. Department of the Interior, Washington, DC.

22. **Kuehne, R. A.** 1975. Evaluation of recovery in a polluted creek after installation of new sewage treatment procedures. U.S. Department of the Interior, Research Report No. 85.

23. **Cox, G. W.** 1972. *Laboratory Manual for General Ecology*, 2nd ed. William C. Brown Co., Dubuque, IA, pp. 174-179.

24. **Jackson, M. L.** 1958. *Soil Chemical Analysis*. Prentice-Hall, Inc., Englewood Cliffs, NJ, pp. 41-49.

25. **Carlisle, V. W. and R. E. Caldwell.** 1964. *A Laboratory Manual for Introductory Soil Science*. John S. Swift Co., Inc., St. Louis, MO, pp. 38-39.

26. **American Public Health Association, American Water Works Association, and Water Pollution Control Federation.** 1975. *Standard Methods for the Examination of Water and Wastewater*, 14th ed. American Public Health Association, Washington, DC.

27. **Analytical Methods for Atomic Absorption Spectrophotometry.** 1973. Perkin-Elmer Co., Norwalk, CT.

28. **Perkins, H. F.** 1970. A rapid method of evaluating the zinc status of coastal plain soils. *Commun. Soil Sci. Plant Anal.* 1:35-42.

29. **Black, C. A.** (ed.). 1965. *Methods of Soil Analysis*. American Society of Agronomy. Agronomy Monograph no. 9. Madison, Wisconsin.

30; **Birge, W. J. and R. A. Cassidy.** 1983. Structure-activity relationships in aquatic toxicology. *Fundam. Appl. Toxicol.* 3:359-368.

31. **Mathis, B. J. and T. F. Cummings.** 1973. Selected metals in sediment, water, and biota in the Illinois River. *J. Water Pollut. Control Fed.* 45:1473-1483.

32. **Gardiner, J.** 1974. The chemistry of cadmium in natural water. II.The adsorption of cadmium on river muds and naturally occurring solids. *Water Res.* 8:157-164.

33. **Peddicord, R. K.** 1980. Direct effects of suspended sediments on aquatic organisms. In R. A. Baker, ed., *Contaminants and Sediments*, Vol. 1. Ann Arbor Science Publishers, Ann Arbor, MI, pp. 501-536.

34. **Cairns, M. A., A. V. Nebeker, J. H. Gakstatter and W. L. Griffis.** 1984. Toxicity of copper-spiked sediments to freshwater invertebrates. *Environ. Toxicol. Chem.* 3:435-445.

35. **Jenne, E. A. and J. M. Zachara.** Factors influencing the sorption of metals. This volume, pp. 83-98.

36. **Birge, W. J.** 1978. Aquatic toxicology of trace elements of coal and fly ash. In J. H. Thorpe and J. W. Gibbons, eds., *Energy and Environmental Stress in Aquatic Systems*. DOE Symposium Series (CONF-771114), Washington, DC, pp. 219-240.

37. **Birge, W. J., J. E. Hudson, J. A. Black and A. G. Westerman.** 1978. Embryo-larval bioassays on inorganic coal elements and in situ biomonitoring of coal-waste effluents. In D. E. Samuel, J. R. Stauffer, C. H. Hocutt and W. T. Mason, eds., *Surface Mining and Fish/Wildlife Needs in the Eastern United States*. FWS/OBS-78/81. Office of Biological Sciences, Fish and Wildlife Service. U.S. Department of the Interior, Washington, DC, pp. 97-104.

38. **Luoma, S. N., D. J. Cain, E. Thomson, C. Johannson, E. A. Jenne and G. W. Bryan.** 1980. Determining the availability of sediment-bound trace metals to aquatic deposit-feeding animals. Open-file Report 80-341. U.S. Geological Survey, Menlo Park, CA.

39. **Brannon, J. M., R. H. Plumb, Jr. and I. Smith, Jr.** 1980. Long-term release of heavy metals from sediments. In R. A. Baker, ed., *Contaminants and Sediments*, Vol. 2. Ann Arbor Science Publishers, Ann Arbor, MI, pp. 221-266.

40. **Prosi, F.** 1979. Bioavailability of heavy metals in different freshwater sediments: Uptake in macrobenthos and biomobilization. In *Management and Control of Heavy Metals in the Environment*. International Conference, London, Sept., 1979 CEP Consultants, Edinburgh, U.K., pp. 288-291.

41. **Malueg, K. W., G. S. Schuytema, D. F. Krawczyk and J. H. Gakstatter.** 1984. Laboratory sediment toxicity tests, sediment chemistry and distribution of benthic macroinvertebrates in sediments from the Keweenaw Waterway, Michigan. *Environ. Toxicol. Chem.* **3**:233-242.

42. **Swartz, R. C., W. A. Daben, K. A. Sercu and J. O. Lamberson.** 1982. Sediment toxicity and the distribution of amphipods in Commencement Bay, Washington, USA. *Marine Pollution Bulletin* **13**:359-364.

43. **Connor, M. S.** 1984. Fish/sediment concentration ratios for organic compounds. *Environ. Sci. Technol.* **18**:31-35.

44. **Birge, W. J., J. A. Black and A. G. Westerman.** 1985. Short-term fish and amphibian embryo-larval tests for determining the effects of toxicant stress on early life stages and estimating chronic values for single compounds and complex effluents. *Environ. Toxicol. Chem.* **4**:807-821.

45. **Horning, W. B. and C. I. Weber** (eds.). 1986. Methods for measuring the chronic toxicity of effluents and receiving waters to freshwater organisms. Environmental Monitoring and Support Laboratory. U.S. Environmental Protection Agency, Cincinnati, Ohio.

46. **Bolton, H. S., R. J. Breteler, B. W. Vigon, J. A. Scanlon and S. L. Clark.** 1985. *National Perspective on Sediment Quality.* U.S. Environmental Protecting Agency, Office of Water. Washington, DC.

47. **NAS-NAE Committee on Water Quality Criteria.** 1973. *Water Quality Criteria 1972.* U.S. Government Printing Office, Washington, DC.

48. **Hamker, J. W. and J. M. Thompson.** 1972. Adsorption. In: Goring, C.A.I. and J. W. Hamaker, eds., *Organic Chemicals in the Soil Environment,* Vol. 1. Marcel-Dekker, N.Y. pp. 49-143.

49. **Birge, W. J. and J. A. Black.** 1981. *In situ* acute/chronic toxicological monitoring of industrial effluents for the NPDES biomonitoring program using fish and amphibian embryo-larval stages as test organisms. OWEP-82-001. U.S. Environmental Protection Agency, Washington, DC.

Chapter 16

Bioavailability of Neutral Lipophilic Organic Chemicals Contained on Sediments: A Review

William J. Adams

INTRODUCTION

The principles for assessing toxicity of chemicals in water to aquatic organisms are widely accepted and are based upon biological effects in relation to chemical exposure as described in the first Pellston Workshop (1). These same principles ought to be applicable to sediment-dwelling organisms. However, the state of the art of how to apply these principles is just emerging.

Society's need for a better understanding of the significance of chemicals sorbed to aqueous sediments and how to assess the hazard of these chemicals has prompted the need for this workshop to review the existing state of knowledge concerning sediment assessment.

To advance the state of the art, a need exists for an approach that can relate concentrations of chemicals measured on sediments to the potential for toxic effects. Approaches have been described for neutral lipophilic organics by Adams et al. (2) and Pavlou and Weston (3) based upon organism exposure to sediment interstitial water. Additional research is needed in this area, as well as with other classes of organics and metals. The problem is complex because of the numerous types of chemicals that have to be assessed and the wide array of sediment types that exist in aquatic ecosystems. Organic carbon appears to be a good data normalizer for different sediment types for neutral lipophilic organic chemicals, but as yet no similar approach has been found for metals. The complexity of the problem is obvious when one considers the effects of sediment Eh and pH on metal speciation.

The fact that overlying water is brought down into the sediments by benthic invertebrates indicates they are exposed to a mixture of interstitial water and surface water. The actual chemical exposure then is a function of the rate of pumping, the K_{oc} of the chemical, the organic carbon (OC) content of the sediment, the concentration of the chemical on the sediment, in the interstitial water and in the overlying surface water, and the feeding habits of the organism.

Feeding habits, including the source of food and feeding rate are potentially important factors controlling tissue residue levels and toxic effects. Feeding rate is a function of the activity of the organism, its metabolic demand for energy, availability of food, oxygen level, and presence of contaminants to name only a few factors. Feeding and life habits (niches) are very diverse among freshwater and marine benthic invertebrates and are usually species specific. Freshwater invertebrate communities are distinct from marine communities in that they contain thousands of species of aquatic insects which often dominate the community assemblages.

A key question concerning the assessment of chemicals sorbed to sediments is how bioavailable are they to aquatic life? I define sediment bioavailability as the extent to which a sediment-sorbed chemical can be toxic to or be accumulated by biological organisms. Techniques that describe the bioavailability of a given chemical will provide the necessary insight for determining its safety.

For the purposes of initiating discussion, a brief summary of bioavailability of lipophilic organic chemicals will be presented as it relates to three areas: (1) feeding habits of freshwater benthic invertebrates, (2) route of exposure, and (3) impact of organic carbon. The traditional measurements of bioavailability, toxicity, and bioconcentration/bioaccumulation are interwoven in the above three points of discussion. For the sake of clarity and consistency, the following terms are defined. Bioconcentration is chemical uptake by an organism from water only. Bioaccumulation is chemical uptake by an organism from both water and diet.

BIOAVAILABILITY AS A FUNCTION OF FEEDING HABITS

To demonstrate the vast array of feeding and life habits, an abbreviated summary of these species traits are presented (Table 16.1). This information serves three key purposes. It shows the wide variety of invertebrate species that are associated with sediments. It demonstrates the diversity of life habits (burrowing versus non-burrowing) that have evolved. And, it points out the many differences in feeding habits. The primary conclusion from this review is the fact that freshwater benthic invertebrates are not sediment ingesters, with the exception of the oligochaetes (aquatic earthworms) and some chironomids that are both filter feeders and occasionally sediment ingesters. In contrast to this, marine burrowing species frequently ingest sediment, especially the polychaetes.

Table 16.1   Feeding and Life Habits of Freshwater Benthic/Macroinvertebrates Associated with Sediments[a]

Taxon	Approximate no. species	Respiratory method	Mode of existence (habitats)	Feeding mechanism (habitats)
Oligochaeta (aquatic earthworms)	100's	diffusion through integument	burrowers	sediment ingestion
Isopod (aquatic sow bugs)	50	abdominal respiratory appendages	sprawlers (cavities in rocks, wood)	scavengers, herbivores (coarse particulate organic matter)
Amphipoda (scuds)	50	abdominal respiratory appendages	swimmers, sprawlers	scavengers, omnivores, chewers
Decapoda	160	blood gills	sprawlers, swimmers retreat makers (terrestrial burrows)	scavengers, omnivores, tearers and chewers
Ephemeroptera (mayflies)				
Dolania	1	tracheal gills	burrowers	predator
Ephemera	7	tracheal gills	burrowers	collectors[b]
Hexagenia	5	tracheal gills	burrowers	collectors
Litobrancha	1	tracheal gills	burrowers	collectors
Pentagenia	2	tracheal gills	(clay) burrowers	collectors
Ephoron	2	tracheal gills	burrowers	collectors
Odonata (dragonflies)				
Petaluridae	2	anal chamber with tracheal gills	burrowers	predators
Cordulegastridae	7		burrowers	predators
Gomphidae (i.e., *Gomphurus, Gomphus*)	86		burrowers	predators
Megaloptera (alderflies)				
Sialis	23	Tracheal gills	burrowers	predators

Table 16.1 (Contd.)

Taxon	Approximate no. species	Respiratory method	Mode of existence (habitats)	Feeding mechanism (habitats)
Trichoptera (caddisflies)				
Phylocentropus	5	Tracheal gills	burrowers (silk tube)	filterers[c]
Coleoptera (beetles)				
Noteridae (burrowing water beetles)	17	Tracheal gills (larvae)	burrowers	predators, collectors
Hydrophilidae (water scavenger beetles)				
Anacaena	3	tracheal gills (larvae)	(silt) burrowers	unknown
Creniitis	11	tracheal gills (larvae)	burrowers (sand & gravel)	unknown
Cymbiodyta	24	tracheal gills (larvae)	burrowers (sand & gravel)	unknown
Paracymus	8	tracheal gills (larvae)	burrowers (sand & gravel)	unknown
Diptera (true flies)				
Psychodidae (moth flies)	62	integument	burrowers	collectors
Ceratopogonidae (biting midges) (i.e., *Stilobezzia*, *Culicoides*)	256	retractile tracheal gills	burrowers burrowers	predators, collectors
Tamyderidae (primitive crane flies)	4	?	burrowers	unknown
Ptychopteridae	16	Atmospheric breathing tube	burrowers	collectors, sediment ingestion
Syrphidae	125	Atmospheric breathing tube	burrowers (some in low D.O. organic substrates)	collectors, predators
Ephydridae	368	caudal tubercle (tracheal)	most are burrowers	collectors, scavengers, petroleum feeders

Table 16.1 (Contd.)

Taxon	Approximate no. species	Respiratory method	Mode of existence (habitats)	Feeding mechanism (habitats)
Tipulidae (crane flies) (i.e., *Tipula, Limnophila, Hexatoma*)	573+	integument, spiricles (atmospheric)	most are burrowers	herbivores, predators
Chironomidae (midges)	500+	blood gills, integument	many are burrowers, tube builders	filterers[d] collectors, sediment ingestion
(i.e., *Chironomus, Polypedilum, Stictochironomus, Parachironomus, Paratendipes, Microtendipes*)				
Pelecypoda (clams)				
Unionidae (freshwater mussels)	500+	blood gills	burrowers	filterers
Sphaeriidae (fingernail clams)	200+	blood gills	burrowers	filterers

[a] Information derived from Merritt and Cummins (4) and Pennak (5).
[b] Detritivores that gather deposited fine particulate organic matter from the surface of the sediment or other objects (plants, rocks, etc.).
[c] Detritivores that gather suspended fine particulate organic matter.
[d] Chironomid larvae are known to exhibit more than one feeding mechanism within a species depending on the habitat of an individual.

The primary food source for most benthic invertebrates is associated with particulates and organic matter of various types. This fact indicates that discriminate feeders will consume particles with higher OC concentrations and higher contaminant concentrations than non-discriminate feeders. Although bioconcentration factors are independent of exposure concentration, tissue chemical concentrations are directly related to the exposure concentration. A key question is whether food with higher OC and contaminate concentrations results in higher tissue chemical concentrations than food with a lower OC and comparatively lower contaminate levels. On the basis of chemical potential or pollutant fugacity, both food sources should result in equal tissue concentrations if the chemical is adsorbed to the food as a function of OC.

To appreciate the extent to which feeding habits can effect bioaccumulation of lipophilic organics, a careful review of the relative importance of the various routes of exposure (food, water, sediment interstitial water, and sediment) is needed, as well as the role of OC in controlling the bioavailability. Both subjects are discussed in the next two sections.

BIOAVAILABILITY AS A FUNCTION OF ROUTE OF EXPOSURE

Route of exposure is often overlooked in various assessment schemes for determining hazard of chemicals in aquatic ecosystems. The most obvious reason for this is that most people simply want to know whether or not a chemical is toxic at a given concentration when it is applied either to the water or the sediment. Occasionally, it is applied to food and fed to fish or larger invertebrates, but this is usually the exception and not the rule. A key route of exposure that is nearly always overlooked is the sediment interstitial water. Since the focus of this workshop is sediments, I would like to discuss the importance of sediment interstitial water to benthic invertebrates. Sediment interstitial water refers to the water that exists between the sediment particles at the bottom of aquatic environments. It is also referred to as pore water.

Experiments that were conducted in our laboratory with kepone and the midge *Chironomus tentans* indicate that the sediment interstitial water concentrations were the same as would be predicted from the K_{oc} for kepone (2). The actual concentration that the midges were exposed to is not known although it would be presumed to be an amount intermediate to the sediment interstitial water and the overlying water concentration. However, the toxicity and bioaccumulation data indicated the midges were exposed to a concentration equal to the sediment interstitial water concentration (Tables 16.2 and 16.3). This will be discussed in detail below. These data also strongly suggest that for kepone and the midge *C. tentans* the primary route of exposure was the interstitial water and not the column water, food, or the ingested sediment per se.

Table 16.2 Comparison of Chronic NOEL-LOEL[a] for Different Routes of
Exposure for the Midge *Chironomus tentans* Exposed to Kepone for 14 days

Type of test	Type of exposure	Flow-thru chronic NOEL-LOEL (ppb)
Flow-thru	Water column	>5.4 <11.8
Water exposure[b]	Interstitial water	>5.4<11.8
Flow-thru		
Sediment exposure[b]	Sediment (1.5% OC)	>3,200<6,900
	Interstitial water	>13.1<29.7
	Water column	>1.8<3.6
Flow-thru		
Sediment exposure[b]	Sediment (12% OC)	>17,100<47,300
	Interstitial water	>8.6<20.7
	Water column	>1.3<2.6
Flow-thru		
Food exposure	Food	<17,900
	Water column	<1.0

[a] NOEL-LOEL are the no observed effect level and the lowest observed effect
level, respectively.

[b] Fourteen day partial life cycle tests. Data were reported by Adams et al. (2).

Table 16.3

Bioaccumulation Factors for Midges Exposed to C-14 Kepone via Water and Sediment Exposure Routes.
(Bioaccumulation factors were calculated for each test by dividing the midge tissue kepone concentration by
interstitial water, water column and sediment exposure concentrations.[a])

| Route of exposure | Type of test | Bioaccumulation factors for each test and each possible exposure route | | |
		Interstitial water	Water column	Sediment
Water	Flow-through	17,600	21,600	600
Sediment (1.5% OC)	Flow-through	5,200	39,600	20.0
	Static	10,800	14,100	70
Sediment (12.3% OC)	Flow-through	5,800	55,100	3.0
	Static	6,800	6,400	5.0

[a] Data were previously reported by Adams et al. [2].

The information shown in Table 16.2 is a summary of four 14-d partial life cycle toxicity tests with the midge *Chironomus tentans* and kepone. The endpoints measured were survival, growth, and bioaccumulation. The first test conducted was a traditional water exposure toxicity test. This was performed to determine the chronic toxicity of kepone in the water to the midge. The lowest observed effect level was 11.8 μg/L. This number formed the basis for comparison with effect levels resulting from sediment interstitial water, sediment, and food. A second study with kepone adsorbed to food and no kepone added to the water or sediment indicates that there was no effect on the midges at concentrations up to 17,900 μg/kg. Two additional studies were performed with kepone sorbed to two sediment types (1.5% and 12% organic carbon), and no kepone spiked on the food or in the water. Flow-through tests were conducted to keep the overlying water (water column) concentrations below the 11.8 μg/L effect level. Kepone concentrations on the sediment ranged from 490 to 92,800 μg/kg. The chronic no effect-effect concentrations based upon the kepone content of the sediments were $> 3,200 < 6,999$ and $> 17,100 < 47,300$ μg/kg for the 1.5% and 12% organic carbon sediments, respectively. The corresponding sediment interstitial water concentrations at these sediment effect levels were $> 13 < 29.7$ and $> 8.6 < 20.7$ μg/L.

The sediment effect concentrations and the organic carbon content of the two sediment types both varied by a factor of 8. The two experiments were designed such that the thermodynamic potential of kepone on the sediment, normalized by organic carbon, remained the same. Hence, the effect levels based on sediment interstitial water concentrations remained essentially the same in both studies, demonstrating the importance of the interstitial water as a route of exposure. The sediment interstitial water concentration effect levels agree reasonably well (differences are factors of 2.5 or less) with the 11.8 μg/L effect level observed when kepone was spiked directly into the water. The kepone concentrations in the water column were always a factor of 3.5 or more less than the chronic water column effect level (11.8 μg/L).

Additional evidence that sediment interstitial water was the primary route of exposure in the sediment tests comes from the bioaccumulation data (Table 16.3). Bioaccumulation factors based upon either the water column concentrations or the sediment concentrations were highly variable with differences exceeding an order of magnitude or more. Bioaccumulation factors based upon the sediment interstitial water concentrations were relatively constant and agree reasonably well with published values (5,000-13,000) for several marine species and with the bioaccumulation factors obtained in the spiked water study.

Additional evidence for the importance of sediment interstitial water as a key route exposure comes from an unpublished study conducted in our laboratory with the midge *Paratanytarsus parthenogenetica* and Compound AW. This is a chemical which is moderately chlorinated and has a water solubility of 60 μg/L. The K_{oc} for this chemical is 1.2×10^5. In contrast to kepone, this chemical is much less water soluble and has a greater affinity for sediment particles. Studies were conducted similar to those reported for kepone except only one sediment type was used (0.9% organic carbon) and full life cycle tests were performed. The results of this study confirm the importance of sediment interstitial water as a principal route of exposure. The chronic no observed effect level (NOEL) and lowest observed effect level (LOEL) for Compound AW in water were $>1.28 < 3.03$ μg/L (Table 16.4). Effects were observed in the sediment exposure study at sediment interstitial water concentrations of $>2.19 < 7.05$ μg/L. This agrees very well with the water exposure effect levels. The overlying water column chemical concentrations were ≤ 0.28 μg/L. The effect levels based upon the concentration of chemical on the sediment were $>27,600 < 59,000$ μg/kg. No effects were observed by feeding this chemical at levels on the food up to 85,500 μg/kg.

Table 16.4   Chronic NOEL-LOEL for Compound AW and the Midge
Paratanytarsus parthenogenetica Based on Water,
Sediment and Food Exposure Routes

Type of exposure[a]	Most sensitive parameter	NOEL-LOEL
Water	Emergence	1.28-3.03 ppb
Sediment (0.9% OC)	Growth and no. eggs/mass	2,760-5,900 ppb
Sediment interstitial water	_[b]	2.19-7.05 ppb
Food	_[c]	>85,500 ppb

[a] Three full life cycle flow through tests (28 d) were performed with measured concentrations.
[b] Sediment interstitial water numbers were obtained from the sediment test at the end of the 28-d chronic test.
[c] Highest concentration tested produced no effects.

Any discussion of sediment interstitial water as a primary route of exposure for benthic invertebrates would not be complete without considering the potential for food to be a source of exposure for bioaccumulation and toxicity of chemical contaminants. Macek et al. (7) and Bruggeman et al. (8) have compared uptake of chemicals by fish from food versus water for several chemicals with varying degrees of lipophilicity. Both studies conclude that water is the principal route of exposure for most lipophilic neutral organic chemicals and that the diet has the potential to contribute to the body residue as the elimination rate ($t_{1/2}$) becomes very large (Table 16.5). Macek et al. (7) indicated that DDT was the only chemical of the eight data sets they reviewed with an elimination half-life large enough (≈ 125 d) that the diet significantly contributed to the total body residue. Chemicals with elimination half-lives of 42 d or less had no more than 14% of the total fish residue attributable to the diet (Table 16.5).

Table 16.5 A Comparison of Time Required by Fish to Eliminate 50 percent of Their Body Chemical Residue ($t_{1/2}$) with Dietary Contribution to Body Residue and with Water Solubility for Several Chemicals

Chemical	Species	Water[a] solubility (ppm)	Dietary contribution (%)[b]	Elimination half-life ($t_{1/2}$, days)[b]
Trichlorobenzene	bluegill	30	6.0	<7
Kepone	oysters	3.0	<0.1	<4
Kepone	sheepshead minnow	3.0	<0.1	<28
Leptophos	bluegill	2.4	1.2	<10
Di-2-ethylhexylphthalate	bluegill	0.6	14	<3
Endrin	catfish	0.024	—[c]	<10
Endrin	fathead minnow	0.024	10	—[c]
Aroclor 1254	grass shrimp	0.010	—[c]	<14
Aroclor 1254	spot	0.010	—[c]	<42
Aroclor 1254	fathead minnow	0.010	—[c]	<42
DDT	lake trout	0.001	—[c]	>125
DDT	rainbow trout	0.001	—[c]	>160
DDT	fathead minnow	0.001	27 to 62	—[c]

[a] Water solubility values were reported by Kenega and Goring (10).
[b] Reported by Macek et al. (7).
[c] No data available.

Bruggeman et al. (8) present a discussion of this subject using a thermo-dynamic modeling approach. They state that the rate of transport of a chemical from water to the storage compartment (i.e., lipid) is equal to the activity gradient between water and the storage compartment (fugacity potential), divided by the sum of the resistance factors. A resistance factor is anything that impedes the rate of transfer and may be defined as the reciprocal of the transfer coefficient. The capacity of the transport compartment (blood) plays an important role in moving a chemical to the storage compartment (lipid). As the octanol-water partition coefficient (log K_{ow}) increases, the resistance decreases, the transfer capacity to storage capacity ratio becomes smaller, and the chemical moves rapidly to the storage compartment. For highly lipophilic chemicals, the transport compartment becomes much less important in terms of the actual residue or capacity and it serves as a facilitator to move the chemical to storage. The capacity of the transport compartment will play an important role after a sudden change in the aqueous concentration such as at the start of a clearance experiment where the deloading of the transport compartment will dominate the flow. This is dependent, however, on the log K_{ow} of the chemical. McKim et al. (9) has shown that for chemicals with a log K_{ow} of 3.0 or greater elimination via the gill significantly decreases.

Aqueous solutions of highly lipophilic chemicals with high activity coeffi-cients will be transported faster to storage than solutions of chemicals with a smaller log K_{ow}. For chemicals with a large log K_{ow}, the gill/water extraction efficiency transfer coefficient is high, and movement to storage is rapid. In this case, the ventilation rate can become the limiting factor for uptake of the chemical. When this happens, elimination is slow and the diet can contribute to the overall body burden of chemical.

The relationship between lipophilicity and the contribution of the diet to the total body burden is nicely demonstrated by the data of Bruggeman et al. (8). They show an increase in bioaccumulation potential with a decrease in water solubility (increase in log K_{ow}) for a series of chlorobiphenyls (Table 16.6). Bioaccumulation is defined in the present paper as body residue levels that are a result of chemical uptake from both water and the diet. The data in Table 16.6 also show that the uptake rate constant increases and the elimination rate constant decreases as a function of increasing log K_{ow}. This inverse relationship between the elimination rate constants and log K_{ow} also clearly indicates that time to equilibrium (steady state) and time to eliminate one half of the body residue increase significantly as the log P increases. This relationship has recently been demonstrated by Mackay and Hughes (11) mathematically and empirically for PCB congeners and for n-alkyl esters of *p*-aminobenzoic acid.

The elimination rate constant (k_2) is, by itself, an important descriptor of the bioaccumulation process of chemicals via the diet. In combination with the intestinal adsorption efficiency and gill/water extraction efficiency,

Table 16.6 Comparison of Uptake and Elimination Rate Constants and Bioaccumulation Potential with Water Solubility and log K_{ow} for Goldfish Exposed to Five Chlorinated Biphenyls (a modification of the data published by Bruggeman et al. (8)).

PCB	Water solubility (μg/L)	Octanol/water partition coefficient (log K_{ow})[a]	Water uptake rate constant k_1 (X10^{-3}d^{-1})	Elimination rate constant k_2 (X10^3d^{-1})	Bioaccumulation factor[b] K_{BA}
2,5-di	190	5.4×10^4	0.92	66	0.28
2',2',5-tri	110	8.4×10^4	0.95	48	0.33
2',4',5-tri	75	1.1×10^5	0.89	21	0.95
2,2',5,5'-tetra	55	1.5×10^5	0.74	15	1.17
2,3',4',5-tetra	22	3.0×10^5	0.42	10	1.60

[a] Calculated from water solubility by the equation of Kenega and Goring (10), log K_{ow} = 4.158-0.800 (log water solubility, mg/L). These values were calculated for illustrative purposes. Actual measurements of log K_{ow} may not show the same agreement with water solubility.
[b] Concentration in fish (extractable lipids)/concentration in food (extractable lipids).
Extractable lipids in fish: 3% of fresh weight (average).
Extractable lipids in food.: 10% of dry weight.

k_2 can realistically estimate the potential for bioaccumulation. Bruggeman et al. (8) show intestinal adsorption efficiencies of 50% or more for five chlorobiphenyls. Recent studies by McKim et al. (9) shows gill/water extraction efficiencies to be 60% for chemicals with a log K_{ow} of 2.8 or greater. These data indicate that uptake efficiencies via the intestines or via the gill are very similar. The relative rates of these two processes are important in determining the impact of diet on total body residue. Bioaccumulation can then be described mathematically as shown by Bruggeman et al. (8) as follows:

$$K_{BA} = \frac{\epsilon \cdot f}{k_2} = \frac{\epsilon}{E} \cdot \frac{f}{R_v/F} \cdot K_L \cdot L \qquad (1)$$

where K_{BA} = bioaccumulation factor, ϵ = intestinal adsorption efficiency, f = feeding rate, k_2 = elimination rate constant, E = gill adsorption efficiency, R_v = ventilation rate, F = fish weight, L = lipid content, and K_L = lipid/water partition coefficient ($k_1/k_2 \cdot L$, where k_1 and k_2 = uptake and elimination rate constants). When K_{BA} is equal to 1.0 or greater, there is evidence to suggest that the diet contributes significantly to the body burden (Table 16.6).

Several summary statements are possible for neutral organic chemicals:

- Elimination rate is inversely related to the lipid content of the organism.
- Body residues should be normalized for lipid content.
- Equilibrium bioconcentration factors are inversely related to the elimination rate constant and directly related to log K_{ow}.
- Bioconcentration factors are independent of exposure concentration.
- Tissue residue levels are directly related to exposure concentration.
- The lipid/water partition coefficient (K_L) should be relatively organism independent.
- The elimination rate constant (k_2) will reflect the ratios between lipid concentration factors for various chemicals.
- k_2 ought to describe the potential for bioaccumulation for chemicals other than neutral organics.

Bioaccumulation can be said to be a function of (1) the ratio of intestinal to gill adsorption efficiencies; (2) the ratio between the weight-specific feeding rate and the weight-specific ventilation rate; and (3) the lipid concentration factor, which can be described as K_L times the lipid content of the fish:

$$BAF = \frac{k_1}{k_2} = \frac{\gamma_w}{\gamma_L} \cdot L = K_L \cdot L \qquad (2)$$

where BAF = bioaccumulation factor and γ_w, γ_L = activity coefficients of the chemical in water and lipid, respectively. This equation points out that K_L is directly related to k_1, and is, therefore, inversely related to k_2. Since the ϵ/E ratio is near unity and the ratio between weight-specific feeding rate and weight-specific ventilation rate, $f/(R_v/F)$, is relatively organism independent and relatively constant for a given organism, the lipid concentration factor is the most important parameter for describing bioaccumulation. In conclusion, as previously stated by Macek et al. (7) and Bruggeman et al. (8), bioaccumulation will only be important for chemicals with a very slow elimination half-life (k_2). These chemicals usually have a log K_{ow} of 5 to 6. It is important to note that most ($\sim 90\%$) chemicals have a log K_{ow} less than 5. Veith (personal communication, 1984) has demonstrated that the mean lop P (i.e., log K_{ow}) value for 19,000 chemicals on the toxic substances inventory is about 2.5 (Fig. 16.1).

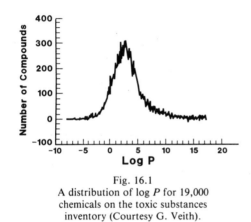

Fig. 16.1
A distribution of log P for 19,000
chemicals on the toxic substances
inventory (Courtesy G. Veith).

Exceptions to the above bioaccumulation rule do occur:

1. When the test chemical impairs the ability of the gill or intestine to carry out normal diffusion processes.

2. When a test chemical has a high log K_{ow} and is readily metabolized, it will behave like a low log K_{ow} chemical, e.g., the overall disappearance rate constant of the parent compound will be large even though the elimination rate constant (k_2) is small for the parent compound.

3. When fish feeding habits change, i.e., when they switch from zooplankters to benthic invertebrates that have been exposed to sediment interstitial water concentrations much higher than the overlying water, to which the zooplankters were exposed. The total mass of chemical the fish are exposed to now has increased and there

is greater potential for the diet to contribute to the body burden, depending on the log K_{ow} of the chemical and the elimination rate constant. This may explain why bottom-feeding fishes are observed to contain tissue residue levels greater than piscivorous fish. This may also explain why PCB levels in ocean fish dramatically increase when they move into harbor areas. Similarly, it has also been reported that dietary PCBs contribute to body residue levels of Lake Michigan lake trout (12).

The model I have described for comparing food versus water as a route of exposure has been developed for fish. However, I believe that it can be conceptually applied to benthic invertebrates as well. This is based on the fact that the adsorption properties of lipids (K_L) are not expected to vary significantly between species.

The importance of sediment interstitial water as a primary route of exposure is supported by the research of several other investigators. Landrum, Eadie and co-workers (13-16) have investigated routes of exposure for several polycyclic aromatic hydrocarbons (PAHs). Their data definitely show that body residues of benthic organisms are not due to the chemical concentrations in the water column. The importance of considering sediment and/or sediment interstitial water concentration in relation to body residue is clear. They conclude that sediment interstitial water or in some cases a mixture of interstitial water and overlying water is generally the source primarily responsible for body residues of PAHs in amphipods and oligochaetes. Additional evidence that sediment interstitial water may be the primary source of body residues has been reported by Muir et al. (17) using the midge *Chironomus tentans* and terbutyrn, flyridone, triphenyl phosphate, trans-permethrin, methoxychlor, and hexachlorobiphenyl.

BIOAVAILABILITY AS A FUNCTION OF ORGANIC CARBON

For neutral lipophilic organic chemicals, the influence of organic carbon in the sediments is perhaps the most important parameter controlling bioavailability. The use of organic carbon (OC) to predict soil adsorption of many chemical types has become widely accepted. The ability of OC to normalize the differences observed in the soil partition coefficients (K_p) for a given chemical with different soils or soil fractions is usually expressed as:

$$K_p = K_{oc} \times F_{oc} = \frac{C^s}{C^w} \tag{3}$$

where K_p = the soil/water partition coefficient, K_{oc} = the organic carbon normalized K_p, F_{oc} = fraction organic carbon = % oc/100, C_s = the concentration of chemical on the sediment phase, and C_W = the chemical concentration in the water phase. This relationship has been investigated by numerous authors (18-21). It has also been demonstrated to be somewhat applicable to chemicals other than neutral organics. Urano et al. (22) have demonstrated linear relationships between OC and concentrations of nonionic and anionic surfactants.

Equation (3) can be used in a number of ways to assess the bioavailability of chemicals in natural surface waters or sediment interstitial waters. The bound versus non-bound (available versus non-available) concentration of a given organic chemical in water as a function of the presence of suspended particles can be estimated from both the water solubility of K_{oc} for the chemical as shown in Table 16.7.

Table 16.7 Estimated Nonbound Fraction of Chemicals in Natural Waters (2)

Chemical	Water[a] solubility, mg/L	Soil partition coefficient,[b] K_{oc}	Calculated non-bound fraction in natural waters,[c] % of total
Hexachlorobiphenyl	0.001	2,700,000	3.6
DDT	0.002	1,500,000	6.3
Methoxychlor	0.003	1,100,000	8.3
DDD	0.005	730,000	12.0
Aroclor 1254	0.010	400,000	20.0
Endrin	0.024	230,000	30.3
Dichlorobiphenyl	0.062	72,000	58.1
Lindane	0.15	44,000	69.4
Trifluralin	0.60	25,000	80.0
t-Butyphenyl diphenyl phosphate	3.2	5,000	95.2
Biphenyl	7.5	1,000	98.0
Parathion	24	800	99.2
Trichlorobenzene	30	500	99.5

[a] Water solubility values previously reported (8).
[b] K_{oc} is calculated from water solubility (23).
[c] Calculation based on the assumption that the natural water contains 250 mg/L of suspended solids (SS), which contain 4.0% organic carbon (this is equivalent to 10 mg/L total organic carbon in the water column).

$$\frac{C_s}{C_w} = K_{oc} \times F_{oc} = K_p \; ; \; K_p \times C_w = C_s \; ; \; C_s \times C_{ss} = C_{WB} \; ;$$

$$C_c + C_w = \text{total} \; \frac{C_w \times 100}{\text{total}} = \text{fraction not bound}$$

where C_s = concentration on the sediment, C_w = concentration in the water column, F_{oc} = fraction of organic carbon, C_{ss} = concentration of suspended solids, C_{WB} = concentration in the water column bound to suspended solids.

These chemicals were chosen to cover a range of water solubilities spanning five orders of magnitude in order to demonstrate the impact of organic carbon on the bioavailability of a chemical. It can be seen that for waters with 250 mg/L suspended solids (4.0% OC) a chemical with a water solubility of 0.062 mg/L ($K_{oc} = 72{,}000$) is approximately 50% non-bound or available for uptake. This would, of course, change as the level of suspended solids and OC change. But, it is clear that the K_{oc} has to be very large to significantly reduce the availability of a chemical in most natural waters since most lakes and streams have suspended solid levels less than 250 mg/L and the OC levels of the suspended solids are usually less than 4.0%. Adams (24) has shown that sediment samples collected across the United States have a mean OC level of 2.0 $\pm$ 1.7% ($N = 187$). A majority (96.3%) of the values occurred within two orders of magnitude (0.1 to 10%).

Eaton et al. (25) have demonstrated a 35% reduction in bioavailability of Kelthane® to minnows in the presence of 65 mg/L suspended clay particles. Organic carbon of the clay particles could not account totally for the reduction in fish tissue levels; nevertheless, this experiment demonstrates the principle that bioavailability can be reduced as a function of OC levels and/or the solids concentration. In the Kelthane experiments the use of suspended clay particles that are very small in particle size and low in OC (< 1.0%) instead of typical soil or river sediments would explain why OC would not account for the reduced bioavailability.

Sources of OC other than the sediments and suspended solids in the water column have the potential to impact the bioavailability of the chemical. One such source is soluble organic carbon both in the water column and in sediment interstitial water. Leversee et al. (26) have demonstrated reduced uptake (2 ×) of benzo(a)pyrene by *Daphnia magna* in the presence of 2 mg/L Aldrich humic acids and naturally occurring humics in stream water. They also demonstrated a small increase in uptake of methylcholanthrene exposed to 0.1-10 µg/L Aldrich humic acids. Adams et al. (27) showed a decrease in the toxicity of di-2-ethylhexyl phthalate to *D. magna* in the presence of 250 mg/L Aldrich fulvic acid. The toxicity of butylbenzyl phthalate, on the other hand, did not change in the presence of 250 mg/L fulvic acid. Most natural surface waters do not contain high levels of dissolved organic carbon (DOC). They are typically in the range of microgram to low milligrams per liter. Small lakes and farm ponds may approach 100 mg/L and some streams originating in coniferous forest may contain 200-300 mg/L. These values are based on personal experience. Sediment interstitial water DOC levels are usually higher than the overlying water and may range up to very high levels in peat bog areas. Landrum et al. (28) reported sediment interstitial water DOC levels in the range of 7-10 mg/L for several Great Lakes sediments.

The existing data on the binding of chemicals to DOC is quite meager. Insufficient data are available to make definitive conclusions. At present, the data suggest the DOC is not nearly as important in controlling bioavailability as is non-dissolved carbon. This may be due to several factors: (1) DOC levels usually occur in the microgram to milligram per liter range whereas OC levels in sediment are typically in the 0.5 to 5.0% range (10^3- to 10^4-fold difference), and (2) the adsorption of organic chemicals by DOC does not appear to be directly related to the presence of the carbon molecule itself, but is dependent on the size of the molecule and the number of binding sites available. It has been shown that naturally occurring humic acids with small particle sizes sorb only 10% of the amount of commercial humic which are thought to contain large particle sizes [28]. Landrum et al. [28] indicate that at low DOC levels (1 to 2 mg/L) the partition coefficient,

$$\frac{\text{g chemical/g DOC}}{\text{g chemical/mL water}} \tag{4}$$

is approximately equal to the log K_{ow} of a given chemical. In summary, we are just beginning to understand the role of DOC in binding organic chemicals. At the present, it appears that DOC partition coefficients are independent of chemical concentration and may decrease slightly as a function of increasing DOC.

In our laboratory, we have demonstrated the importance of the sediment organic carbon concentration in relation to the acute and chronic toxicity of sediments spiked with kepone to midges [2, 29]. The data indicate that the sediment interstitial water concentrations remained quite constant over a wide range of sediment kepone concentrations as a function of the OC (Fig. 16.2). The toxicity in both the acute and chronic studies was directly related to the interstitial water, which was a function of the sediment OC levels and not the sediment concentrations per se. This conclusion was based on the toxicity and bioaccumulation results as well as the life habits of the midge.

The importance of OC in natural sediments is further demonstrated by the relationship between diphenyl oxide (DPO) sediment concentrations and sediment OC (Fig. 16.3). These data were recently obtained from a 2-acre site on a large river in southeastern United States. Both the concentrations of DPO and OC span approximately three orders of magnitude and show a moderate degree of correlation. The results point out the variability in sediment concentrations of chemicals that can exist in fairly confined areas as a function of OC and indicates sample to sample variability can be very high. Clearly, organic carbon levels need to be measured when assessing the significance of lipophilic organics on sediments.

Fugacity is a concept that has traditionally been used in gas low calculations involving equilibrium and movement of gases, but is extendable to other areas. Chemical fugacity can be defined as the tendency of a chemical to leave one compartment and move to another. Fugacity and absolute activity are both used to replace chemical potential in thermodynamic equations. Fugacity is usually expressed in units of pressure (pascals) and

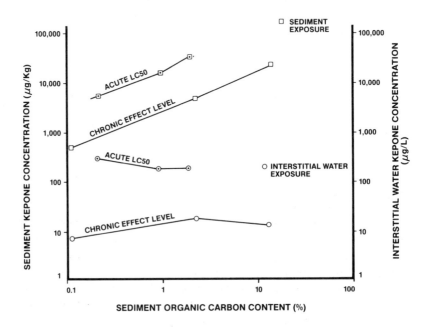

Fig. 16.2
Comparison of acute (LC$_{50}$) and chronic (geometric mean
NOEL-LOEL) values with kepone concentrations on the
sediment and in the sediment interstitial water as a
function of sediment organic carbon levels.

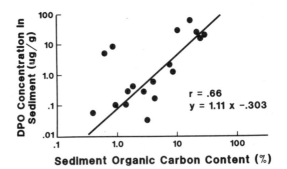

Fig. 16.3
Regression of diphenyl oxide (DPO)
concentrations on sediment with the organic
carbon content of the sediment.

absolute activity is expressed by the activity coefficient (γ). The absolute activity or effective concentration of a chemical determines the extent to which it influences or acts in determining the equilibrium properties of the solution. The activity (a) can be expressed as

$$a = \gamma C \qquad (5)$$

where γ = the activity coefficient and C = the molar concentration. Activity coefficients for chemicals in solutions are usually less than 1.0 and approach 1.0 as the solution becomes more dilute. Activity coefficients can be used to quantify pollutant affinity for a given compartment of an environmental model. Since activity coefficients generally do not vary by more than a factor of 10 from chemical to chemical, they are useful in first approximations designed to estimate chemical partitioning of a group of chemicals such as lipophilic organic chemicals.

The concepts of chemical potential or pollutant fugacity are important for assessing the significance of chemicals sorbed to sediments because they can be used to describe the partitioning of a given chemical between a source (sediment) and a sink (midge) at equilibrium. Mackay and co-workers (30, 31) have demonstrated the usefulness of using fugacity capacity as a means of describing chemical partitioning between ecosystem components (air, soil, water, sediment and biota) on a worldwide basis. These same concepts can be used to describe chemical partitioning between sediment, water, and biota at specific sites or in laboratory test systems.

The utility of the fugacity of chemical activity approach can be demonstrated by means of the following summary statements reported by Karickhoff (32):

- Chemicals will move across phase boundaries from higher to lower fugacity until the fugacities are equal in each phase. This movement is independent of chemical concentration and dependent on fugacity capacity.

- Chemical activity coefficients describe the fraction of total chemical present in a source that is actively involved in movement. An activity coefficient for a given chemical on sediment OC is largely independent of sediment source, and the activity coefficient for a chemical in organism lipid is independent of the organism. Therefore, it can be shown that source (sediment) and sink (organism) specificity for a chemical or chemical fugacity between source and sink is directly related to the activity coefficients for the chemicals in the source and sink.

- To assess the potential of organisms to accumulate hydrophobic organics from sediments, it can be assumed that the dominant fraction controlling distribution on the sediment particles is the organic carbon, and similarly the lipid content of the organism. For a first approximation, the bio-uptake of lipophilic organics from sediment ought to be a function of the OC content of the sediment and the lipid content of the organism. This relationship can be expressed as:

$$\text{BAF}_L = \frac{\gamma oc}{\gamma L} \tag{6}$$

where BAF_L = bioaccumulation factor for organism lipid or distribution coefficient of the chemical between the organismic lipid (γ_L) and the sediment OC (γ_{oc}) as a function of the respective activity coefficients (γ).

● For lipophilic organics, large variations (more than 10 ×) in γ_{oc} or γ_L are not expected from chemical to chemical. Any change in the γ_{oc} would also be expected in the γ_L for a given chemical. Thus, bioaccumulation is expected to be controlled by the organism lipid to sediment OC ratio and should be chemical independent as related to log P.

● The actual concentration of chemical in the organism lipid and the sediment OC may vary some from chemical to chemical, but generally should be around 1.0 and should not exceed 10.

The last three statements are theoretical postulates that need to be tested. This process can be described conceptually by the following equations for passive uptake of chemicals from sediment by aquatic organisms. Continual exposure to a constant chemical source (sediment) will result in a steady-state condition analogous to thermodynamic equilibrium, provided the organism does not die, as shown:

$$C_O = (\text{BAF})C_s \tag{7}$$

where C_O = concentration of chemical in the organism (midge), C_s = chemical concentration in the source (sediment), and BAF = bioaccumulation factor for the organism. If biological uptake is only by diffusion (passive uptake), then the BAFs for different sources (water, sediment, sediment interstitial water) should be interrelated by the relative fugacity capacity or activity of the chemical in each source. This can be described as:

$$(\text{BAF})_s = \frac{(\text{BAF})_w}{K_p} = \frac{(\text{BAF})_w}{K_{oc} f_{oc}} \tag{8}$$

where s = source, w = water, K_p = sediment/water partition coefficient defined as $K_{oc} \times f_{oc} = K_p$. For lipophilic organics sorbed to a source such as sediment, bioaccumulation can be expressed as:

$$(BAF)_s f_{oc} = (BAF)_{oc} = \frac{(BAF)_w}{K_{oc}} \tag{9}$$

Thus, bioaccumulation is related to the organic carbon content of the source and the partitioning between the source and the water. This assumes that uptake is from the water; however, the equation could be written for uptake via food if so desired. The above bioaccumulation factor (BAF_{oc}) should be applicable for most sediment types. If it is assumed that pollutant fugacities of natural organic matter approximate fugacity capacities of lipid in the midge tissue, then $(BAF)_{oc} \simeq 1.0$, as previously stated. A similar set of equations could also be written for active uptake or to describe biological effects.

The data of Adams et al. (2) for midges and kepone demonstrates this relationship between organism and sediment concentrations of kepone. The mean midge $(BAF)_w$ based on interstitial water concentrations (Table 16.2) = 9,240 and the average K_{oc} for kepone = 16,000. The midge $(BCF)_{oc}$ = 9,240/16,000 = 0.6. This agrees fairly well with the theoretical value of 1.0, recognizing that the theoretical and actual values could be orders of magnitude apart. Eadie et al. (14, 15) have shown the concentrations for sediments and organisms in the sediments to be close to unity for five polycyclic aromatic hydrocarbons (PAH) with amphipods and oligochaetes. The values ranged from 1.0 to 3.4 when the PAH analyses were based on bulk sediments. Measurements of PAHs on fine grain sediments ($< 60\ \mu$m), which closely resembles the ingestible size fraction for these organisms, provided values close to unity (0.4 to 1.9). Assuming the midge lipid content to be similar to the organic carbon content of the sediments, then the ratio of the concentrations in both phases would be expected to be near 1.0.

SUMMARY

Bioavailability of chemicals sorbed to settled sediments and suspended sediments is a critical aspect of the hazard assessment process for these chemicals. Principles of hazard assessment for chemicals in water based on measurement of exposure and effects ought to be applicable to sediments as well. The challenge before us is to define accurately what constitutes exposure and what is biologically available for benthic invertebrates.

As a summary of this review of bioavailability, it is concluded that the following principles are applicable for the sediment assessment process:

- Organic carbon is the primary controlling factor of the bioavailability of neutral organic chemicals sorbed to sediment (suspended or settled).

- Bioavailability is affected by both DOC and particulate OC. Particulate OC is the dominant factor.

- Water, especially sediment interstitial water is the primary route of uptake for most benthic invertebrates and most lipophilic chemicals. Dietary uptake can contribute to body residues for chemicals with a log P in the range of 5 to 7; however, this is not true for all chemicals in this range because of metabolism and other considerations.

- Benthic invertebrate food sources, i.e., algae, bacteria, detritus, and fine grain sediments are no better accumulators of organic chemicals than invertebrates or fish when normalized for OC ($BAF_{OC} \simeq$). The implications of this are (a) in ecosystems which are at thermodynamic equilibrium for a given chemical, fish and invertebrates have equal exposure concentrations and will have, theoretically, equal tissue concentrations on a lipid normalized basis; and (b) in ecosystems in which the sediment's contaminant concentrations are not in equilibrium with the water column, benthic invertebrates are likely to have higher lipid normalized tissue residues than fish because sediment interstitial water concentrations are higher than overlying surface water concentrations. In this situation, benthic invertebrates may present a possible important dietary source of contaminants for predators depending on the log P of the chemical, elimination rate of the chemical by the predator, and feeding rate of the predator.

- For hazard assessment purposes, the aquatic ecosystem can conceptually be thought of as a system with increasing sediment concentrations from the top of the water column to the bottom of the biotic zone in the settled sediments. Sediment concentrations in typical ecosystems range from 1.0 mg suspended solids per liter of water to 1.0 kg of sediment per liter of water, i.e., bottom sediments where the sediment to water ratio is approximately 1:1. Using this conceptual framework, all water in the system is sediment interstitial water, and exposure concentrations or bioavailability can be calculated on the basis of partitioning to organic carbon. Calculations using this approach have been made by Staples et al. (33). This concept is a tool intended for assessment purposes. There will be situations and chemicals for which it will not apply, but as a general guideline ought to be applicable to most lipophilic chemicals.

REFERENCES

1. **Cairns, J. Jr., K. L. Dickson and A. W. Maki** (eds.). 1978. *Estimating the Hazard of Chemical Substances to Aquatic Life.* ASTM STP 657. American Society for Testing and Materials, Philadelphia, PA.

2. **Adams, W. J., R. A. Kimerle and R. G. Mosher.** 19895. Aquatic safety assessment of chemicals sorbed to sediments. In R. D. Cardwell, R. Purdy and R. C. Bahner, eds., *Aquatic Toxicology and Hazard Assessment: Seventh Symposium.* ASTM STP 854. American Society for Testing and Materials, Philadelphia, PA, pp. 429-453.

3. **Pavlou, S. P. and D. P. Weston.** 1984. Initial evaluation of alternatives for development of sediment related criteria for toxic contaminants in marine water (Puget Sound). Phase II: Development and testing the sediment-water equilibrium partitioning approach. Prepared by JRB Associates for the U.S. Environmental Protection Agency under EPA Contract No. 68-01-6388.

4. **Merritt, R. W. and. K. W. Cumins.** 1978. *An Introduction to the Aquatic Insects of North America.* Kendall/Hunt Publishing Company, Dubuque, IA.

5. **Pennak, R. W.** 1953. *Fresh-water Invertebrates of the United States.* The Ronald Press Company, New York.

6. **Bahner, L. H., A. J. Wilson, J. M. Sheppard, J. M. Pattrick, L. R. Goodman and G. E. Walsh.** 1977. Kepone bioconcentration, accumulation, loss, and transfer through estuarine food chains. *Chesapeake Science* **18**:299-308.

7. **Macek, K. J., S. Petrocelli and B. H. Sleight.** 1979. Considerations in assessing the potential for, and significance of, biomagnification of chemical residues in aquatic food chains. In L. L. Marking and R. A. Kimerle, eds., *Aquatic Toxicology: Second Symposium.* ASTM STP 67. American Society for Testing and Materials, Philadelphia, PA, pp. 251-268.

8. **Bruggeman, W. A., L. B. J. M. Martron, D. Kooiman and O. Hutzinger.** 1981. Accumulation and elimination kinetics of di-, tri-, and tetrachlorobiphenyls by goldfish after dietary and aqueous exposure. *Chemosphere* **10**:811-832.

9. **McKim, J., P. Schmieder and G. Veith.** 1984. Adsorption dynamics of organic chemical transport across trout gills as related to octanol/water partition coefficient. *Environ. Toxicol. and Chem.* (in press).

10. **Kenega, E. E. and C. A. I. Goring.** 1980. Relationship between water solubility, soil sorption, octanol-water partitioning, and concentration of chemicals in biota. In J. G. Eaton, P. R. Parrish and A. C. Hendricks, eds., *Aquatic Toxicology: Third Symposium.* ASTM STP 707, American Society for Testing and Materials, Philadelphia, PA, pp. 78-115.

11. **Mackay, D. and A. I. Hughes.** 1984. Three parameter equation describing the uptake of organic compounds by fish. *Environ. Sci. Technol.* **18**:439-444.

12. **Weinger, D.** 1978. Accumulation of PCBs by lake trout in Lake Michigan. Ph.D. dissertation, University of Wisconsin, Madison.

13. **Landrum, P. F., B. J. Eadie, W. R. Faust, N. R. Morehead and M. J. McCormick.** 1985. Role of sediment in the bioaccumulation of benzo(a)pyrene by the amphipod, *Pontoporeia hoyi.* In M. W. Cooke and A. J. Dennis, eds., Eighth International Symposium on Polynuclear Aromatic Hydrocarbons; Mechanisms, Methods, and Metabolism, Battelle Press, Columbus, OH, pp. 799-812.

14. **Eadie, B. J., W. R. Faust, P. F. Landrum, N. R. Morehead, W. S. Gardner and T. Nalepa.** 1983. Bioconcentration of PAH by some benthic organisms of the Great Lakes. In M. W. Cooke and A. J. Dennis, eds., *Polynuclear Aromatic Hydrocarbons,* Seventh International Symposium on Formation, Metabolism and Measurement, Battelle Press, Columbus, OH, pp. 437-449.

15. **Eadie, B. J., W. R. Faust, P. F. Landrum and N. R. Morehead.** 1985. Factors affecting bioconcentration of PAH by the dominant benthic organisms of the Great Lakes. Eighth International Symposium on Polynuclear Aromatic Hydrdocarbons, Battelle Press, Columbus, OH, pp. 363-378.

16. **Landrum, P. F. and D. Scavia.** 1983. Influence of sediment on anthracene uptake, depuration, and biotransformation by the amphipod *Hyalella aztecca. Can. J. Fish. Aquat. Sci.* **40**:298-305.

17. **Muir, D. C. G., B. E. Townsend and W. L. Lockhart.** 1983. Bioavailability of six organic chemicals to *Chironomus tentans* larvae in sediment and water. *Environ. Toxicol. Chem.* 2:269-281.

18. **Karickhoff, S. W., D. S. Brown and T. A. Scott.** 1979. Sorption of hydrophobic pollutants on natural sediments. *Water Res.* 13:241-248.

19. **Karickhoff, S. W.** 1981. Semi-empirical estimation of sorption of hydrophobic pollutants on natural sediments and soils. *Chemosphere* 10:833-846.

20. **Hassett, J. J., J. Means, W. L. Banwart and S. G. Wood.** 1980. Sorption properties of sediments and energy-related pollutants. EPA-600/3-80-041. U.S. Environmental Protection Agency, Washington, DC.

21. **Chiou, C. T., P. E. Porter and D. W. Schmeddling.** 1983. Partition equilibria of nonionic organic compounds between soil organic matter and water. *Environ. Sci. Technol.* 17:227-231.

22. **Urano, K., M. Saito and C. Murata.** 1984. Adsorption of surfactants on sediments. *Chemosphere* 13:293-300.

23. **Mill, T., D. M. Hendry, W. E. Mabey and D. J. Johnson.** 1980. Laboratory protocols for evaluating fate of organic chemical in air and water. EPA-600/3-80-069. U.S. Environmental Protection Agency, Washington, DC.

24. **Adams, W. J.** 1986. Levels of organic carbon in aquatic sediments of the United States. Monsanto Environmental Sciences Report No. ESC-EAG-86-10, MSL-5413, St. Louis, Mo.

25. **Eaton, J. E., V. R. Mattson, L. H. Mueller and D.K. Tanner.** 1983. Effects of suspended clay on bioconcentration of Kelthane in fathead minnows. *Arch. Environ. Contam. Toxicol.* 12:439-445.

26. **Leversee, G. J., P. F. Landrum, J. P. Geisy and T. Fannin.** 1983. Humic acids reduce bioaccumulation of some polycyclic aromatic hydrocarbons. *Can. J. Fish. Aquat. Sci.* 40:63-69.

27. **Adams, W. J., W. E. Gledhill and S. W. Landvatter.** 1978. Acute toxicity of di-2-ethylhexyl phthalate (DEHP) to *Daphnia magna* in the presence of fulvic acid. Monsanto Environmental Sciences Report ES-78-SS-14, St. Louis, MO.

28. **Landrum, P. F., S. R. Nihart, B. J. Eadie and W. S. Gardner.** 1984. Reverse phase separation method for determining pollutant binding to Aldrich humic acid and dissolved organic carbon of natural waters. *Environ. Sci. Technol.* 18:187-192.

29. **Ziegenfuss, P. S., W. J. Renaudette and W. J. Adams.** 1984. Methodology for assessing the acute toxicity of chemicals sorbed to sediments. Monsanto Environmental Sciences Report ESC-EAG-78-SS-10, St. Louis, MO.

30. **Mackay, D.** 1979. Finding fugacity feasible. *Environ. Sci. Technol.* 13:1218-1223.

31. **Mackay, D. and S. Paterson.** 1981. Calculating fugacity. *Environ. Sci. Technol.* 15:1006-1014.

32. **Karickhoff, S. W.** 1983. Bioconcentration potential: Sediment sources. Unpublished Report, U.S. Environmental Protection Agency, Athens, GA.

33. **Staples, C. A., K. L. Dickson, J. H. Rodgers, Jr. and F. Y. Saleh.** 1985. A model for predicting the influence of suspended sediments on bioavailability of neutral organic chemicals in the water compartment. In R. D. Cardwell, R. Purdy and R. C. Bahner, eds., *Aquatic Toxicology and Hazard Assessment: Seventh Symposium.* ASTM STP 854. American Society for Testing and Materials, Philadelphia, PA, pp. 417-428.

Chapter 17

Bioavailability of Sediment-Bound Chemicals to Aquatic Organisms— Some Theory, Evidence and Research Needs

John H. Rodgers, Jr., Kenneth L. Dickson, Farida Y. Saleh and Charles A. Staples

INTRODUCTION

Assessing the persistence and compartmentalization of potentially toxic chemicals released both accidentally and purposefully to aquatic systems requires consideration of their interactions with sediments. For some chemicals, both suspended and consolidated sediments may serve as major storage reservoirs. Sorption of chemicals to sediments may play a major role in transport, activity, and persistence.

Considerable recent and historical data have suggested that some chemicals are not bioavailable or bioactive when sorbed to seston or settled sediments. The purpose of this paper is to review some of these data and to suggest a research strategy to improve our predictive capability of the roles that suspended and settled solids play in regulating the bioavailability of recently released chemicals in aquatic systems.

If a compound is "bioavailable," then the compound, over a range of concentrations, should elicit some response from an organism. The response is a function of exposure concentration and duration of exposure. The important participants in this interaction are the organisms, the chemical, and the particular sorbing materials. Since (by definition) the bioavailability of a chemical is measured by some response, a wide variety of single species and community responses may be used as indicators. Perhaps most frequently, the following questions are asked:

1. Does the presence of suspended or settled solids alter the toxicity of a chemical? (i.e. is that chemical less bioavailable to cause acute and/or chronic toxicity?)

2. Does the presence of suspended or settled solids alter bioconcentration of a chemical?

3. Does the presence of suspended or settled solids alter biodegradation, biotransformation, or transformation of a chemical?

This paper primarily addresses the part of the first question dealing with acute toxicity. Only a little discussion is directed toward the second question since this is the principal topic of another paper (by William Adams) in this publication. Finally, the third question of biotransformation and sorption is addressed briefly and some research needs are presented.

ACUTE TOXICITY AS A MEASURE OF BIOAVAILABILITY OF SORBED CHEMICALS—SOME THEORY AND CONSIDERATIONS

A Mass Balance Model for-Predicting Aqueous-Phase Chemical Concentrations in the Presence of Solids

A basic hypothesis or premise that may spawn strategies for research on bioavailability is that chemicals that are particulate-bound are not bioavailable. Based on evidence, we may reject or fail to reject hypotheses generated from this premise. Early in our research, the need for a tool that would allow rapid predictions of particulate-phase chemical concentrations and aqueous-phase chemical concentrations became apparent. Details of the model that evolved for organic chemicals have been published and only a brief summary is provided here (1).

The model is based upon simple equilibrium partitioning concepts and assumes rapid attainment of equilibrium. The model predicts how much of the total chemical concentration of a neutral organic chemical is distributed between the particulate-bound phase (C_p) and the dissolved phase (C_d). The model relates a chemical's sorption coefficient (K_p) to the organic content and concentration of suspended solids or particulates in water to develop a prediction of the percentages of the chemical that will be particulate bound and dissolved.

Similar to several other environmental fate models, this partitioning model is based on mass balance and fugacity principles. Sediment sorption tests, for example, are simply laboratory analogs based on these equilibrium concepts for the "real world" situation of a chemical introduced into a sediment-water system. The model reflects these analogs and the implied partitioning into dissolved test chemical (C_d) and particulate-bound chemical

(C_p) phases. Total chemical concentration (C_t) does not change except due to kinetic fate (degradation processes). The model describes the equilibrium distribution of total chemical in terms of a particulate-to-total chemical ratio (C_p/C_t).

The model is developed as follows:

If $\qquad$ $C_t = C_p + C_d$ $\qquad\qquad\qquad$ (1)

where C_t, C_p and C_d are total, particulate-bound and dissolved chemical (mass/vol)

and $\qquad$ $C_s = C_p/S$ $\qquad\qquad\qquad\qquad$ (2)

where C_s = sorbed chemical (mass/mass) and S = suspended solids concentration (mass/vol)

then $\qquad$ $F = (C_p/C_t)100$ $\qquad\qquad\qquad$ (3)

where F = the fraction of C_t that is particulate-bound.

Since, at equilibrium,

$$K_p = C_s/C_d \qquad\qquad\qquad (4)$$

where C_p = partition coefficient, which is either directly measured or calculated from

$$K = (K_{oc}) (\text{o.c.}) \qquad\qquad\qquad (5)$$

where K_{oc} = sediment sorption coefficient and o.c. is the sediment organic content,

then $\qquad$ $K_p = C_p/(S\,C_d)$ $\qquad\qquad\qquad$ (6)

or $\qquad$ $C_p = K_p/S\,C_d$ $\qquad\qquad\qquad$ (7)

Equation (7) shows that for every 1 unit mass per volume (i.e., 1 mg/L) of C_d there are K_p times C_d times S units mass per mass (i.e., mg/kg) of particulate-bound chemical. Therefore, knowledge of K_p and S enable the ratio F or (C_p/C_t) 100 to be determined for some mass input of a chemical into a system.

To illustrate how the model works, consider the following. Given a K_{oc} of 10,000 and suspended solids organic carbon content of 1%, $K_p = 100$ (Fig. 17.1). Since $K_p = C_s/C_d$, where C_s and C_d are suspended solids and water concentrations, respectively, a K_p of 100 means that for every 1 mg/L aqueous chemical there are 100 mg/kg particulate-bound chemical. For a

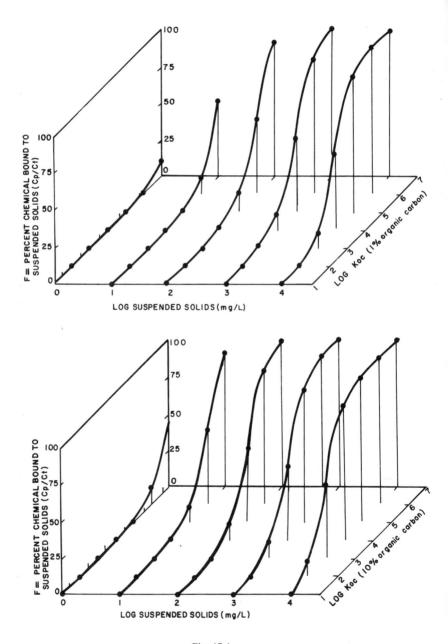

Fig. 17.1
Fraction of test chemical expected to be particulate bound at
1% and 10% organic carbon content of suspended solids.

water column suspended solids concentration of 10 mg/L (0.00001 kg/L) there are hence 100 mg/kg times 0.00001 kg of particulate-bound chemical per liter solution or 0.001 mg/L. Since $0.001/1.001$ (C_p/C_t) is 0.001 this represents only 0.1% shift of total chemical from dissolved to particulate phase. Next consider a K_{oc} of 1,000,000 (10^6) and a suspended solids organic carbon content of 10% (Fig. 17.1). K_p is then 100,000. So, for every 1 mg/L dissolved chemical, there are 100,000 mg/kg particulate-bound chemical. A suspended solids concentration of 1,000 mg/L or 0.001 kg/L would have then 100 mg of particulate-bound chemical for each mg/L aqueous chemical. The particulate to total chemical ratio would be 100/101, i.e., 99.01% of total chemical would be adsorbed to the suspended solids.

Some Considerations Required for Design of Acute Toxicity Tests of the Model

The model provided a tool that could be used to predict the reduction in aqueous-phase concentrations of organic chemicals as a result of the presence of suspended sediments. Initial calculations facilitated by the model demonstrated that experimental designs using acute toxicity bioassays to test the limits of the model would not be trivial. Chemicals were tested that extensively sorb, as well as chemicals that were very soluble and did not appreciably sorb. These experiments were limited to petroleum refinery effluent chemicals.

In order to test the model vigorously, chemicals were used that both sorb and were acutely toxic. One rapidly becomes aware of the solubility, sorption, and acute toxicity trade-offs after more detailed examinations. Most water-soluble chemicals do not appreciably sorb. Most water-insoluble chemicals sorb extensively but are acutely toxic at concentrations in excess of their water solubilities.

Identification and selection of test sediments was an unanticipated and extensive activity. We anticipated that the organic carbon content would be one of the driving variables regulating the amount of chemical sorbed and thus the amount bioavailable. Careful examination of published information and other sources revealed that the organic carbon content of suspended or bed sediments in "natural" situations does not vary an order of magnitude. Therefore, the sediment organic carbon axis of the model was compressed to a range from 0 to 2% (2) and was not as suitable an independent variable for prediction of bioavailable (in solution) test chemical as initially anticipated.

A more important factor was suspended solids concentrations that can range from 10 to 10,000 mg/L in aquatic systems (2). Estimation of sorption partition coefficients for potential test chemicals was hampered by laboratory measures of partitioning potential. Sorption tests with chemicals that partition to solids almost totally (i.e., high K_p) are difficult to conduct because of inaccuracies in determining the aqueous concentration of relatively water-insoluble chemicals. Derivation of accurate K_p values at low solids levels (< 5 mg/L) is also difficult because these are usually determined by difference between total and centrifuged sample analyses. Sorption tests with water-soluble chemicals are difficult because chemicals that do not sorb appreciably are also difficult to analyze in the solid phase. In sorption tests, a few percent error in analytical analyses could cause an order of magnitude error in the observed or estimated K_p. Analyses of aqueous- and sorbed-phase test chemical concentrations are required to validate the cause-effect relationships that are observed in toxicity tests.

In addition to the solubility/acute toxicity considerations, another factor had to be recognized in choosing test chemicals. The shape of the dose-response curve for a candidate chemical and test organism proved to be an important consideration. The response of the organism per unit dose (concentration) cannot readily be predicted from LC_{50} values. In order to measure reduction in toxicity due to chemicals removed from solution by sorption, the decrease in soluble-phase chemical (dose) must yield a sufficiently large reduction in toxicity (response) to be measurable. Given the inherent variability in toxicity tests and a manageable number of test animals, any reduction in toxicity would only be observed if the dose-response curve for a test chemical was relatively steep (Fig. 17.2).

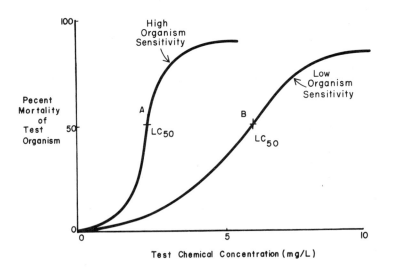

Fig. 17.2
Representative dose-response curves from toxicity tests.

An important factor was the unit response in organism mortality per unit reduction in aqueous-phase concentrations. Certainly, dose-response curves with slopes ≥ 1 (curve A-Fig. 17.2) would be more likely to yield measurable differences if the test chemical concentration in the aqueous phase was reduced by some percentage.

Finally, several mechanical and other problems had to be solved. An obvious problem was mechanical abrasion and the requirement to keep the suspended solids "suspended." A further consideration was control mortality or mortality resulting from suspended solids alone.

A Preliminary Test of the Bioavailability Hypothesis

Based on the hypothesis that particulate-bound chemicals are not bio-available, two comprehensive experiments were conducted. Details of these experiments may be found in the Institute of Applied Sciences' Final Report to the American Petroleum Institute [3]. The strategy used involved conducting standard acute toxicity tests to determine the shape of the dose-response curve. In subsequent experiments, animals were used to test for bioavailable test chemical after the chemicals had some contact time with suspended solids. The endpoint used in these bioavailability experiments was the acute mortality of the test animals: daphnia (*Daphnia magna* (Strauss)) and fathead minnows (*Pimephales promelas* (Rafinesque)). The test animals were cultured according to recommended protocols [4, 5]. Toxicity tests and bioavailability experiments were conducted with animals cultured using the same procedures.

Acenaphthene (Eastman Kodak, CAS #93-32-9) and 1, 2, 4, 5-tetrachlorobenzene (Aldrich, CAS #95-94-3) were selected as test chemicals (Table 17.1) for acute toxicity tests and subsequent bioavailability experiments.

Table 17.1 Physical and Chemical Characteristics of Test Chemicals

Characteristics	Test chemicals	
	Acenaphthene[a]	1, 2, 4, 5-Tetrachlorobenzene[b]
Molecular weight	154.2	215.9
Molecular formula	$C_{12}H_{10}$	$C_6H_2Cl_4$
Water solubility at 25°C (mg/L)	3.42	0.60
Vapor pressure (Torr at 25°C)	10^{-2} to 10^{-3}	4.89×10^{-2}
Low K_{ow}	4.33	4.93

[a] Source - [6]
[b] Source - [7]

After extraction of hexane acenaphthene was analyzed by spectro-fluorometry. After extraction with hexane tetrachlorobenzene was analyzed by gas chromatography (GC) with electron capture detection. Analytical quality control included regular calibration using fresh standards, replicate analyses of samples, and careful mass balance calculations.

Three types of suspended solids were selected to be used for the bioavailability and sorption experiments. Sediments were obtained from Pat Mayse Lake in Lamar Country, Texas; Roselawn Cemetery Pond in Denton County, Texas; and montmorillonite clay from the Department of Geology at the University of Missouri in St. Louis, Missouri (Table 17.2). Water used in these experiments was U.S. EPA reconstituted hard water.

Both sorption experiments and acute toxicity tests were conducted with the two test chemicals, acenaphthene and 1,2,4,5-tetrachlorobenzene, prior to the bioavailability experiments. Sorption experiments were designed to simulate the bioavailability experiments and to provide some estimate of the expected removal of test chemicals from the water by suspended solids. For acenaphthene, 11%, 20% and 10% of the total concentrations (1.40 mg/L) were removed by 1,000 mg/L of Roselawn Pond solids, Pat Mayse Lake solids, amd montmorillonite clay, respectively. The sorption model predicted 17%, 9% and 0% respectively. For tetrachlorobenzene (0.5 mg/L), the percent sorbed to 1,000 mg/L of the test solids ranged from 33% to 0.5%. The sorption model predicted 26%, 13% and 0% of the tetrachlorobenzene would be sorbed by suspended solids from Roselawn Pond, Pat Mayse Lake and montmorillonite clay. Tetrachlorobenzene was very unstable in the sorption experiments with a half-life of 5 to 8 h.

The acute toxicity tests with the test chemicals showed that daphnia and fathead minnows did not precipitously respond to acenaphthene and tetrachlorobenzene (Fig. 17.3). The 48-h LC_{50} for daphnia and acenaphthene was 0.85 mg/L, and the 96-h LC_{50} for fathead minnows was 1.08 mg/L. For tetrachlorobenzene, the 8-h LC_{50} for daphnia was 0.33 mg/L, and the 24-h LC_{50} for fathead minnows was 0.06 mg/L.

Table 17.2   Physical-Chemical Characteristics of Sediments Used in Bioavailability and Sorption Experiments

	Roselawn Pond sediment	Pat Mayse sediment	Montmorillonite clay
pH at 25°C	8.04	7.36	7.90
Particle size μm	<180	<180	<180
Particle size distribution			
% Clay	67	66	100
% Silt	33	22	0
% Sand	0	12	0
Cation exchange meq/100 g	43.3	16.8	65
Percent organic carbon	1.636 ± 0.001[a]	0.695 ± 0.016[a]	0
Percent acid extractable carbon[b]	0.470	0.450	—
Percent dry weight at 100°C	31	50	100
Percent volatile solids at 550°C	10.29 ± 0.35[a]	8.64 ± 0.03[a]	—

[a] Standard deviation of replicate measurement.
[b] Extractable carbon compounds at pH 2 under N^2.

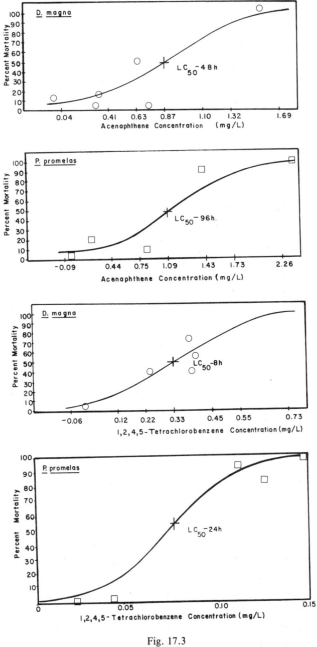

Fig. 17.3
Acute toxicity curves for acenaphthene and
1,2,4,5-tetrachlorobenzene to *Daphnia magna*
and *Pimephales promelas.*

Bioavailability Experiments with Acenaphthene and Tetrachlorobenzene

Bioavailability experiments were conducted using three sources of sediments (suspended solids) to explore the role of suspended solids in regulating the bioavailability of acenaphthene and 1,2,4,5-tetrachlorobenzene to daphnia and fathead minnows. These experiments were conducted in a series of replicated glass columns (Fig. 17.4) with fish and daphnia contained in chambers (Fig. 17.5).

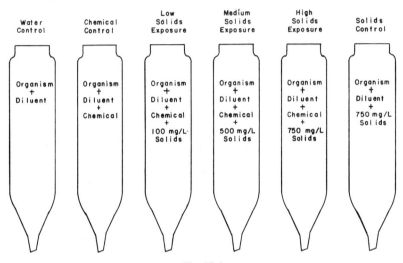

Fig. 17.4
Glass columns (see Fig. 17.5) representing the experimental
design for the bioavailability experiments.

Each fish or daphnia chamber contained 10 to 12 or 20 to 30 organisms, respectively. Target suspended solids concentrations were 0, 100, 500 and 750 mg/L. The glass columns were designated as follows:

1. Water control—a unit to test for organism survival in the system; contained water + organisms.
2. Chemical control—contained test chemical and no added solids.
3. Low solids exposure—100 mg/L target concentration of solids + test chemical.
4. Medium solids exposure—500 mg/L target concentration of solids + test chemical.
5. High solids exposure—750 mg/L target concentration of solids + test chemical.
6. Solids control—a unit to test for organism responses to high solids levels; 750 mg/L target concentration of solids.

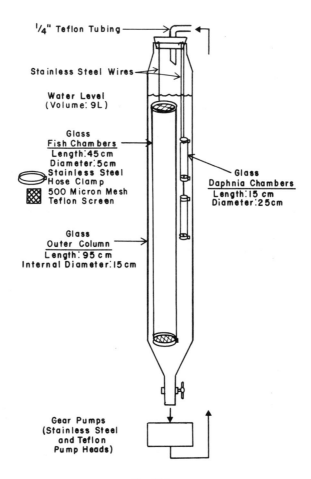

¼" Teflon Tubing

Stainless Steel Wires

Water Level
(Volume: 9L)

Glass
Fish Chambers
Length: 45 cm
Diameter: 5 cm
Stainless Steel
Hose Clamp
500 Micron Mesh
Teflon Screen

Glass
Daphnia Chambers
Length: 15 cm
Diameter: 2.5 cm

Glass
Outer Column
Length: 95 cm
Internal Diameter: 15 cm

Gear Pumps
(Stainless Steel
and Teflon
Pump Heads)

Fig. 17.5
Exposure system used in bioavailability experiments
with acenaphthene and 1,2,4,5-tetrachlorobenzene
and *Daphnia magna* and *Pimephales promelas*.

The same mass of test chemical was introduced to each chamber that received chemicals. For acenaphthene, the target concentration was 2.0 mg/L, and for tetrachlorobenzene the target concentration was 0.6 mg/L. Renewal of the test solutions (test chemical + solids) was required in order to maintain test chemical concentrations. The test chemicals were allowed a 24-h (acenaphthene) or 15-min (tetrachlorobenzene) contact time with the solids prior to introduction of the test organisms. Animals were observed at least every 24 h. Daphnia were removed after 42 to 48 h and fish were exposed for 33 to 96 h.

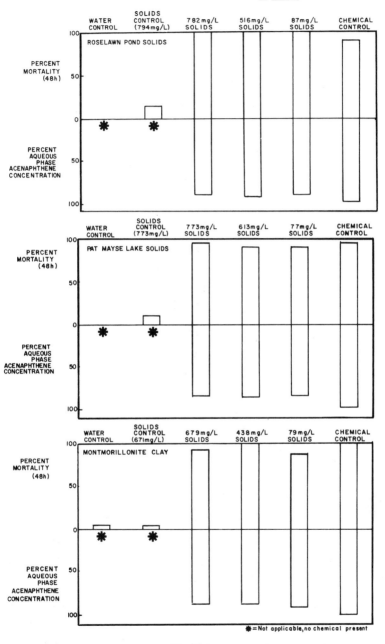

Fig. 17.6
Results of bioavailability experiment with
acenaphthene and *Daphnia magna*.

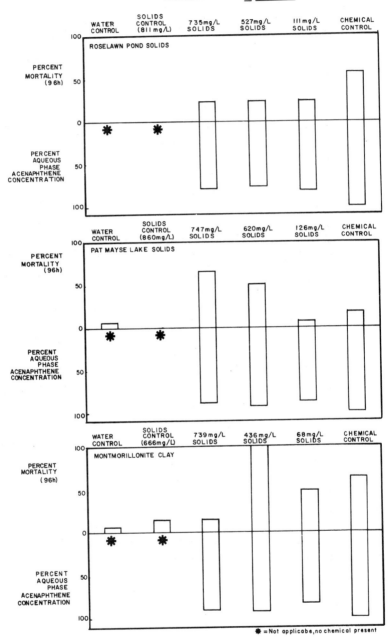

Fig. 17.7
Results of bioavailability experiments with
acenaphthene and *Pimephales promelas*.

The following discussion summarizes the results of the bioavailability experiments with acenaphthene (Figs. 17.6 and 17.7):

1. There was no obvious decrease in daphnia mortality due to acenaphthene in the presence of up to 750 mg/L solids although a measurable decrease in fathead minnow mortality was observed.

2. Mortality in the daphnia controls was as expected: $\leqslant 10\%$ in solids and water controls and $\leqslant 90\text{-}100\%$ in the chemical control ($\simeq 0$ mg/L suspended solids). Mortality in the fathead minnow controls was more variable. Water and solids controls had $< 15\%$ mortality. Chemical controls mortality ranged from 20 to 70%.

3. Mass balances ranged from 87 to 126%. Observed sediment partition coefficients ranged from 222 to 474, which compares well with the sorption model prediction of 211.

4. The model predicted 17% sorbed acenaphthene at 750 mg/L suspended solids. The mean percent sorbed acenaphthene observed was 20.6% and 19.9% at 48 and 96 h, respectively.

5. Acenaphthene was very unstable in this experimental apparatus, and the exposure concentration had a coefficient of variation of about 50% during the experiment.

6. Sorption and bioassay results with montmorillonite clay were similar to results with sediments containing organic carbon.

7. Based on previously conducted acute toxicity tests, daphnia were expected to show a greater response. However, fathead minnows were more responsive.

8. A 20% reduction in aqueous-phase chemical concentration was very difficult to detect accurately with the inherent variability in the bioassays.

Static renewal bioavailability experiments were also conducted using three types of solids and 1,2,4,5-tetrachlorobenzene. Tetrachlorobenzene solutions and solids mixtures were renewed every 8 to 12 h. Initial contact time between solids and chemicals in solution was reduced to 15 min instead of 24 h used with acenaphthene. This was necessary to minimize the loss of tetrachlorobenzene prior to the start of the experiment. A near-saturated solution was required since acute toxicity concentrations were a large fraction of tetrachlorobenzene's solubility in water.

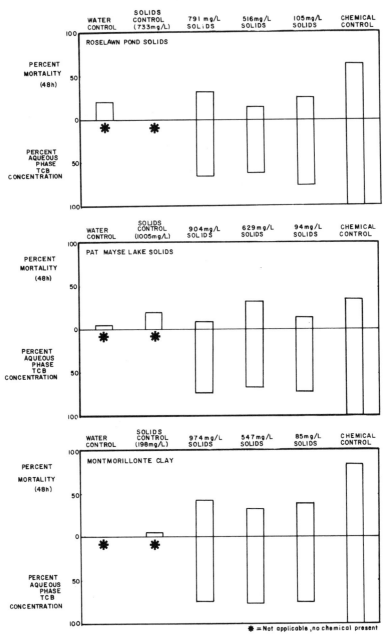

Fig. 17.8
Results of bioavailability experiment with
tetrachlorobenzene and *Daphnia magna*.

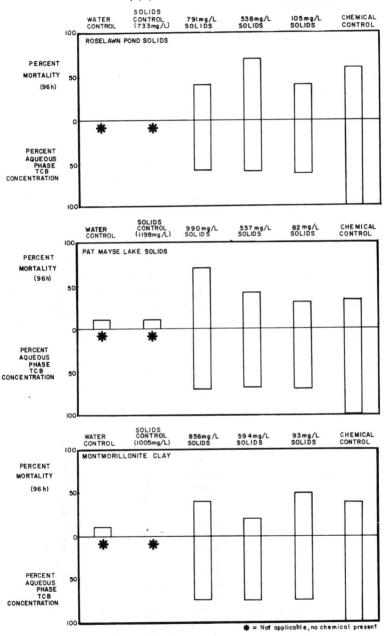

Fig. 17.9
Results of bioavailability experiment with
tetrachlorobenzene and *Pimephales promelas*.

The following discussion summarizes the results of the bioavailability experiments with tetrachlorobenzene (Figs. 17.8 and 17.9):

1. For all three sources of suspended solids, there was an apparent decrease in daphnia mortality in the presence of solids. The trends were not uniform with increasing solids concentrations.

2. There was no apparent decrease in fathead minnow mortality as a result of the presence of solids in any of the tetrachlorobenzene bioavailability experiments.

3. Analytical mass balances ranged from 57 to 90%.

4. Up to 27% of the initial aqueous tetrachlorobenzene was sorbed by montmorillonite clay. The model predicted 0% as it would for all organic free solids.

5. Roselawn Pond and Pat Mayse Lake suspended solids sorbed up to 56% and 50% of the aqueous tetrachlorobenzene, respectively, with model predictions of 26% and 13%.

6. The exposure concentration of tetrachlorobenzene varied by about 50% because of rapid degradation and requirements for frequent renewal.

7. Based on previously conducted acute toxicity tests and dose-response curves, fathead minnows should have been more sensitive than daphnia to tetrachlorobenzene. However, the contrary was observed.

BIOAVAILABILITY FOR BIOCONCENTRATION

A discussion of the availability of sorbed chemicals for bioconcentration is presented by William Adams elsewhere in this publication. A variety of questions on this topic remain to be answered:

1. Are only chemicals that are dissolved in interstitial waters of sediments bioavailable? Can we assume that particulate-bound chemicals are not bioavailable?

2. Can we generalize from one animal to another regardless of feeding habits and modes of exposure? A basic ecological principle is that there is a wide variety of feeding mechanisms and habits. That organisms can bioaccumulate and biomagnify certain chemicals is well-known. Sorbed chemicals may be available only to organisms that ingest sediments and/or have acidic digestive systems.

4. What is the depth of "the well mixed layer" of sediments and how is this depth defined? Does bioturbation play a significant role?

5. Perhaps even more basic, what is the significance of a bioconcentrated chemical? At what body burden is an organism significantly affected?

6. Can bioconcentration be regulated by desorption rates from particulates?
7. Can chemicals that do not sorb bioconcentrate to levels that are harmful?
8. Can predictive capabilities be developed for metals as well as organics? To date, much of our research interests have been devoted to organic compounds (8).

Although further research is needed to define the limits of extrapolation, a number of studies support the hypothesis that sorbed chemicals are not bioavailable. Rossi (9) and Roesijadi et al. (10, 11) found that sediment-bound organic chemicals (e.g., PAHs) were not bioconcentrated or bioaccumulated by benthic organisms. Marine polychaete worms ingested sediments containing naphthalenes but did not measurably bioaccumulate more chemical than could be accounted for by bioconcentration alone (9). Similar results were observed for sipunculid worms exposed to sediments contaminated with Prudhoe Bay crude oil (12). Roesijadi et al. (11) reported that sediment-feeding clams placed in heavily oil-contaminated sediment did not bioaccumulate differently than those held in water without sediment. The authors inferred that the clams bioaccumulated chemicals from interstitial waters. Neff (13) indicated that polynuclear aromatic hydrocarbons (PAHs) accumulation from sediments may be generally attributed to uptake of PAHs desorbed from sediment particles into interstitial waters. Halter and Johnson (14) found PCBs in fathead minnows higher for fish that were observed "mouthing" contaminated sediments than in fatheads separated from the sediments. They postulated that PCBs were desorbing in the fathead's buccal cavity.

An extensive study of the adverse effects caused by open-water disposal of dredged material demonstrated that chemical contaminants associated with natural sediments are generally unavailable to affect aquatic life adversely (15). Lee and Mariani (15) suggested that at the edge of the mixing zone of dredged-sediment disposal, the contaminants present are generally unavailable because they are particulate bound. They suggested that water quality criteria based on total chemical (C_t) are inappropriate.

BIOAVAILABILITY FOR BIODEGRADATION

Biotransformation of organic chemicals may be altered by the presence of seston or hydrosoils. The possible consequences of the presence of particulates may be that biodegradation rates are significantly retarded or enhanced or not significantly altered.

Biotransformation rates of diquat (9,10-dihydro-8a,10a-diazoniaphenanthrene ion) are significantly reduced in the presence of suspended solids (16). Sorbed diquat is neither bioavailable for degradation nor for phytotoxicity.

Sorption of cationic diquat occurs at negatively charged sites on both clay minerals and organic matter (17). The attraction of diquat to clays and organic matter is strong and essentially irreversible. Consequently, diquat concentrations must be increased several times to achieve a level of bioactivity in waters containing suspended materials, and the herbicide may persist and accumulate in sediments (18). Resistance of diquat and similarly strongly sorbed chemicals to degradation may be due to inaccessibility to enzymes, and the biotransformation rate may be regulated by the rate of desorption. Steen et al. (19) suggested that sorption of toxics such as chlorpropham and di-n-butyl phthalate were unavailable for biodegradation when sorbed.

Another possibility is that the presence of suspended or sedimented particulates could enhance degradation rates. The particulates may provide a different microbial consortium with unique degradative capabilities (20). The concentration of microbes is generally increased in sediments relative to overlying waters (20), and this may lead to a faster degradation rate. Sorption may bring the sediments or solids and associated microbes into contact and may locally increase the concentration of a chemical to a level sufficient to induce or depress degradative enzyme systems (20). Finally, suspended particulates or sediments may provide inorganic nutrients necessary to maintain degradation rates of organics (21). Increased degradation rates of lindane, some organophosphorus insecticides (21), acenaphthene (this study) and the herbicides endothall (22) and 2,4-D (23) in the presence of sediments have been reported.

Another possibility that must be considered is that solids do not alter biotransformation rates in aquatic systems. Biotransformation rates of some relatively soluble chemicals such as 1,2,4,5-tetrachlorobenzene were not significantly altered by the presence of up to 750 mg/L suspended sediments (4).

It is apparent that the role of suspended solids in biodegradation cannot be simply explained. Considerable theoretical development and research would contribute greatly to our understanding in this important area.

CONCLUSIONS AND SUGGESTIONS FOR FURTHER RESEARCH

Suspended solids and bed sediments can alter the bioavailability of chemicals to aquatic organisms. The direction and degree of alteration are not easily predicted, although some useful tools are rapidly being developed to resolve this situation. Many of these developments will directly impact potential modifications of National Water Quality Criteria for chemicals via the guidelines issued by EPA in its *Water Quality Standards Handbook* (24). Some specific conclusions are:

1. Many chemicals do not have high sorption coefficients ($K_p = {} < 10^4$) and are unlikely to sorb sufficiently to suspended or bed solids to markedly affect their bioavailability.

2. Chemicals with high sorption coefficients usually do not exhibit acute toxicity at concentrations below their solubility. This has direct implications for derivation of site-specific water quality criteria.

3. Sorption of chemicals to suspended sediment is not simply related to organic carbon content. The relatively narrow range of organic carbon content (1-10%) of most sediments would probably not contribute greatly to observed variability in sorption coefficients.

4. A critical factor in determining whether or not suspended solids will change bioavailability of a chemical for acute toxicity is the shape of the dose-response curve for an organism and the test chemical. The dose-response must be sufficiently precipitous to yield a measurable reduction in toxicity as a result of test chemical sorption.

5. If an organism is not sufficiently sensitive to a test chemical and only a small percentage of the test chemical is removed from solution by sorption to suspended solids, no reduction in toxicity will be observed.

6. Bioavailability of chemicals for acute toxicity may be species specific and/or a function of the mode or route of exposure.

Clearly, there is considerable room for research on bioavailability of chemicals. A better understanding is needed of the role of suspended sediments in regulating bioavailability of pesticides and metals for bio-concentration. It is apparent that chronic toxicity and bioconcentration are major concerns for chemicals that sorb appreciably. The notion that sorption is an "equilibrium process" should be carefully reevaluated. Research needs to be conducted on sorption-desorption kinetics and hysteresis. Our understanding of bioavailability of metals would be furthered by simple models to predict their partitioning (25). Such a model would be a useful decision support tool. And last, but probably most important, there is a serious need for carefully designed laboratory and field studies to validate the notions incorporated in these models. Through wide use without sufficient testing, the ideas encompassed by the models run the risk of becoming indefensible "facts."

REFERENCES

1. **Staples, C. A., K. L. Dickson, J. H. Rodgers, Jr. and F. Y. Saleh.** 1984. A model for predicting the influence of suspended sediments on bioavailability of neutral organic chemicals in the water compartment. Proceedings Seventh Annual Symposium on Aquatic Toxicology, Milwaukee, WI.

2. **Wetzel, R. G.** 1975. *Limnology.* Holt, Rinehart, and Winston, Philadelphia, PA.

3. **Dickson, K. L., J. H. Rodgers, Jr., F. Saleh, C. Staples, D. Wilcox, S. Hall, S. Rafferty and A. Entezami.** 1984. An assessment of the role of suspended solids in regulating the bioavailability of petroleum refinery effluent chemicals in the aquatic environment. Report submitted to American Petroleum Institute, Washington, D.C. by the Institute of Applied Sciences, North Texas State University.

4. **U.S. Environmental Protection Agency.** 1975. Methods for acute toxicity tests with fish, macroinvertebrates, and amphibians. EPA-660/3-75-009. U.S. EPA, Washington, D.C.

5. **U.S. Environmental Protection Agency.** 1979. Interim NPDES compliance biomonitoring inspection manual. U.S. EPA Enforcement Division, Office of Water Enforcement, Washington, D.C.

6. **U.S. Environmental Protection Agency.** 1979. Water related environmental fate of 129 priority pollutants. EPA-44014-79-029a. U.S. EPA, Washington, D.C.

7. **CRC.** 1980. *Handbook of Chemistry and Physics.* CRC Press, Inc., Boca Raton, FL.

8. **Weber, J. B.** 1972. Interaction of organic pesticides with particulate matter in aquatic and soil systems. In R. F. Gould, ed., *Fate of Organic Pesticides in the Aquatic Environment.* American Chemical Society, Washington, D.C., pp. 55-120.

9. **Rossi, S. S.** 1977. Bioavailability of petroleum hydrocarbons from water, sediments, and detritus to the marine annelid, *Neanthes arenaceodentata.* Proceedings 1977 Oil Spill Conference. American Petroleum Institute, Washington, D.C., pp. 621-626.

10. **Roesijadi, G., D. L. Woodruff and J. W. Anderson.** 1978. Bioavailability of naphthalenes for marine sediment artificially contaminated with Prudhoe Bay crude oil, *Environ. Poll.* **15**:223-229.

11. **Roesijadi, G., J. W. Anderson and J. W. Blaylock.** 1978. Uptake of hydrocarbons from marine sediments contaminated with Prudhoe Bay crude oil: Influence of feeding type of test species on availability of polycyclic aromatic hydrocarbons. *J. Fish Res. Board. Can.* **35**:608-614.

12. **Anderson, J. W., L. J. Moore, J. W. Blaylock, D. L. Wodruff and S. L. Keissa.** 1977. Bioavailability of sediment-sorbed naphthalenes to the sipuculid worm, *Phascolosoma agassizzi.* In D. A. Wolfe, ed., *Fate and Effects of Petroleum Hydrocarbons in Marine Ecosystems and Organisms.* Pergamon Press, New York, pp. 275-285.

13. **Neff, J. M.** 1979. *Polycyclic Aromatic Hydrocarbons in the Aquatic Environment.* Applied Science Publ. Ltd., London.

14. **Halter, M. T. and H. E. Johnson.** 1977. A model system to study the desorption and biological availability of PCB in hydrosoils. In F. L. Mayer and J. L. Hamelink, eds., *Aquatic Toxicology and Hazard Evaluation.* ASTM STP 634. American Society for Testing and Materials, Philadelphia, PA, pp. 178-195.

15. **Lee, G. F. and G. M. Mariani.** 1977. Evaluation of the significance of waterway sediment-associated contaminants on water quality at the dredged material disposal site. In F. L. Mayer and J. L.Hamelink, eds., *Aquatic Toxicology and Hazard Evaluation.* ASTM STP 634. American Society for Testing and Materials, Philadelphia, PA, pp. 196-213.

16. **Hance, R. J.** 1970. Influence of sorption on the decomposition of pesticides. In *Sorption and Transport Processes in Soils.* Monograph No. 37. Society of Chemical Industry, pp. 92-104.

17. **Knight, B. A. G., J. Coutts and T. E. Tomlinson.** 1970. Sorption of ionized pesticides by soil. In *Sorption and Transport Processes in Soils.* Monograph No. 37. Society of Chemistry Industry, pp. 54-62.

18. **Bowmer, K. H.** 1982. Aggregates of particulate matter and aufwuchs on *Elodea canadensis* irrigation waters, and inactivation of diquat. *Aust. J. Mar. Freshwater Res.* **33**:589-593.

19. **Steen, W. C., D. F. Paris and G. L. Baughman.** 1980. Effects of sediment sorption on microbial degradation of toxic substances. In R. A. Baker, ed., *Contaminants and Sediments,* Vol. 1—Fate and Transport Case Studies, Modeling, Toxicity. Ann Arbor Science Publishers, Ann Arbor, MI, pp. 477-482.

20. **Marshall, K. C.** 1980. Microorganisms and interfaces. *Bioscience* **30**:246-249.

21. **Pionke, H. B. and G. Chesters.** 1973. Pesticide-sediment-water interactions. *J. Environ. Qual.* **2**:29-45.

22. **Simsiman, G. V. and G. Chesters.** 1975. Persistence of endothall in the aquatic environment. *Water, Air, and Soil Poll.* **4**:399-413.

23. **Nesbitt, J. H. and J. R. Watson.** 1980. Degradation of the herbicide, 2,4-D in river water. II. The role of suspended sediments, nutrients, and water temperature. *Water Res.* **14**:1689-1694.

24. **U.S. Environmental Protection Agency.** 1983. Water quality standards handbook. Office of Water Regulations, Criteria and Standards, Washington, D.C.

25. **Allen, H. E., R. H. Hall and T. D. Brisbin.** 1980. Metal speciation: Effects on aquatic toxicity. *Environ. Sci. Technol.* **14**:441-442.

Chapter 18
Synopsis of Discussion Session 3:

Biological Effects, Bioaccumulation, and Ecotoxicology of Sediment- Associated Chemicals

J. Anderson (*Chairman*), W. Birge, J. Gentile, J. Lake, J. Rodgers, Jr. and R. Swartz

INTRODUCTION

Biological effects of sediment-associated chemicals include such responses as mortality, bioconcentration, reproductive impairments, pathological effects, alterations in distribution or behavior of aquatic species, and biochemical or physiological modifications that affect various biological processes. However, considerations in this report will be limited primarily to those responses that can serve as clearly detectable endpoints in sediment toxicity tests, suitable for use in hazard assessment and regulation of sediment contamination. Since exposure to sediment-associated chemicals is habitat-related, consideration will be given to freshwater and marine ecosystems and to organisms in the water column, organisms or life cycle stages that dwell on or near the sediment surface, and benthic species that inhabit the sediment compartment.

The objective of this session is to summarize the available literature, the papers presented at this workshop, and the group discussions concerning the following topics:

1. Considerations for proper collection, storage and testing of sediments as related to the specific regulatory or environmental concerns.

2. Interpretation of the chemical analyses of sediments in light of

degradation potential and alterations in the environment that may change the sorbed contaminants to more or less available forms.
3. Factors to consider in assessing the bioavailability of contaminants in sediments including routes of transfer.
4. Recommended organisms, both freshwater and marine, for use in determining bioaccumulation from sediments.
5. Tiers of biological effects testing to evaluate the toxicity of sediments.
6. Additional research needs.

SEDIMENT COLLECTION AND HANDLING

It is evident that many types of sediment collections from aquatic environments have taken place that bear little relationship to one another. Some of this variability is justified since the questions posed regarding the characteristics of the sediments have been driven by different regulatory mandates. To determine the potential toxicity of sediments in the vicinity of an industrial or municipal outfall, it is logical to be concerned with relatively small samples of surface sediments (top 2 cm) along a gradient. Enough additional sediment should be collected in this effort to supply the analytical chemists with samples for characterization of bulk sediment analyses and toxics including polyaromatic hydrocarbons (PAHs), PCBs, trace metals, and other priority pollutants. If only this top layer is tested for biological responses, then it should probably be the only depth analyzed for chemical characteristics. This type of limited testing may indeed be sufficient for some types of field investigations, and several types of grabs can be used to supply samples.

The problems with sediment handling begin to get complex when there is a need to ask questions such as: (1) Are the number of stations to be sampled on a cruise limited by the speed at which tests can be conducted using "fresh" (1-week-old) sediments? (2) Is it possible that below 2 cm there is a very contaminated layer of sediment? (3) How can equivocal result be reexamined by either the same test or another test without a back-up sediment sample? (4) Now that the screening tests indicate a problem, how can chronic or community-level tests be conducted to verify the findings? (5) If a particular toxicant not originally looked for in the sample may have been present, could it have been responsible for the effects observed?

Suggested approaches that could be used to provide answers to the above questions all possess trade-offs in expense of collection (including collecting gear, vessel, and time), speed of producing final analyses, and potential for toxicant alteration within the sediment matrix. The only means of determining the depth of sediment containing the highest level of contaminants is coring and production of a depth profile. It is generally best to use large-gauge cores such that only the central portion is sampled for chemical analyses. Transition zones can frequently be identified (such as native marine sediment) and compared to a few key depths to reduce the number of analyses. These chemical characterizations can then serve as the guide for biological effects testing. It is clear that the depth of coring should be guided by the overall objective of the program, and only deep dredging would warrant a major effort.

Once the profile of sediment contaminant has been obtained, it may be justified to go back to stations to collect only surface (2 cm) samples of a number not exceeding the limit to biological testing within a week of collection. It has been recommended that any storage take place at 4°C, and it would be best if no oxygen were present over the sealed samples. Zero air space or nitrogen will slow the microbial degradation of petroleum hydrocarbons. There has been strong objection to either drying or freezing sediment samples before the magnitude of variation in chemical characteristics and biological effects is documented. There is a very important trade-off regarding freezing of sediment for future chemical or biological analyses and the possible need to collect more of the "same" material. Variability resulting from freezing must be compared to the variability produced by attempting to later occupy the same field station and obtain a "second" sample. Since there is little chance the same sediment can be obtained in a second collection, the best alternative is probably sequential tests for a given reliable biological response with a large sediment sample. This approach should probably be used if sediment is either maintained at 4°C or frozen. Chemical analyses should also be conducted at intervals during the storage, but alterations may not be as readily detected by chemical means as through bioassay tests.

CHEMICAL ANALYSIS OF SEDIMENTS AND HAZARD ASSESSMENT

Duration of Impacts from Contamination Sediments

There has been an attempt to classify sediment scheduled for dredging and disposal on a basis of the content of PAHs, PCBs, and trace metals, plus the toxicity they produce in short-term tests. It should be recognized that, during removal of the material and discharge to a deep water site, there will be considerable release of the more toxic and water soluble mono- and diaromatic hydrocarbons. Furthermore, a major portion of these aromatics that are present in the aerobic layer of discharge material will degrade over a period of weeks and months. More soluble species of trace metals contained in the pore water will likely be released during dredging and disposal resulting in short-term higher concentrations of available metals in the water column. Higher molecular weight PAHs and chlorinated organics (PCBs) will not likely exert an acute toxicity, but they will be available from the sediments for tissue accumulation for longer periods.

Motile marine species that ingest discharged material from time to time will have the capability of metabolizing PAHs. Estimates of effects and potential bioaccumulation of PAHs from discharged material should consider this capability of either slow release of PAHs (to bivalves) or metabolism (to higher organisms). Chlorinated organics may be available for uptake to the benthos and potential transfer to higher trophic levels. Concern for new chemicals should begin with a careful evaluation of the length of time they will be available on sediments before microbial and infaunal biodegradation reduces contamination levels.

Sediment Analysis

Bulk chemical analyses usually provide a worst-case estimate of available toxic materials. Included in the assessment of these data are the factors known or expected to control the availability of the contaminants. Chemical analysis of sediments may or may not be useful in hazard assessment and subsequent decisions regarding actions such as sediment transfer or isolation. Bulk chemical analyses that reveal relatively high levels of potential toxicants may indeed point to potential problems with contaminated sediments. On the other hand, bulk chemical analyses (including grain size, sulfides, TOC and % volatiles) that show intermediate concentrations of potential toxicants yield no guidance regarding the potential or realized toxicity of that contaminated sediment. Finally, relatively low concentrations of potential toxicants determined by bulk chemical analysis may give a "clean bill of health" to a sediment and allow a wide range of transfer ("disposal") options. It is the sediments with intermediate concentrations of potential toxicants that require bioassays to measure the bioavailability of those contaminants.

Bioavailability is not an "all-or-none" phenomenon. Under one set of environmental conditions or overlying water conditions, a sediment may not show any toxicity. If dissolved oxygen or pH conditions change in the overlying water, the sediment may prove to be potently toxic. Two points can be made: (1) the sediments should be tested with the range of excursions of environmental conditions expected in the overlying water, and (2) sediments cannot be divorced from their overlying water.

Ultimately, only organisms can answer the question of bioavailability of a compound. The futility of bulk chemical analysis has been clearly demonstrated with metals and phosphorus. Mere presence does not mean that a material is bioavailable. Measurable acute and chronic effects and bioconcentration are examples of responses of organisms to "bioavailable" contaminants. Bioassays and bioconcentration studies are currently required by legislation for contaminated sediments that are dredged and discharged. Simply stated, we do not currently know how to determine chemically whether or not materials are bioavailable. With considerable research, chemical analyses may be developed that more accurately predict bioavailability of toxicants in contaminated sediments.

BIOAVAILABILITY

This workshop committee defined bioavailable toxicant as the total concentration of a sediment-released toxicant to which aquatic organisms are exposed. The point at which the degree of bioavailability results in measurable biological responses (e.g., mortality, bioconcentration) was termed the "bioavailable effect level."

Two conditions are essential to measurements of the bioavailability of sediment-bound toxicants. First, determinations should be made on a site-specific basis. Second, the quantification of the bioavailability of sediment-sorbed chemicals should include consideration of all avenues of exposure. Organisms in the water column are exposed primarily to toxicants released from the sediment back into the overlying water. However, benthic species, as well as eggs and larvae of fish and other organisms that spawn on sediments, may receive exposure via three avenues: (1) the ingestion of contaminated sediment particles, (2) toxicants in the sediment interstitial water, and (3) toxicants released back into the water column. Thus, calculations of bioavailability will be more complex for organisms that inhabit the sediment or frequent the water/sediment interface. Exposure to interstitial water may be particularly important for such organisms and may vary depending upon the specific characteristics of the toxicants and various environmental factors.

In assessing bioavailability, sediment-associated toxicants may sequester into four different compartments: (1) the sediment solid phase, (2) the sediment interstitial water, (3) the overlying water column, and (4) the biomass. Factors that affect sorption/desorption kinetics governing bioavailability include chemical/physical characteristics of the toxicant, structure and chemistry of the sediment and interstitial water, chemistry and hydrology of the water column, and structure, density, and habitat of the biomass.

A variety of questions may be asked regarding bioavailability of sediment-bound toxicants. These questions may be loosely grouped as historic questions, present or immediate questions, and future questions. Dredged-material removal and dumping is an example of a source of inquiry that is principally historic. The sediments and toxic chemicals have had sufficient contact time that unique behavior would be expected. For example, desorption or release of toxic compounds from these sediments would be different than release from recently exposed sediments. Present or immediate questions arise when a particular toxicant is found to be accumulating in the sediment of an effluent's mixing zone. What is the role of suspended sediments in mitigating acute toxicity as described by the Site-Specific Water

Quality Criteria? An example of a future question might be what sort of tests or decision criteria are appropriate for a premanufacture notice chemical that appears to have potential to persist in sediments. Each of these questions has some common threads, as well as some unique considerations. For proper interpretation of experimental design and data, appropriate consideration should be given to the question or questions being addressed. If this is done, considerable confusion and misinterpretation can be avoided.

Summary of Bioavailability Considerations

1. Organic carbon is the primary controlling factor of bioavailability of hydrophobic organic chemicals sorbed to sediments (suspended and/or settled). The clay content of sediments can also be important in controlling sorption of toxicants.

2. Bioavailability is affected by both DOC and TOC. Particulate or TOC is the dominant factor, and DOC is less important for the water column. DOC is important for the interstitial water.

3. Dietary uptake of chemicals can be important for chemicals with a log K_{ow} of 5 to 7 or a half-life $\geq$ 50 d.

4. Chemical concentrations of neutral organics in fish, invertebrates, sediments, and detritus theoretically can be normalized by the use of organic carbon.

5. Interstitial water appears to be the primary source of chemical exposure for many benthic organisms.

6. For hazard assessment purposes, use of interstitial water chemical concentrations provides a means of calculating safety factors for benthic invertebrates. The practical (field) application of these analyses requires further research and testing.

BIOACCUMULATION

Marine Organisms

The first concept that should be made very clear is that organisms that are the most suitable for bioassays have generally been used least in testing for bioaccumulation of contaminants from sediments. It is clear that small sensitive species are not available for chemical analyses either because they cannot tolerate the toxicants long enough to exhibit a high body burden or because their biomass is so small that large numbers are needed for complete chemical characterization. The latter factor can be overcome by larger sample sizes and/or more sophisticated analytical techniques. The reason there are few, if any, good correlations between body burdens and toxicity responses may be that tests for bioaccumulation have been conducted with very tolerant species. Exceptions to this pattern include mussels that are now being tested for "scope for growth" and other effects [1].

The habits of several marine organisms determine their utility in bioassays. Flat fish, which do not purposely ingest sediments, live in close association with the substrate, and they have demonstrated both high body burdens and histopathological effects. Oysters, mussels, and some species of clams receive contaminants from suspended particles during filter feeding, but other clam species feed on fine detritus particles on or in surface sediments. In environments characterized by high suspended loads or frequent resuspension of bottom sediments, we would expect contaminants present on these particles to be accumulated by filter feeders. Once contaminated particles have been deposited, filter feeders would not be expected to serve as good indicators of bioaccumulation. Detritivorous clams, as *Macoma* sp, have been shown to accumulate much higher levels of contaminants from sediments, while placed side by side with filter-feeding clams (2, 3). Detritivores have, in general, also been found to be more sensitive to contaminants sorbed to sediments.

Table 18.1 provides a brief summary of the usefulness of groups of marine species for demonstrating bioaccumulation from suspended and/or deposited sediments containing different classes of toxicants. Marine species have not shown a great deal of accumulation from exposures to sediments contaminated with heavy metals, with the exception of Hg, Cd, and Cu. Accumulation of PAH from sediments has primarily been shown by species that do not possess high capabilities for PAH degradation (bivalves). If sampled soon after exposure, fish may contain elevated levels of PAH in the liver and gall bladder. Chlorinated hydrocarbons including DDT, DDE, PCBs, and kepone appear to be accumulated by all groups of organisms listed as a result of partitioning into lipids and slow turnover rates in the body.

There are surprisingly few data concerning the levels of specific toxic materials in organisms at the time they exhibited an adverse response to sediment exposure. Studies with marine fish and PCBs showed that concentrations of 0.1 to 7 μ g/g in fertilized fish eggs was correlated with reduced survival of offspring (4, 5). Field experiments with oiled sediment and clams showed that concentrations of about 1 to 8 μ g/g of aromatic hydrocarbons in the whole animals correlated with either reduced growth over several months (*Protothaca*) or reduced condition index (*Macoma*) after 1 month (2, 6, 7). In a 6-month field exposure, *Macoma* condition index was reduced at a tissue content of about 75 μ g Cu/g (7). There are probably some data on the tissue concentration of Hg in marine organisms associated with an adverse response, but in general we are not aware of papers on other metals.

Table 18.1   Bioaccumulation in Marine Species

Classes of toxicants	Polychaetes (deposit feeders)	Clams ff	Clams df	Mussels ff	Oysters ff	Fish	Shrimp	Amphipods
Trace metals	−[a]	−	−	+[c]	+	−	−	−
PAH	(−)[b]	+	++	+	+	+	−	−
PCB	++	+	++	+	++	++	++	+
Other chlorinated organics	++	+	++	+	++	++	++	+

(Crustaceans: Shrimp, Amphipods)

[a] No strong evidence.
[b] Probably negative because of metabolism (AHH).
[c] Probably from suspension (mussel data on Hg).
ff = filter feeder.
df = detritus feeder, feeding on fine particles on surface or in interstitial spaces (with interstitial water).

Freshwater Organisms

In freshwaters, sediment test protocols have been developed independently and for a variety of purposes. To address questions regarding non-suspended sediments, both elutriate and solid-phase tests have been used. One problem in evaluating the appropriateness of these tests is that the tests are not generally used routinely from lab to lab. The tests arise and are used by a laboratory to answer some question or to fill some data requirement. Another laboratory may use a similar exposure apparatus but with a different species, or they may make modifications as required by their unique situation. Exposure times also vary based on the data pursued. There is serious need for detailed synthesis and standardization of approaches for sediment tests for non-suspended sediments in freshwater. These needs are important for both bioconcentration and toxicity studies.

A summary of some tests used for determining bioconcentration of contaminants from the aqueous phase and bioaccumulation of sediment-associated contaminants is shown in Table 18.2. The bioconcentration test is well established, relatively cheap, and produces results for accumulation that are well correlated with structure activity relationships. This test, however, does not consider contaminant interactions with sediments.

The presence of sediments confounds predictions of bioaccumulation because of variability in the process of adsorption and desorption of contaminants on sediments. In addition, bulk sediment concentrations may have little relationship to bioavailability. This latter difficulty may be overcome if appropriate sediment properties for use as normalizing factors for bioavailability can be identified and tested. For most non-polar organic compounds, normalization of concentrations to total organic carbon appears appropriate, but proper normalizing factors for metals and polar organics are not known.

In general, appropriately scaled and operated microcosms will probably provide the most useful bioaccumulation data; however, the exposure and consequent inability to test numerous sediments has resulted in the development of screening tests that are simpler and more rapid (Table 18.2). Monitoring that uses both introduced organisms and indigenous organisms has demonstrated environmental accumulations of organic and inorganic contaminants. In many monitoring studies, insufficient exposure data is obtained, and a quantitative link between accumulated residues and exposure concentrations cannot be established. In these cases, generalization of bioaccumulation findings is precluded.

Sediment test protocols used by the Corvallis Environmental Research Laboratory include elutriate tests and solid-phase tests. Both use 4:1 water to sediment ratios with the elutriate tested in the absence of the settled sediment; after elution, settling, and the centrifugation, the solid-phase test is conducted with the water gently added over the sediment. Both tests are

Table 18.2   Tests and Exposure Systems for Determining/Predicting Bioaccumulation

Procedure	Outputs	Applicable areas	Limitations	Measurements needed	Compounds	Present state of use development	Exposure systems
Correlative models							
Bioconcentration	Steady state concentration in organism. Bioconcentration Factor (BCF)	Predict concentration of contaminants in laboratory organisms. Predict concentration of contaminants in organisms from field. Application to uptake from elutriates of sediments.	Limited to uptake from dissolved phase. No sediment present.	(a) Log P or aqueous solubility of compound. (b) Concentration of contaminants in aqueous phase. (c) K_I and K_2	Organics	Well established ASTM	Laboratory aquaria (usually a flow-through test)
Screening Tests							
DMRP Bioaccumulation	Concentration in organisms after 10-day uptake.	Access bioavailability of contaminants associated with dredged materials in 10 days.	Not a steady-state exposure. Test organism may be inappropriate.	Concentrations of contaminants in organisms and in sediments.	Organics and Inorganics	In Use DMRP U.S. Army Corps of Engineers	Laboratory aquaria
Distributional Approach	Estimation of maximum concentration of contaminant that can be accumulated by a specific organisms from a specific waste.	Screening of dredged materials and other mixed wastes for the maximum concentration of contaminants that can be bioaccumulated.	Does not predict concentration of contaminants in field exposed organisms. Makes numerous assumptions regarding exchange and distribution of contaminants.	Concentration of contaminants TOC of waste. Lipids of target organisms.	Organics	Being tested (8)	Testing model— Laboratory exposure systems. Suspensions of test material—filter feeders and fish. Bedded sediments— infaunal organisms.
Macoma Exposure Test	Steady state concentrations in organisms.	Uptake of contaminants from natural and spiked sediments.	Feeding habits of organism result in maximal exposure to surface	Concentration of contaminants in organisms.	Organics and Inorganics	In Use	Laboratory aquaria

Table 18.2   Tests and Exposure Systems for Determining/Predicting Bioaccumulation

Procedure	Outputs	Applicable areas	Limitations	Measurements needed	Compounds	Present state of use development	Exposure systems
Screening Tests (Continued) Embryo-Larval Test	Concentration in organisms.	Solid phase waste. Natural sediments.	Metals reach steady state. Organics may not reach steady state.	Concentration in sediment, water column, and tissues.	Organics and Inorganics	Used for metals being tested for organics (9)	Petri dishes (state and flow-through exposures).
Microcosms	Distributions of contaminants in "more realistic" environments including organisms.	Solid phase waste. Possibility for close simulation of natural system.	(a) Microcosm may not adequately mimic environment. (b) Exposure	Contaminant concentration in desired compartments (i.e., organisms)	Organics and Inorganics	In Use	
Monitoring Introduced organisms	Concentrations in organisms.	Can be deployed to monitor various environmental exposures.	Difficult to determine exposure concentrations	Concentrations in organisms.	Organics and Inorganics	In Use	
Indigenous organisms	Concentrations in organisms.	Where organisms are available they can be utilized.	Difficult to determine exposure concentrations	Concentrations in organisms.	Organics and Inorganics	In Use	

conducted in 1-L beakers. Organisms used in both protocols include *Daphnia magna*, the amphipods *Hyallela azteca*, and *Gammarus* sp., and the midge *Chironomous tentans*. The test endpoint is routinely survival, but studies have been conducted with young of *Daphnia* and *Hyallela*, and chronic tests are feasible with these species. Other test procedures are being evaluated for the worm *Lumbricus* sp.

Tests have also been conducted using the Prater-Anderson static recycling apparatus (10) in which a large volume of water is repeatedly passed over a quantity of sediment. Test organisms included the burrowing mayfly *Hexagenia limbata* (10) to provide bioturbation and intimate sediment exposure and caged *Daphnia magna* to evaluate water-column toxicity. All tests are of 10-d duration except for acute tests with *Daphnia*, which last 48 h.

Residue-Effects Linkages

A general concern for data collected in bioaccumulation tests is that the accumulated residues found in organisms have not been related to effects. In reviewing the exposure systems and tests available for determining bioaccumulation of sediment-bound contaminants, documentations of two major considerations should be emphasized: (1) where possible, contaminant-specific effect observations should be incorporated into bioaccumulation tests, or (2) if this is not possible, general organism responses (i.e., behavioral, metabolic) should be recorded. The latter approach will not allow residue-effects linkages to be made unequivocally, but it may help identify relationships for examination in future accumulation tests. As appropriate endpoints are identified by these or other methods, field studies should be undertaken to determine the correspondence between residues and effects in field organisms.

Summary of Bioaccumulation Findings

In some instances bioaccumulation can be defined as a function of the ratio of intestinal to gill adsorption efficiencies or the ratio between the weight-specific feeding rate and the weight-specific ventilation rate. Bioaccumulation may be expressed as the lipid concentration (BAL) factor, which can be described as K times the lipid fraction of the organism weight:

$$\text{BAF} = \frac{K^1}{K^2} = \frac{r^W}{r^l} \ X^L - K_L \times L \qquad (1)$$

The following summary statements apply to neutral organics:

1. Elimination is inversely related to the lipid content of the organism and the capacity for degradation.
2. Body residues should be normalized for lipid content.
3. Equilibrium bioconcentration factors (BCF) are inversely related to the elimination rate constant and directly related to Log K_{ow} (and K_l).
4. BCFs are independent of exposure concentration.
5. Tissue residue levels are directly related to exposure concentration.
6. The lipid/water partition coefficient (K_L) should predict bioaccumulation, be organism independent (except for degradation capacity), and chemical dependent.
7. The elimination rate constant (K_2) will reflect the ratios between lipid concentration factors for various chemicals.
8. The elimination rate constant (K_2) ought to describe the potential for accumulation via the diet for chemicals other than neutral organics.
9. Biomagnification or, more specifically, dietary accumulation of neutral organics will only be important for chemicals with a large Log K_{ow} (i.e., 5-7) or for chemicals with a very slow elimination rate (K_2), that is a very large elimination half-life ($\geq$ 50 d).

TESTING PROCEDURES FOR DETERMINING BIOLOGICAL EFFECTS OF SEDIMENTS

In order to implement the regulations in a manner consistent with the criteria, the Environmental Protection Agency and the Army Corps of Engineers prepared technical guidance in the form of an implementation manual (11). The manual specified which test procedures were to be followed in collecting the necessary information to be used in making a disposal decision that is consistent with the regulatory criteria. Among the procedures detailed in the manual were those for chemically characterizing the proposed dredged material, for determining the acute toxicity of liquid, suspended particulate, and solid phases, for estimating the potential for contaminant bioaccumulation, and for describing the initial mixing during disposal.

These procedures represented a balance between the technical state of the art and what was routinely implemented at the time of preparation. The recommended test methods were chosen to provide technical information that was consistent with the criteria specified in the regulations. Because the technical state of the art is a dynamic process, it was recognized from the outset that the implementation manual would require both augmentation and revision on a periodic basis.

Several years of operational experience with the sediment test methods and the interpretive guidance provided in the manual has led to the identification of the following technical issues:

● The toxicity test methods are limited to measuring only mortality.

● The experimental design is limited to static testing.

● The species commonly used for these tests were not among the most sensitive nor are the sensitive life stages included.

● The bioaccumulation test method is static, the 10-d exposure probably does not allow for steady-state, and the species selected are often inappropriate.

● The use of an elutriate for the liquid and suspended phases presents interpretation difficulties.

In order to address the above deficiencies, the adoption of a tiered approach to the biological testing of contaminated sediments is recommended, which, when coupled with environmental exposure information, permits the determination of risk and margins of safety (Table 18.3).

Table 18.3   Recommended Data to be Acquired Prior to
Initiation of Biological Testing and
Considerations for Testing with Freshwater Species

Tier I

—Estimates of toxicity from structure activity data.
—Existing toxicological data.
—Examination of existing field ecology data.

(1)   Primarily site specific.

(2)   Chemical and physical characterization of sediment, water, and compound.

(3)   Calculated sediment sorption capacity.

(4)   Calculated desorption rate, leading to estimates of toxicant concentrations in water column.

(5)   Available toxicity data (e.g., acute LC_{50}).

(6)   Since toxicity is likely to be largely chronic, application factor of 10 to 50 could be applied to LC_{50}s.

(7)   Exposure and effects data should be linked together in developing criteria on whether to go on to Tier II (testing).

(8)   Possible approach to criteria:

(a)   High sediment sorption capacity and/or high release rate and high toxicity _____ Tier II.

(b)   Low sorption to sediment and/or slow desorption rate and toxicity _____ no further testing.

(c)   Intermediate results treated case-by-case using best scientific judgement.

Test Species Selection

Test species should be selected on the basis of their relative sensitivity to sediment contaminants and their ecological importance. Non-indigenous species should be used only when they are at least as sensitive as dominant species in the habitat likely to be contaminated. The sensitivity of a species is often related to its phylogenetic position and relation to the substrate. Among marine invertebrates, crustaceans are usually more sensitive than molluscs or polychaetes in acute lethality tests. Phoxocephalid amphipods, ampeliscid amphipods, and mysid shrimp are recommended for lethal tests. Taxa that are likely to survive the sediment exposure (i.e., polychaetes, bivalves, oligochaetes, and fish) are recommended for behavioral or other sublethal response tests. Experimental populations or the entire macro-benthic assemblage is recommended for study of longer-term responses such as recruitment, succession, and population growth (Table 18.4). Within higher taxa, sensitivity is usually related to the degree of contact between sediment and organism:

Most sensitive

1. Infaunal burrowers that respire interstitial water.
2. Infauna that extend tubes, burrows, or body points into overlying water.
3. Epifauna that are in near continuous contact with sediment.
4. Demersal fauna that occasionally contact sediment.

Least sensitive

5. Pelagic biota.

Interspecies comparisons of sensitivity to sediment contaminants provide the best method for test species selection. The distribution of benthic organisms in the field along pollution gradients will demonstrate the range of sensitivities for organisms and can also indicate appropriate test species.

Since many benthic species are affected by changes in natural geochemical properties of sediment, the sensitivity of bioassay species to interstitial water salinity, sediment particle size distribution, and organic environment must be documented. The statistical precision of the species' response to sediment contamination must also be known in relation to sample size and replication.

Table 18.4   Summary of Methods and Response Parameters for Taxa Used in Sediment Testing

A = method is available; R = method is recommended; NA = method is not available

Taxa	Lethality	Tier II responses Behavioral	Genotoxicity	Embryo/Larval development	References
Marine Taxa					
Amphipods	A,R	A,R	NA	NA	15-23
Mysid	A,R	NA	NA	NA	22-24
Polychaetes	A	A,R	A,R	NA	15,22,25-28
Bivalves	NA	A	NA	A,R	15,27,29-34
Oligochaetes	NA	NA	NA	NA	—
Nematodes	NA	NA	NA	NA	—
Caridien shrimp	A	NA	NA	NA	20,27,29,30,35-37
Cumaceans	NA,R	NA	NA	NA	15
Echinoderms	NA,R	NA	NA	NA	30
Fish	A	A	A,R[a]	A,R	23,27,38-44
Freshwater Taxa[b]					
Amphibians					
Gastrophryne carolinensis	A,R	NA	NA	A,R	45
Rana pipiens	A,R	NA	NA	A,R	45,46
Xenopus laevis	A,R	NA	NA	A,R	45,46
Fish					
Carassius auratus	A,R	NA	NA	A,R	45,46
Micropterus salmoides	A,R	NA	NA	A,R	45-46
Salmo gairdneri	A,R	NA	NA	A,R	45
Pimephales promelas	NA,R	NA	NA	NA,R	10
Epibenthic Invertebrates					
Cladocerans					
Daphnia magna	A,R	NA	NA	NA	10,47,48
Daphnia pulex	NA,R	NA	NA	NA	—
Ceriodaphnia reticulata	NA,R	NA	NA	NA,R	—

Table 18.4 (Contd.)

Taxa	Tier II responses				References
	Lethality	Behavioral	Genotoxicity	Embryo/Larval development	
Epibenthic Invertebrates (Continued)					
Amphipods					
Hyallela azteca	A,R	NA	NA	NA	47
Gammarus sp.	A,R	NA	NA	NA	47
Pelycypods					
Sphaerium	A	A	NA	NA	—
Unionidae	A	A	NA	NA	—
Corbicula	A	A	NA	A,R	—
Gastropods					
Inoperculate snails	A	NA	NA	A,R	—
Insects					
Odonates	A	NA	NA	A,R	—
Crustaceans					
Procambarus sp.	A	A	NA	A,R	—
Benthic Invertebrates					
Insects					
Ephemeroptera					
Hexagenia	A,R	NA,R	NA	NA	10,48
Chironomidae					
Chironomus sp.	A,R	NA,R	NA	NA	47
Syrphidae					
Rat-tail maggot	A	NA	NA	NA	—
Oligochaetes	A,R	NA	NA	NA	—
Plants (rooted, submerged)	NA,R	NA	NA,R	NA	—
Microbial communities					
(e.g., sediment O_2 demand)	A,R	NA	A?	NA	—

a Solvent extraction of sediment should not be used.

b (1) Emphasis should be placed on bioconcentration and chronicity for freshwater organisms inhabiting the water column, down to the sediment surface.

(2) Delete acute tests for the above organisms and go directly to the three short chronic tests developed for complex effluents.

Test Material

Sediment bioassays have been applied to material collected from potentially contaminated field sites, and to unpolluted sediment to which toxic materials have been experimentally added. Proper collection and handling of field sediment samples can minimize changes in contaminant bioavailabiilty. Methods of spiking sediment that simulate natural processes of contamination have not yet been properly validated. The use of spiking to predict contaminant effects must therefore be coupled to the application of the data. Spiking usually does not allow for the long-term partitioning of chemicals between interstitial water and particulates and may best represent conditions of recent discharge. The exact method of spiking and incubation period prior to exposure should always be documented. Particulate and interstitial water concentrations of contaminants should be determined at the initiation and termination of bioassays.

Tiered Testing Approach

The tiered approach for sediment toxicity testing is generally organized along increasing levels of biological organization from single species to population, and ultimately community tests. In addition, within the individual species level of organization, the responses are tiered from short-term (e.g., lethality, behavior, embryo, genotoxic) through long-term exposures that measure a range of functional (biochemical, physiological) responses to the most integrative and predictive (growth and reproduction) responses. The tiered approach offers the opportunity for decision making at a variety of steps that eliminate unnecessary testing and focus resources on the important problem. This approach, however, required "decision criteria" that are based upon the integration of both toxicity and environmental exposure data. This presents a unique problem in that there is no broad toxicological data base that includes measures of short-term and chronic responses from which safety margins can be calculated. In view of this, we suggest using the guidance given at previous Pellston Workshops for decision criteria for those cases where empirical data are absent [12-14].

The first tier (Table 18.3) is designed to utilize existing information from a variety of sources to identify the magnitude and types of contaminants of concern: the physical (solubility) and chemical (sorption/desorption) properties of the contaminants and sediment (grain size, organic carbon). These data are then used to make predictions of potential toxicity and bioaccumulation and are compared with laboratory toxicity data and field ecology data. The preliminary estimates are then compared to exposure data such as mass input and contaminant bioavailability.

The second tier (Table 18.4) represents the initial level of biological testing. Here the acquisition of acute toxicity data with an amphipod (*Rhepoxinius* and/or *Ampelisca*) and a mysid (*Mysidopsis*) is mandated.

Behavioral data can be collected from the amphipod acute tests, and larval development tests with proven sensitive species are also recommended. Genotoxic tests are recommended when the chemical characterization indicates the presence of known or potential carcinogens. However, these tests must not be carried out with solvent extracts of sediments as this precludes the operation of the normal process of bioavailability and transport across cell membranes.

The third tier (Table 18.5) represents more functional and integrative responses. Measures of reproduction and growth are mandatory on at least one and preferably two sensitive species (amphipod and mysid) with pathological, physiological, and biochemical responses measured for complimentation. Predictions of population response from chronic reproductive tests using demographic techniques is strongly recommended.

The fourth tier (Table 18.6) represents population and community level tests. Population tests have not been widely applied to contaminated sediments and are recommended with a cautionary note. Recruitment tests, on the other hand, have been widely used for a variety of sediments and drilling fluids and are strongly recommended. Community tests are recommended in only limited cases due to complexity and cost.

Table 18.5   Summary of Methods and Response Parameters for Taxa Used in Sediment Testing
A = method is available; R = method is recommended; NA = method is not available

| Taxa | Tier III Responses[a] | | | | | | References |
	Bio-chemical	Physio-logical	Path-ology	Growth	Repro-duction	Population models	
Marine Taxa							
Amphipods	NA	NA	A,R	A,R	A,R	A,R	21,22,49
Mysid	NA	A	A	A,R	A,R	A,R	22,24,49
Polychaetes	A,R	A,R	A,R	A	NA	NA	28,49-54
Bivalves	A,R	A,R	A	A,R	NA	NA	1,49,54
Oligochaetes	NA	NA	A	NA	NA	NA	43
Nematodes	NA	NA	A	NA	NA	A	56
Caridien shrimp	NA	NA	A	NA	NA	NA	—
Cumaceans	NA	NA	NA	NA	NA	NA	—
Echinoderms	NA	NA	NA	NA	NA	NA	—
Fish	A	A	A,R	A,R	A	NA	38,49,56
Freshwater Taxa							
Amphibians							
Gastrophryne carolinensis	NA	NA	NA	NA	A,R	NA	45
Rana pipiens	NA	NA	NA	NA	A,R	NA	45,46
Xenopus laevis	NA	NA	NA	NA	A,R	NA	45,46
Fish							
Carassius auratus	NA	NA	NA	NA	A,R	NA	45,46
Micropterus salmoides	NA	NA	NA	NA	A,R	NA	45,46
Salmo gairdneri	NA	NA	NA	NA	A,R	NA	45
Pimephales promelas	NA	NA	NA	NA	A,R	NA	—
Epibenthic Invertebrates							
Cladocerans							
Daphnia magna	NA	NA	NA	NA	A,R	NA	47
Daphnia pulex	NA	NA	NA	NA	A,R	NA	—
Ceriodaphnia reticulata	NA	NA	NA	NA	A,R	NA	—

Table 18.5 (Contd.)

Taxa	Tier III Responses[a]						References
	Bio-chemical	Physio-logical	Path-ology	Growth	Repro-duction	Population models	
Amphipods							
Hyallela azteca	NA	NA	NA	A	A	NA?	47
Gammarus sp.	NA	NA	NA	A	A	NA?	47
Pelycypods							
Sphaerium	NA	NA	NA	A	NA	NA	—
Unionidae	NA	NA	NA	A	NA	NA	—
Corbicula	NA	NA	NA	A	NA	NA	—
Gastropods							
Inoperculate snails	NA	NA	NA	A	NA	NA	—
Insects							
Odonates	NA	NA	NA	A	NA	NA	—
Crustaceans							
Procambarus sp.	NA	NA	NA	A	NA	NA	—
Benthic Invertebrates							
Insects							
Ephemeroptera							
Hexagenia	NA	NA	NA	A	NA	NA	—
Chironomidae							
Chironomus sp.	NA	NA	NA	A	NA	NA	47,57,58
Syrphidae							
Rat-tail maggot	NA	NA	NA	NA	NA	NA	—
Oligochaetes	NA	NA	NA	A	NA	NA	—
Plants (rooted, submerged)	NA	A	NA	A	NA	NA	—
Microbial communities	NA	A	NA	NA	NA	NA	—

[a] Stress flexibility and list tool by which problem sediments can be approached on a site-specific case-by-case basis.

Table 18.6   Summary of Methods and Response Parameters for Taxa Used in Sediment Testing
A = method is available; R = method is recommended; NA = method is not recommended; N/AP = not applicable.

Taxa	Tier IV responses			
	Population	Recruitment	Community (S/F)	References
Marine Taxa				
Amphipods	A,R	N/AP	N/AP	22
Mysid	A,R	N/AP	N/AP	22
Polychaetes	NA	N/AP	N/AP	44
Bivalves	A	N/AP	N/AP	—
Oligochaetes	NA	N/AP	N/AP	—
Nematodes	A,R	N/AP	N/AP	55
Caridien shrimp	A	N/AP	N/AP	—
Cumaceans	A	N/AP	N/AP	—
Echinoderms	A	N/AP	N/AP	—
Fish	A	N/AP	N/AP	—
Macrobenthic assemblages	NA	A,R	A,R	59-71
Freshwater Taxa				
Epibenthic Invertebrates				
Amphipods				
Hyallela azteca	A,R	N/AP	N/AP	47
Gammarus sp.	A,R	N/AP	N/AP	47
Pelycypods				
Sphaerium	NA	N/AP	N/AP	—
Unionidae	N/A	N/AP	N/AP	—
Corbicula	A	N/AP	N/AP	—

Table 18.6 *(Contd.)*

Taxa	Tier IV responses			
	Population	Recruitment	Community (S/F)	References
Freshwater Taxa (Continued)				
Epibenthic Invertebrates (Continued)				
Gastropods				
Inoperculate snails	NA	N/AP	N/AP	—
Insects				
Odonates	NA	N/AP	N/AP	—
Crustaceans				
Procambarus sp.	A	N/AP	N/AP	—
Benthic Invertebrates				
Insects				
Ephemeroptera	NA	N/AP	N/AP	—
Hexagenia	NA	N/AP	N/AP	—
Chironomidae				
Chironomus sp.	NA	N/AP	N/AP	—
Syrphidae				
Rat-tail maggot	NA	N/AP	N/AP	—
Oligochaetes	NA	N/AP	N/AP	—
Plants (rooted, submerged)	A	NA,R	NA,R	—
Microbial communities				
(e.g., S.O.D.)	A	NA,R	NA,R	—
Benthic communities	NA	A,R	A,R	—

Freshwater Sediment Tests

Relatively few sediment bioassays have been performed with freshwater organisms. Scuds and the midge are now being used in 10-d lethality tests on sediment and sediment elutriate (10, 47). The same investigators are also using 48-h acute tests with *Daphnia magna* (10). Other research involves the use of a freshwater recycling system in 96-h sediment tests with fathead minnows, daphnias, and mayfly nymphs. In still other cases, 14-d lethality tests with sediment have been performed with midges (57, 58). In addition, short-term embryo-larval tests with fish and amphibians have been used in the monitoring of natural sediments and in sediment-spiked investigations. Exposure ranged from 7 to 8 d with warm water species and was held to 20 d in tests with the rainbow trout.

Caution should be taken in equating biological effects produced by a sediment-released toxicant, compared to effects by the same toxicant when added directly to water in a standard toxicity test. There are some data which indicate that at the same analyzed concentrations, a given toxicant may produce significantly different levels of biological effects depending on whether it comes from sediment release or is added directly to the test water from the reagent bottle. These differences may be the result of organic or inorganic complexes with the toxic agent, rendering it less bioavailable. There may have also been incomplete incubation and transformation of the added chemical into appropriate sediment components.

With respect to freshwater species that primarily inhabit the water column, significant cases of acute toxic effects have been encountered infrequently. This is due primarily to the small concentrations of contaminants released per unit time from bottom sediments. In fish, the predominant effect on adult stages generally has been the bioconcentration of sediment-released toxicants. Tissue residues of such toxicants as PCBs, kepone, and mercury sometimes exceed human health standards. However, it remains unclear if or to what extent tissue uptake of sediment-released contaminants impact natural fish populations, and this consideration, at the present time, limits the usefulness of bioconcentration as a test endpoint in sediment hazard assessment.

Concerning freshwater fish and amphibians, the critical target sites of sediment-associated chemicals are the sensitive life stages (e.g., embryos, early larvae), particularly those of species which spawn directly on the bottom sediment surface. Constraints in sediment toxicity testing preclude use of standard chronic tests with fish. However, three short chronic tests, recently proposed (72) for use with complex effluents, appear adaptable and suitable for use with sediments. One has been used successfully with sediments, and the other two can easily be adapted for either sediments or sediment elutriates. These tests include a short-term embryo-larval procedure used with fish and amphibians and a 7-d larval test conducted with the fathead minnow. The former test requires no feeding and has been used successfully in direct sediment monitoring and in sediment-spiked bioassays. Eggs are deposited on the sediment surface and covered with uncontaminated water. Exposure is initiated at fertilization and continued through 4 d past hatching, giving exposure periods of 6 to 8 d for such species as the bluegill sunfish (*Lepomis macrochirus*), largemouth bass (*Micropterus salmoides*), or fathead minnow (*Pimephales promelas*). The endpoints include frequences of mortality and terata (deformed organisms). The 7-d larval method has not as yet been used in sediment tests, but it appears easily adaptable for such use. It is initiated with day-old minnow larval, and the primary response is growth impairment.

Both tests can be used to express acute toxicity to sensitive life stages by calculating LC_{50} values either for the full exposure period or the first 96 h. In addition, the toxicity threshold (LC_1) or no observed effect (NOEC) values determined in these tests indicate conditions likely to produce chronic toxicity, and they provide a sensitive measurement of the "bioavailable effect level" for sediment-released toxicants. These tests can be used (1) to monitor natural sediments and dredged material for toxic effects, (2) to quantify toxicity in spiked sediment, and (3) to evaluate actual or calculated relationships among sediment sorption capacity, desorption rate and biological effects.

Sediment tests with invertebrate species that primarily inhabit the water column have been restricted largely to *Daphnia magna*, involving exposure periods of 48 to 96 h. Such tests have been performed on both solid sediments and the elutriate. However, as toxic effects of sediment-released chemicals are more likely to be chronic than acute, it is recommended that consideration be given to modifying the 7-d *Ceriodaphnia* sp. test for use with sediments or sediment elutriates.

RESEARCH NEEDS

- Comparison of the sensitivity of species and response endpoints used in sediment toxicity bioassays.
- Evaluation of sediment spiking methods and storage methods in relation to contaminant partitioning.
- Field validation of sediment toxicological methods in relation to the distribution of contaminants and macrobenthic communities along pollution gradients.
- Assessment of the influence of multiple contaminant interactions on sediment toxicity.
- Relative contribution of particulate-bound and interstitial water contaminants to sediment toxicity.
- Development of chronic test methods for sensitive infaunal species.
- Documented relationships between bioconcentration (tissue content)

and other biological responses (physiological, behavioral, genetic, etc.).

● Development of standardized sediment toxicity testing methods for use with freshwater species.

● Identification of additional suitable freshwater vertebrate and invertebrate species for sediment toxicity testing.

In addition to the above items, researchers need to review the three short chronic tests developed for complex effluents for use in freshwater sediment toxicity testing (72). Plants and attached microbial communities should be investigated for response to contaminated sediments. Using spiked sediments and development of bioconcentration and toxicity data is needed for an appropriate selection of inorganic and organic "benchmark" chemicals to evaluate relationships between exposure to sediment contaminants and biological effects. Comparison of toxicity data, bulk chemical sediment analyses, and interstitial water analyses on field collected sediments with varying content of the classes of toxicants would be helpful.

The role of sediments in altering the bioavailability of organics is relatively unexplored; furthermore, protocols for test methods are widely unpublished and not standardized nor compared. We should increase efforts to fit new field study findings, particularly on sediments with high content of neutral organic contaminants, into models such as Pavlou and Weston (73) have developed. Sediment effects testing (e.g., bioconcentration, toxicity) should be used in evaluating predictive models. The bottom line is that we need validation of the models already developed (e.g., mathematical, animal, laboratory, etc.) so that the role of current discharges in creating future problems will receive equal emphasis to historic problems.

REFERENCES

1. **Nelson, W. G., D. Black and D. Phelps.** 1984. A report on the utility of the scope for growth index to assess the physiological impact of Black Rock Harbor suspended sediment on the blue mussel, *Mytilus edulis*. Technical Report D-84-XX, prepared by U.S. Environmental Protection Agency, Narrangansett, RI, for the U.S. Army Engineer Waterways Experiment Station, Vicksburg, MS.

2. **Augenfeld, J.M., J. W. Anderson, D. L. Woodruff and J. L. Webster.** 1980. Effects of Prudhoe Bay crude oil contaminated sediments on *Protothaca staminea* (Mollusca: Pelecypoda): Hydrocarbon content, condition index, free amino acid level. *Mar. Environ. Res.* **4**:135-143.

3. **Augenfeld, J. M., R. G. Riley, B. L. Thomas and J. W. Anderson.** 1982. The fate of polyaromatic hydrocarbons in an intertidal sediments exposure system: Bioavailability to *Macoma inquinata* (Mollusca: Pelecypoda) and *Abarenicola pacifica* (Annelida: Polychaeta). *Mar. Environ. Res.* **7**:31-50.

4. **Von Westernhaten, H., H. Rosenthal, V. Dethlefsen, W. Ernst, U. Harms and P. D. Hansen.** 1981. Bioaccumulating substances and reproductive success in Baltic flounder, *Platichthys flesus*. ICES Report, C.M. 1981/E: 4, Marine Environmental Quality Committee.

5. **Hansen, D. J., Schimmel, S. C. and J. Forester.** 1983. Aroclor 1254 in eggs of sheepshead minnow: Effect on fertilization success and survival of embryos and fry. Proceedings from the 27th Annual Conference of the Southeastern Association of Game and Fish Commissioners.

6. **Anderson, J. W., R. G. Riley, S. L. Kiesser, B. L. Thomas and G. W. Fellingham.** 1983. Natural weathering of oil in marine sediments: Tissue contamination and growth of the clam, *Protothaca staminea. Can. J. Fish. Aquatic Sc.* **40**(2):70-77.

7. **Roesijadi, G., J. S. Young, A. S. Drum and J. M. Gurtisen.** 1984. Behavior of trace metals in *Mytilus edulis* during a reciprocal transplant field experiment. *Marine Ecology Progress Series* (in press).

8. **Lake, J. L., N. Rubinstein and S. Pavignano.** Prediction bioaccumulation: Development of a simple partitioning model for use as a screening tool for regulating ocean disposal of wastes. This volume, pp. 151-166.

9. **Birge, W., J. A. Black and B. Ramey.** 1981. The reproductive toxicology of aquatic contaminants. In *Hazard Assessment of Chemicals*, Vol. 1, Academic Press, pp. 58-115.

10. **Prater, B. L. and M. A. Anderson.** 1977. A 96-hour sediment bioassay of Duluth and Superior Harbor Basins (Minnesota) using *Hexagenia limbata, Asellus communis, Daphnia magna,* and *Pimephales promelas* as test organisms. *Bull. Environ. Contam. Toxicol.* **18**(2):159-169.

11. **U.S. Environmental Protection Agency & U.S. Army Corps of Engineers (EPA/USACE).** 1977. The ecological evaluation of proposed discharge of dredged material into ocean waters; Implementation Manual for Section 103 of PL 92-532. Environmental Effects Lab, USACE, Waterways Experiment Station, Vicksburg, MS.

12. **Cairns, J., Jr., K. L. Dickson and A. W. Maki** (eds.). 1978. *Estimating the Hazard of Chemical Substances to Aquatic Life.* ASTM STP 657. American Society for Testing and Materials, Philadelphia, PA.

13. **Dickson, K. L., A. W. Maki and J. Cairns, Jr.** (eds.). 1979. *Analyzing the Hazard Evaluation Process.* American Fisheries Society, Washington, DC.

14. **Bergman, H. L., R. A. Kimerle and A. W. Maki** (eds.). 1985. *Environmental Hazard Assessment of Effluents.* Pergamon Press, Elmsford, NY.

15. **Swartz, R. C., W. A. DeBen and F. A. Cole.** 1979. A bioassay for toxicity of sediments to marine macrobenthos. *Water Pollut. Control Fed.* **51**:944-950.

16. **Swartz, R. C., W. A. DeBen, K. A. Sercu and J. O. Lamberson.** 1982. Sediment toxicity and the distribution of amphipods in Commencement Bay, Washington, D.C. *Mar. Pollut. Bull.* **13**:359-364.

17. **Swartz, R. C., D. W. Schults, G. R. Ditsworth and W. A. DeBen.** 1984. Toxicity of sewage sludge to *Rhepoxynius abronius,* a marine benthic amphipod. *Arch. Environ. Contam. Toxicol.* **13**:207-216.

18. **Swartz, R. C., D. W. Schults, G. R. Ditsworth, W. A. DeBen and F. A. Cole.** 1984. Sediment toxicity, contamination, and macrobenthic communities near a large sewage outfall. In T. P. Boyle, ed., *Validation and Predictability of Laboratory Methods for Assessing the Fate and Effects of Contaminants in Aquatic Ecosystems.* ASTM STP 865. American Society for Testing and Materials, Philadelphia, PA (in press).

19. **Swartz, R. C., W. A. DeBen, J. K. P. Jones, J. O. Lamberson and F. A. Cole.** 1985. Phoxocephalid amphipod bioassay for marine sediment toxicity. In R. D. Cardwell, R. Purdy and R. C. Bahner, eds., *Aquatic Toxicology and Hazard Assessment: Seventh Symposium;* ASTM STP 854. American Society for Testing and Materials, Philadelphia, PA, pp. 284-306.

20. **Shuba, P. J., H. E. Tatem and J. H. Carroll.** 1978. Biological assessment methods to predict the impact of open-water disposal of dredged material. Technical Report D-78-50. U.S. Army Engineer Waterways Experiment Station, Vicksburg, MS.

21. **Scott, K. N., P. P. Yevich and W. S. Boothman.** Toxicological methods using the benthic amphipod *Ampelisca abdita* Mills. (Submitted to *Aquat. Toxicol.*).

22. **Gentile, J. H., K. J. Scott, S. Lussier and M. Redmond.** 1984. Application of population responses for evaluating the effects of dredged material. Technical Report D-84-XX, prepared by U.S. Environmental Protection Agency, Narrangansett, RI, for the U.S. Army Engineer Waterways Experiment Station, Vicksburg, MS.

23. **Rogerson, P., S. Schimmel and G. Hoffman.** 1984. Chemical and biological characterization of Black Rock Harbor dredged material. Technical Report D-84-XX, prepared by U.S. Environmental Protection Agency, Narrangansett, RI, for the U.S. Army Engineer Waterways Experiment Station, Vicksburg, MS.

24. **Nimmo, D. R., T. L. Hamaker, E. Matthews and W. T. Young.** 1982. The long-term effects of suspended particulates on survival and reproduction of the mysid shrimp, *Mysidopsis bahia,* in the laboratory. In G. F. Mayer ed., *Ecological Stress and the New York Bight: Science and Management.* Estuarine Research Federation, Columbia, SC, pp. 413-422.

25. **McLeese, D. W., L. E. Burridge and J. Van Dinter.** 1982. Toxicities of five organochlorine compounds in water and sediment to *Nereis virens. Bull. Environ. Contam. Toxicol.* 28:216-210.

26. **Rubinstein, N.I.** 1979. A benthic bioassay using time-lapse photography to measure the effect of toxicants on the feeding behavior of lugworms (Polychaete: Arenicolidae), In W. B. Vernberg, A. Calabrese, F. P. Thurberg, F. J. Vernberg, eds., *Marine Pollution: Functional Responses.* Academic Press, New York, pp. 341-351.

27. **Mohlenberg, F. and T. Kiorboe.** 1983. Burrowing and avoidance behaviour in marine organisms exposed to pesticide-contaminated sediment. *Mar. Pollut. Bull.* 14:57-60.

28. **Pesch, G., C. Mueller, C. Pesch, J. Heltshe and P. S. Schauer.** 1984. The application of sister chromatid exchange in marine polychaetes to Black Rock Harbor sediment. Technical Report D-84-XX, prepared by U.S. Environmental Protection Agency, Narrangansett, RI, for the U.S. Army Engineer Waterways Experiment Station, Vicksburg, MS.

29. **Peddicord, R. K.** 1980. Direct effects of suspended sediments on aquatic organisms. In R. A. Baker, ed., *Contaminants and Sediments,* Vol. 1. Ann Arbor Science Publishers, Ann Arbor, MI, pp. 501-536.

30. **Chang, B. D. and C. D. Levings.** 1975. Laboratory experiments on the effects of ocean dumping on benthic invertebrates. I. Choice test with solid wastes. Fisheries and Marine Service Technical Report No. 637, West Vancouver, BC.

31. **Olla, B. L. and A. J. Bejda.** 1983. Effects of oiled sediment on the burrowing behavior of the hard clam, *Mercenaria mercenaria. Mar. Environ. Res.* 9:183-193.

32. **Phelps, H. L., J. T. Hardy, W. H. Pearson and C. W. Apts.** 1983. Clam burrowing behavior: Inhibition by copper-enriched sediments. *Mar. Pollut. Bull.* 14:452-455.

33. **Pearson, W. H., D. L. Woodruff, P. C. Sugarman and B. L. Olla.** 1981. Effects of oiled sediment on predation on the littleneck clam, *Protothaca staminea,* by the Dungeness crab, *Cancer magister. Estuarine Coastal and Shelf Sci.* 13:445-454.

34. **Chapman, P. M. and. J. D. Morgan.** 1983. Sediment bioassays with oyster larvae. *Bull. Environ. Contam. Toxicol.* 31:438-444.

35. **McLeese, D. W. and C. D. Metcalfe.** 1980. Toxicities of eight organochlorine compounds in sediment and seawater *Crangon septemspinosa. Bull. Environ. Contam. Toxicol.* 25:921-928.

36. **Arden, III, R. W. and R. J. Young, Jr.** 1982. Open ocean disposal of materials dredged from a highly industrialized estuary: An evaluation of potential lethal effects. *Arch. Environ. Contam. Toxicol.* 11:567-576.

37. **Lee, G. F. and G. M. Mariani.** 1976. Evaluation of the significance of water sediment-associated contaminants on water quality at the dredged material disposal site. In F. L. Mayer and J. L. Hamelink, eds., *Aquatic Toxicology and Hazard Evaluation.* ASTM STP 634. American Society for Testing and Materials, Philadelphia, PA, pp. 196-213.

38. **American Society for Testing and Materials.** 1984. Proposed new practice for conducting fish early life-stage toxicity tests. Draft #8. ASTM Committee E-47.01.

39. **Hoss, D. E., L. C. Coston and W. E. Schaaf.** 1974. Effects of sea water extracts of sediment from Charleston Harbor, S.C, on larval estuarine fishes. *Estuarine Coastal Mar. Sci.* 2:323-328.

40. **Tsai, C., J. Welch, K. Chang, J. Shaefer and L. E. Cronin.** 1979. Bioassay of Baltimore Harbor sediments. *Estuaries* 2:141-153.

41. **Rubinstein, N. I., W. T. Gilliam and N. R. Gregory.** 1983. Evaluation of three fish species as bioassay organisms for dredged material testing. EPA-600/X-83-062. U.S Environmental Protection Agency, Gulf Breeze, FL.

42. **Pearson, W. H., D. L. Woodruff, P. C. Sugarman and B. O. Olla.** 1984. The burrowing behavior of sand lance, *Ammodytes hexapterus*: Effects of oil-contaminated sediment. *Mar. Environ. Res.* 11:17-32.

43. **Chapman, P. M., G. A. Vigers, M. A. Farrell, R. N. Dexter, E. A. Quinlan, R. M. Kocan and M. L. Landolt.** 1982. Survey of biological effects of toxicants upon Puget Sound biota. I. Broad-scale toxicity survey. NOAA Technical Memorandum OMPA-25. Boulder, CO.

44. **Chapman, P. M., D. R. Munday, J. Morgan, R. Fink, R. M. Kocan, M. L. Landolt and R. N. Dexter.** 1983. Survey of biological effects of toxicants upon Puget Sound biota. II. Tests of reproductive impairment. NOAA Technical Report NOS 102 OMS 1. Rockville, MD.

45. **Birge, W. J., P. C. Francis, J. A. Black and A. G. Westerman.** Toxicity of sediment-associated chemicals to freshwater organisms: Biomonitoring procedures. This volume, pp. 199-218.

46. **Francis, P. C., W. J. Birge and J. A. Black.** 1984. Effects of cadmium-enriched sediment on fish and amphibian embryo-larval stages. *Ecotox. Environ. Safety* 8:378-387.

47. **Nebeker, A. V., M. A. Cairns, J. H. Gakstatter, K. W. Malueg, G. S. Schuytema and D. F. Krawczyk.** Biological methods for determining toxicity of contaminated freshwater sediments to invertebrates. *Environ. Tox. Chem.* (in press).

48. **Prater, B. L. and M. A. Anderson.** 1977. A 96-hour bioassay of Otter Creek, Ohio. *J. Water Pollut. Control Fed.* 49(19):2099-2106.

49. **Yevich, P., C. A. Yevich, K. J. Scott, M. Redmond, D. Black, P. S. Schauer and C. E. Pesch.** 1984. Histopathological effects of Black Rock Harbor dredged sediments on marine organisms. Technical Report D-84-XX, prepared by U.S. Environmental Protection Agency, Narrangansett, RI, for the U.S. Army Engineer Waterways Experiment Station Vicksburg, MS.

50. **Lee, R. F., S. C. Singer, K. R. Tenore, W. S. Gardner and R. M. Philpot.** 1979. Detoxification system in polychaete worms: Importance in the degradation of sediment hydrocarbons. In W. B. Vernberg, A. Calabrese, F. P.Thurbeg, F. J. Vernberg, eds., *Marine Pollution: Functional Reponses*. Academic Press, New York, pp. 23-27.

51. **DeCoursey, P. J. and W. B. Vernberg,** 1975. The effect of dredging in a polluted estuary on the physiology of larval zooplankton. *Water Res.* 9:149-154.

52. **Johns, D. M. and R. Gutjahr-Gobell.** 1984. The impact of dredged material on the bioenergetics of the polychaete *Nephtys incisa*. In F. J. Vernberg, A. Calabrese, F. Thurberg and W. B. Vernberg, eds. *Pollution and Physiology of Marine Organisms.* Academic Press, New York (in press).

53. **Johns, D. M., R. Gutjahr-Gobell and P. S. Schauer.** 1984. Use of bioenergetics to investigate the impact of dredged material on benthic species: A study with polychaetes and Black Rock Harbor material. Technical Report D-84-XXC, prepared by U.S. Environmental Protection Agency, Narrangansett, RI, for the U.S. Army Engineer Waterways Experiment Station, Vicksburg, MS.

54. **Zaroogian, G. E., C. Pesch, P. Schauer and D. Black.** 1984. Evaluation of adenylate energy charge as a test for stress in *Mytilus edulis* and *Nephtys incisa* treated with dredged material. Technical Report D-84-XX, prepared by U.S. Environmental Protection Agency, Narrangansett, RI, for the U.S. Army Engineer Waterways Experiment Station, Vicksburg, MS.

55. **Tietjen, J. H. and J. J. Lee.** 1984. The use of free-living nematodes as a bioassay for estuarine sediments. *Mar. Environ. Res.* 11:233-251.

56. **Kehoe, D. M.** 1983. Effects of Grays Harbor estuary sediment on the osmoregulatory ability of coho salmon smolts (*Oncorhynchus kisutch*). *Bull. Environ. Contam. Toxicol.* 30:522-529.

57. **Wentsel, R., A. McIntosh and G. Atchison.** 1977. Sublethal effects of heavy metal contaminated sediment on midge larvae (*Chironomus tentans*). *Hydrobiologia* 57(3):195-196.

58. **Wentsel, R., A. McIntosh and W. P. McCafferty.** 1978. Emergence of the midge *Chironomus tentans* when exposed to heavy metal contaminated sediment. *Hydrobiologia* **57**(3):195-196.

59. **MacKenzie, C. L. Jr., D. Radosh and R. N. Reid.** 1983. Reduced numbers of early stages of Mollusca and Amphipoda on experimental sediments containing sewage sludge. *Coastal Ocean Pollution Assessment News* **3**:7-8.

60. **Hansen, D. J. and M. E. Tagatz.** 1980. A laboratory test for assessing impacts of substances on developing communities of benthic estuarine organisms. In J. G. Eaton, P. R. Parrish and A. C. Hendricks, eds., *Aquatic Toxicology.* ASTM STP 707. American Society for Testing and Materials, Philadelphia, PA, pp. 40-57.

61. **Hansen, D. J.** 1974. Aroclor 1254: Effect on composition of developing estuarine animal communities. *Estuarine Coastal Mar. Sci.* **7**:401-407.

62. **Tagatz, M. E. and M. Tobia.** 1978. Effects of barite (BaSO) on development of estuarine communities. *Estuarine Coastal Mar. Sci.* **7**:401-407.

63. **Vanderhorst, J. R., J. W. Blaylock, P. Wilkinson, M. Wilkinson and G. Fellingham.** 1980. Effects of experimental oiling on recovery of Strait of Juan De Fuca intertidal habitats. EPA-600/7-81-088. U.S. Environmental Protection Agency, Washington, DC.

64. **Young, D. K. and M. W. Young.** 1978. Regulation of species densities of seagrass-associated macrobenthos: Evidence from field experiments in the Indian River estuary, Florida. *Mar. Res.* **36**:569-593.

65. **Eleftheriou, A., D. C. Moore, D. J. Basford and M. R. Robertson.** 1982. Underwater experiments on the effects of sewage sludge on a marine ecosystem. *Neth. Sea Res.* **16**:465-473.

66. **Perez, K. T.** 1983. Environmental assessment of a phthalate ester di(2-ethylhexyl) phthalate (DEHP), derived from a marine microcosm. In W. E. Bishop, R. D. Cardwell and B. B. Heidolph, eds., *Aquatic Toxicology and Hazard Assessment.* ASTM STP 801. American Society for Testing and Materials, Philadelphia, PA, pp. 180-191.

67. **Emgren, R., G. A. Vargo, J. F. Grassle, J. P. Grassle, D. R. Heinle, G. Langlois and S. L. Vargo.** 1980. Trophic interactions in experimental marine ecosystems perturbed by oil. In J. P. Giesy, Jr., ed., *Microcosms in Ecological Research.* U.S. Technical Information Center, U.S. Department of Energy, Symposium Series 52, Washington, DC, pp. 779-800.

68. **Grassle, J. F., R. Elmgren and J. P. Grassle.** 1981. Response of benthic communities in MERL experimental ecosystems to low level chronic additions of No. 2 fuel oil. *Mar. Environ. Res.* **4**:279-297.

69. **Oviatt, C., J. Frithsen, J. Gearing and P. Gearing.** 1982. Low chronic additions of No. 2 fuel oil: Chemical behavior, biological impact and recovery in a simulated estuarine environment. *Mar. Ecol. Prog. Ser.* **9**:121-136.

70. **Oviatt, C. A., M. E. Q. Pilson, S. W. Nixon, J. B. Frithsen, D. T. Rudnick, J. R. Kelly, J. F. Grassle and J. P. Grassle.** 1984. Recovery of a polluted estuarine system: A mesocosm experiment. *Mar. Ecol. Prog. Ser.* **16**:203-217.

71. **Kuiper, J., P. De Wilde and W. Wolff.** 1984. Effects of an oil spill in outdoor model tidal flat ecosystems. *Mar. Pollut. Bull.* **15**:102-106.

72. **Dickson, K. L.** 1982. Research needs in aquatic toxicology and hazard assessment: A sojourner's perspective. In J. G. Pearson, R. B. Foster and W. E. Bishop, eds., *Aquatic Toxicology and Hazard Assessment: Fifth Symposium.* ASTM STP 766. American Society for Testing and Materials Philadelphia, PA, pp. 9-14.

73. **Pavlou, S. P. and D. P. Weston.** 1983. Initial evaluation of alternatives for development of sediment related criteria for toxic contaminants in marine water (Puget Sound)-Phase I: Development of conceptual framework. Final Report, Work Assignment No. 42 of EPA Contract NO. 68-01-6388. Water Quality Branch, EPA Region X, Seattle, WA.

74. **McFarland, V. A.** 1984. Activity based evaluation of potential bioaccumulation for sediments. In R. L. Montgomery and J. W. Leach, eds., *Dredging and Dredged Material Disposal, Vol. I.* American Society of Civil Engineers, York, pp. 461-467.

Session 4

Case Studies

Chapter 19

Interactions of Spilled Oil with Suspended Materials and Sediments in Aquatic Systems

Douglas A. Wolfe

INTRODUCTION

Approximately 3 million metric tons of oil are released directly into the world oceans [1] each year as a result of operational discharges and accidental spills during production, transportation, refining, and use (consumption) of crude oil and its refined products. Much smaller amounts enter the ocean annually from coastal and riverine runoff, atmospheric aerosols, and fallout and from natural seeps. These petroleum materials undergo transport and transformation through a multiplicity of pathways once they enter the ocean, including dissolution, evaporation, particulate and sediment interactions, photo-oxidation, and biodegradation [1,2]. These pathways and interactions are illustrated conceptually in Figure 19.1. The exact nature and composition of transport pathways effective for any given petroleum spill will vary considerably and will depend on the type of oil spilled, the precise location and timing of the spill, and local meteorological and geographic conditions, such as winds, waves, currents, etc. Since spills of oil directly onto the ocean are relatively short-lived, episodic events that are highly variable with respect to oil composition, timing, and location, different pathways are operative for different spills or operational discharges. In any given situation, however, some oil constituents are likely to undergo transformations to a particulate state, or to become associated with suspended particulate materials or sediment. This paper reviews the various potential roles of suspended particulate and sediments in influencing the fate of spilled oil in aquatic environments.

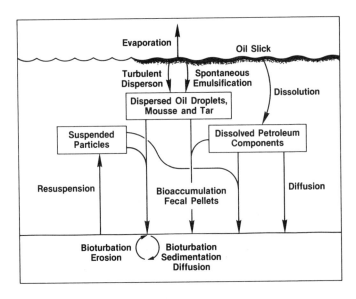

Fig. 19.1
Potential interactions of spilled oil with suspended particulate
material and sediments (redrawn and modified from ref. 24).

OIL INTERACTIONS WITH SUSPENDED MATERIALS

A variety of physical and chemical processes are operative in oil-particulate interactions. As a result of turbulent mixing, suspended particulates may become physically entrapped on the sticky surfaces of floating oil as it undergoes weathering. Oil may exhibit hydrophobic associations with surfaces of suspended particulates, or interact with natural dissolved materials to form micellar aggregations. Some compounds of petroleum origin may undergo ionic exchange or adsorption onto charged particle surfaces. Though most of these processes probably occur simultaneously to some degree, they are discussed separately in the following sections with supporting observations from laboratory results and pertinent oil spills.

Entrainment of Suspended Particulates and Sinking of Floating Oil

When oil is spilled in areas of high suspended particulate concentration, there is potential for mixing with entrainment of relatively dense particulate matter into the oil slick leading to settling of the oil (Fig. 19.2). This phenomenon has rarely been observed in spill instances, primarily because few spills have occurred in areas of very high suspended particulate concentrations, such as those receiving pulsed riverine discharges.

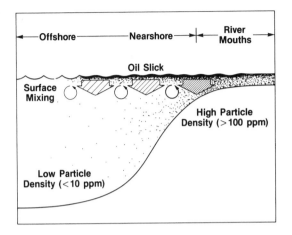

Fig. 19.2
Sorption and sinking of spilled oil
(modified from ref. 3).

After the Santa Barbara Channel blowout in January 1969, however, a large portion of the floating oil slick was carried into the sediment-rich plumes of the Ventura and Santa Clara Rivers, where the oil disappeared from the surface within a few meters of the river plume edges (4). Through this association with suspended sediments, the oil was transported to the bottom where it constituted up to 1.4% (w/w) of the dry weight of the bottom sediments (5). Drake et al. (6) estimated that about 40×10^6 tonnes of suspended particulate materials were introduced into the Santa Barbara Channel, mainly from the Santa Clara River, during the winter rains of 1969. This high flux of sedimenting particulates provided an effective vehicle for downward transport of floating oil in the area.

At the IXTOC-I spill, Patton et al. (7) found that the surfaces of mousse pancakes had a very high density as a result of embedded particulate material. The observed density of the surface was greater than 1.022 g/cm, suggesting that small pancakes or flakes of the surface would be highly susceptible to sinking.

Sorption of Dissolved Oil into Suspended Particulates

Materials of petroleum origin are generally non-polar and very poorly soluble in water (8). Highest solubilities are found for small compounds of low molecular weight and for those with aromatic or olefinic character. There is a strong tendency, therefore, for most petroleum compounds to become associated with other organic phases or phase boundaries such as particulate surfaces or the air-sea interface. This tendency of non-polar substances to become associated with each other has been called hydrophobic bonding (9). As the more soluble (and volatile) materials disappear from an oil slick, the heavier residual compounds partition increasingly with particles. This partitioning of higher molecular-weight compounds into particulates, while lower molecular-weight materials are entering the water column, has been observed in the University of Rhode Island Marine Ecosystem Research Lab (MERL) tanks using Number 2 fuel oil (10), in simulated spills (11) and in coastal waters near a natural oil seep in California (12). In all cases, there was a general partitioning of aromatic materials into the soluble phase as contrasted with aliphatic materials. Smaller, more soluble aromatics partition most strongly into the aqueous phase. Aliphatic materials also undergo a phase fractionation, with larger molecules preferentially being associated with particulates. The aqueous phase thus becomes enriched with aromatics and low molecular-weight aliphatics, while the particulate phase becomes enriched in high molecular-weight aromatics and aliphatics.

With increasing temperature or as a result of increased interaction with dissolved organic matter, hydrocarbons become more soluble and correspondingly smaller amounts associate with suspended particulate matter. In sorption experiments with hydrocarbons and bentonite clay in saline solutions, Meyers and Quinn (13) found increasing association with particulates as the solubility of the hydrocarbons decreased from aromatics to n-alkanes. In addition, Meyers and Oas (14) found that particulate association increased linearly with concentration for n-alkanes. The degree of association increased with increasing chain length, and isoalkanes were adsorbed more effectively than n-alkanes of the same carbon number.

For most systems, the sediment/water partition coefficient (K_p) is constant over a broad range of aqueous concentrations. Thus for any compound, at equilibrium:

$$K_p = \frac{\text{ppb (dry) in solid phase}}{\text{ppb in water}} \tag{1}$$

Measurement of K_p is very difficult for oil and its constituents. Dispersed oil droplets are usually present and interfere with the analysis. Karickhoff et al. (15) showed that the partition coefficients for hydrophobic contaminants on any single particle size fraction was related both to the octanol-water partition coefficient and to the organic carbon content of the sediment:

$$K_p = 0.63 \, K_{ow} \, (OC) \tag{2}$$

where K_{ow} = octanol-water partition coefficient,
OC = organic carbon content of the sediments.

For polynuclear aromatic hydrocarbons, Means et al. [16] found essentially this same relationship, but with a proportionality constant of 0.48 instead of 0.63. For any compound, therefore, the equilibrium partitioning onto natural sediments is directly related to the organic carbon content of the sediment. The proportionality is highly constant for a wide variety of tested sediments [16, 17], indicating that the partitioning does not depend on specific adsorption characteristics of the organic material. The binding of organic solutes by soil organic components may be more analogous to solvation by an organic solvent.

It is useful to normalize the partition coefficients (K_p) for the organic content of the sediment (OC). This gives a new coefficient, designated K_{oc}, representing partitioning of materials into the organic content of the sediment:

$$K_{OC} = \frac{K_p}{OC} \tag{3}$$

Representative values for K_{OC} and K_{OW} are shown in Table 19.1 for a number of petroleum constituents.

Table 19.1   Measured Partition Coefficients, K_{oc} and K_{ow}, for Selected Petroleum Constituents [17]

Compound	K_{ow}	K_{oc}
Benzene	130	60
Naphthalene	2,300	870
Phenanthrene	37,000	12,000
Anthracene	35,000	16,000
9-methylanthracene	120,000	51,000
Pyrene	150,000	68,000
Tetracene	790,000	650,000
7,12-dimenthylbenz(a)anthracene	950,000	220,000
Dibenz(a,b)anthracene	3,200,000	1,700,000
3-methylcholanthrene	2,600,000	1,200,000
Dibenzothiophene	24,000	11,000

The presence of dissolved organic material may also affect the solubility of petroleum hydrocarbons in the soluble phase. Boehm and Quinn [18] found that dissolved humic-like organic materials in seawater promoted the solubility of certain saturated hydrocarbons (n-alkanes and isoprenoids), but not of polycyclic aromatic hydrocarbons. This solubilization was attributed to micellar formation through intermolecular association of hydrocarbons and the surface-active humic substances. In a subsequent paper, Boehm and Quinn [19] distinguished this solubilized oil as colloidal micelles less than 1 μm in diameter from truly dissolved oil which is incorporated into the water structure. Leversee et al. [20] examined the effects of humic acids on 6-h bioaccumulation of six aromatic hydrocarbons by *Daphnia magna*. They found no effect of humics on uptake of naphthalene, anthracene, dimethyl-benzanthracene or dibenzanthracene, but methylcholanthrene uptake was enhanced, and benzo(a)pyrene uptake was reduced. These latter results are not readily interpretable in terms of humic effects on either solubility or partitioning of polynuclear aromatics.

Measurements of dissolved and particulate hydrocarbons in natural waters, especially after a spill, are seriously impeded by the presence of undissolved, dispersed oil in the water column. As a result, water samples are usually not fractionated, and when they are, one cannot be sure that the "soluble" fraction represents truly dissolved material. Those few instances where dissolved and particulate fractions have been analyzed separately in seawater, however, at least approximate the partition coefficients expected on the basis of experimental results (Table 19.2).

Table 19.2   Hydrocarbons in Aqueous and Particulate Fractions of Filtered Seawater Samples from Puget Sound[a] (23)

Compound	Concentrations		K_p
	Dissolved, ng/L	Particulate µg/g	
Naphthalene	43 ± 31	0.09 ± 0.14	2,100
2-methylnaphthalene	14 ± 7	0.02 ± 0.01	1,400
1-methylnaphthalene	9 ± 4	0.05 ± 0.07	5,600
2,6-dimethylnaphthalene	2 ± 2	0.04 ± 0.04	20,000
1,3-dimethylnaphthalene	3 ± 1	0.03 ± 0.04	10,000
2,3-dimethylnaphthalene	1 ± 2	0.01 ± 0.01	10,000
2,3,6-trimethlynaphthalene	6 ± 2	0.12 ± 0.23	20,000
Fluorene	<4	0.03 ± 0.03	>7,500
Phenanthrene	10 ± 8	0.23 ± 0.16	23,000
Anthracene	<1	0.10 ± 0.13	>100,000
Fluoranthene	<1	0.93 ± 1.17	—
Pyrene	9 ± 5	1.21 ± 1.32	130,000
1-methylpyrene	<1	<0.9	—
Chrysene	<1	<0.03	—
Benzo(a)anthracene	<1	<0.04	—
Benzo(a)pyrene	No data	<0.03	—
Perylene	No data	<0.04	—

[a] Data are means from nine samples collected in July 1979 from several locations. Suspended load generally ranged between 1-9 mg/L.

Boehm (21) reported particulate and "dissolved" bulk hydrocarbon concentrations for Atlantic Ocean shelf waters over Georges Bank. If one assumes particulate concentrations of 10 mg/L for surface waters in this area off the New England coast, then the composite K_p for all hydrocarbons would be in the order of 10^4 to 10^5. Payne et al. (22) found a high proportion of water column hydrocarbons associated with particulates during the IXTOC blowout in the Gulf of Mexico, but particulate concentrations were not measured, and K_p cannot be estimated. The organic content of the particulate material was not measured in any of these instances, precluding direct comparison with experimental results in Table 19.1.

Particle Formation by Weathering Oil

As petroleum undergoes weathering, several processes occur simultaneously to incorporate the petroleum into the water column in a particulate state. Except under carefully controlled experimental circumstances, these processes cannot readily be distinguished from each other nor from associations of bulk or dissolved oil with naturally occurring suspended particulate materials. The principal mechanisms of formation of particulate oil are turbulent dispersion of oil droplets, spontaneous emulsification, and mousse-tarball formation.

Turbulence from winds and waves drives small droplets into the water column (Fig. 19.3). These droplets are then convected away from the slick by local currents and eddy diffusivity. Under actual spill conditions, the amount of oil entering the water column through this process may exceed that entering through dissolution (24). However, smaller droplets have much larger surface area-to-volume ratios such that dissolution of soluble components from the smaller droplets is enhanced (25).

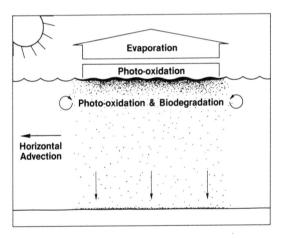

Fig. 19.3
Entrainment and settling of photo-oxidized
(and microbially oxidized) oil particles in
water column (modified from 3).

Payne et al. (24) observed the formation of small micelles, 1 to 10 μm in diameter, in the aqueous phase when crude oil was gently layered over water, and postulated that some form of spontaneous emulsion (26) was taking place at the petroleum/water interface. Compared to turbulent dispersion under spill conditions, spontaneous emulsification is unlikely to account for significant transport of oil into the water column.

Under actual spill circumstances in the marine environment, the oil concentrations in the water column may far exceed solubility limits in sea water, and it is operationally difficult to separate or distinguish the different forms that are present. During the *Amoco Cadiz* spill, concentrations in excess of 1,000 μg/L were detected by a towed underwater fluorometer in a region of high turbulence just inshore of the wreck at the mouth of Aber Wrac'h [27]. Oil concentrations well offshore were in the range of 3 to 20 μg/L, while nearshore concentrations ranged from 2 to 200 μg/L. Concentrations inside the Abers were 30-500 μg/L during this period [28]. These concentrations were generally comparable to those found at other surface spills, for example 30 μg/L at Ekofisk [29], 450 μg/L at Argo Merchant [30], 45 μg/L at Arrow [31]. At the IXTOC-I subsurface blowout in the Bay of Campeche, by contrast, up to 7,000 μg/L were measured in the water column near the wellhead. These high concentrations were postulated to result predominately from oil droplets larger than 1 μm, because UV spectra corresponded closely to spectra obtained from whole oil [32]. At distances of about 25 to 40 km from the IXTOC-I wellhead, lower concentrations (5-20 μg/L) were observed and the UV fluorescence spectra showed these to consist mainly of soluble naphthalenic compounds. Filtration of these samples did not affect either the total concentration or spectral characteristics of the oil. At one station with 416 μg/L oil at 6-m depth, however, about three-fourths of the oil was retained on the glass fiber filter, and the spectrum of the filtered sample was depleted in the three- to five-ring region compared to the original sample [32].

A water-in-oil emulsion or "mousse" forms rapidly in floating oil slicks. Mousse may contain 30 to 83% (v/v) water and has a semi-solid gel-like consistency. Mousse formation is accompanied by evaporation and dissolution, especially of aromatics and lower molecular weight alkanes, and oxidation of the surface layer (Fig. 19.3). Together these processes lead to the formation of tar balls [33]. Degradation of the surface of tar balls produces small dense particles that may sink [34]. At the IXTOC-I spill site, photochemical and evaporative processes caused skinning-over of mousse patches [7]. Wind-driven turbulence promoted cracking and flaking of the skin and drove the resultant particles into the water column. Relative to parent mousse, these flakes were depleted in both saturated and aromatic hydrocarbons, and enriched in polar compounds, probably as a result of photo-oxidation of the skin. Patton et al. [7] suggested that mousse pancakes may undergo successive and alternating stages of skinning, breakup through flaking, and dissolution and diffusion of unweathered constituents from the newly exposed mousse until a new skin develops.

Fecal Pellet Transport of Oil

Petroleum dispersed in the water column may be ingested by zooplankton and eliminated as sedimenting fecal pellets (Fig. 19.4).

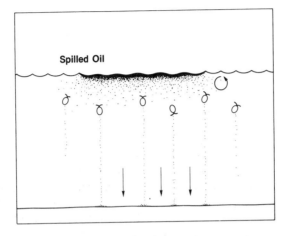

Fig. 19.4
Settling of oil by fecal material after ingestion
and excretion by zooplankton (modified from 3).

After the ARROW incident in Chedabucto Bay, Nova Scotia, copepod feces contained up to 7% (w/w) oil (35). Although dispersed oil droplets exhibit neutral or positive buoyance, the oil rapidly sinks when it is pelletized into fecal matter. Parker (36) also found oil droplets in the guts of copepods and barnacle larvae and in their fecal pellets, and suggested that the oil passed through the gut largely unaltered. Contaminated zooplankton were observed also after the *Argo Merchant* spill (30), the *Ekofisk* blowout (37) and the *Tsesis* spill (38). In all cases oil particles were found adhering to the feeding appendages or in the alimentary tract.

Parker et al. (39) estimated that one copepod (*Calanus finmarchicus*) could ingest up to 0.15 mg of oil per day. At that rate, zooplankton could contribute substantially to the sedimentation of dispersed oil. From sediment trap data, Johansson (38) reported that 30 to 60 mg of oil were sedimented daily per square meter in the vicinity of the *Tsesis* 1 week after the spill. Although the zooplankton biomass was low at this season (~ 1 g wet weight/m²), sedimentation by fecal pellets remains a plausible explanation for the observation (40).

OIL INTERACTIONS WITH SEDIMENTS

Oil is introduced into subtidal sediments through direct settling from the water column, as discussed in the preceding sections and by seaward transport of intertidal sediments that have been oiled by spills in coastal waters. Oil incorporated into sediments, whether inter- or subtidal, may be subjected to sequential burial and erosion, resuspension from turbulence in the bottom boundary layers or from bioturbation and/or desorption into overlying waters. These processes are discussed in the following sections.

Burial and Erosion of Oil in the Intertidal Zone

The geomorphology and sediment characteristics of the coast strongly influence the amount of oil that may come ashore in a spill and its persistence in the intertidal zone. Based on observations at numerous spills, an index of coastal oil spill vulnerability has been developed (41, 42), and is currently being used for spill contingency planning in the United States. The vulnerability index (Table 19.3) characterizes ten types of coastal environments according to steepness of beach profile, composition, and grain size of substrate, and relates these characteristics to probability of impingement and persistence of oil in a spill situation. The index characterizes rocky headlands as least vulnerable and salt marshes as most vulnerable (Table 19.3).

Steep rocky beaches tend to deflect the wave energy and hold floating oil away from the shore. Fine sandy beaches are only modestly vulnerable because the oil does not penetrate readily into the beach. Generally, beach vulnerability increases as the grainsize of the sediment increases on the beach. Oil can penetrate readily into cobble, gravel, or coarse sand and may percolate successively with the tidal water table down to hard substrates (42, 43, 44). In sheltered tidal flats and salt marshes, oil penetrates into anoxic muds facilitated by animal burrows and interstitial water movements where it may persist for long periods.

At the *Amoco Cadiz* spill, oil came ashore initially during a period of maximum beach erosion, depositing oil on to the lower strata of the beach. Subsequent tides and waves deposited alternating layers of clean sand and relatively fresh oil in the beach (Fig. 19.5). In some areas, as many as five separate layers were visible in the beach, with the deepest buried under nearly 1 m of sand. Much of this oil was reexposed during subsequent winter erosion, but in some cases the buried oil appears to have migrated deeper into the beach and stabilized at the beach water table (43, 44).

Oil was transported to the bottom at the *Amoco Cadiz* scene while surface slicks were still extensive, probably through a combination of sediment entrainment, adsorption of oil onto particulates, and sinking of weathered oil. In addition, however, oil was carried seaward off the beaches, some 2 to 6 weeks after initial deposition (28). Erosion or scouring of oiled beach sediments into nearshore sedimentary areas could also be a significant route of entry for oil into shallow benthic environments (Fig. 19.5).

Table 19.3   Relation between Shoreline Type and Vulnerability
to Oil Deposition and Persistence (40)

Vulnerability Index	Shoreline type	Comments
1	Exposed rocky headlands	Wave reflection keeps most of the oil off-shore.
2	Eroding wave-cut platforms	Wave swept. Most oil removed by natural processes within weeks.
3	Fine-grained sand beaches	Oil doesn't penetrate into the sediment, facilitating mechanical removal. Otherwise, oil may persist several months.
4	Coarse-grained sand beaches	Oil may sink and/or be buried rapidly. Under moderate to high energy conditions, oil will be removed naturally within months from most of the beachface.
5	Exposed, compacted tidal flats	Moist oil does not adhere to, nor penetrate into, the compacted tidal flat.
6	Mixed sand and gravel beaches	Oil may undergo rapid penetration and burial. Under moderate to low energy conditions, oil may persist for years.
7	Gravel beaches	Same as above. A solid asphalt pavement may form under heavy oil accumulations.
8	Sheltered rocky coasts	Areas of reduced wave action. Oil may persist for many years.
9	Sheltered tidal flats	Areas of great biologic activity and low wave energy. Oil may persist for years.
10	Salt marshes and mangroves	Most productive of aquatic environments. Oil may persist for years. These areas should receive priority protection by using booms or oil sorbent materials.

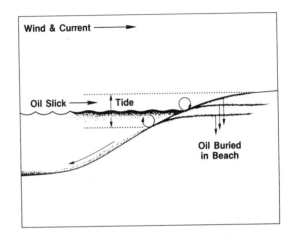

Fig. 19.5
Mixing and burial of spilled oil in intertidal
beach sediment during accretionary periods. In
the surf zone, oil may mix with sediments and
subsequently be transported seaward by bottom
currents. Tidal pumping can transport the
beached oil downward to the water table.

Resuspension and Dissolution of Once-settled Oiled Sediments

Once deposited on the bottom, oiled sediments may be buried by further deposition and biodegraded, resuspended by bottom currents or storm-induced turbulence, or subjected to physical mixing as a result of bioturbation. Resuspension and bioturbation promote the contact of the sediments with interstitial and overlying waters and facilitate the dissolution of soluble hydrocarbon components from the sediments. These mixing processes also maintain the oxygenation of sediments and thereby promote the microbial biodegradation of deposited oil.

At the *Amoco Cadiz* spill, sediment oil concentrations decreased much more rapidly at offshore and intertidal stations subjected to wave action than at protected subtidal stations in the Abers (Fig. 19.6). Cores taken at Ile Grande 1 year after the spill showed penetration of *Amoco Cadiz* oil to depths of about 15 cm. Little mixing had occurred in this area as evidenced by the high concentrations of oil in the upper layers of the sediment and decreasing concentrations with depth, in contrast with the distribution of hydrocarbons not from the spill (Fig. 19.7).

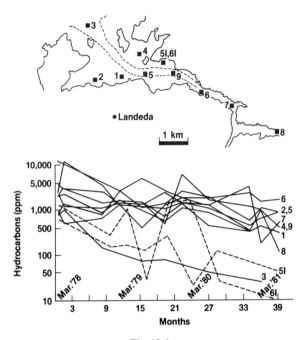

Fig. 19.6
Hydrocarbon concentrations in Aber Wrac'h sediments
from March 1978 through June 1981. Samples at
station 1 through 9 were analyzed by infrared
spectroscopy: samples at stations 51 and 61
(intertidal) were analyzed by GC (28).

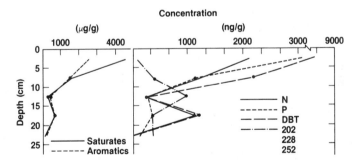

Fig. 19.7
Depth profiles for hydrocarbon components from a
representative core taken at Ile Grande in March 1979 (N = total
naphthalene; P = total phenanthrene; DBT = total dibenzo-
thiophene; 202, 228 and 252 indicate summations of polynuclear
hydrocarbons (not attributable to *Amoco Cadiz*) of molecular
weights 202, 228 and 252 (28)).

In heavily oiled areas, oil constituents may be highly mobile in interstitial waters. On the beaches at St. Michel-en-Greve, a light oil sheen was usually evident on the water that seeped into footprints as one walked on the beaches 6 weeks after the *Amoco Cadiz* spill (after the surficial oil had been effectively removed from the beach by cleanup crews and by tidal action). Interstitial water that seeped into shallow (3-6 cm) holes scooped in this fine sand beach contained 170 to 180 μg/L of petroleum hydrocarbons, measured by UV fluorescence in comparison to a reference *Amoco Cadiz* mousse sample. During this same period, surface waters from nearby areas contained only 2.2 to 14 μg/L (R. C. Clark, Jr. and D. A. Woolfe, unpublished observations). Surficial sediments subject to mixing should therefore be rapidly cleansed of soluble oil constituents.

PERSISTENCE OF SPILLED OIL IN MARINE SYSTEMS

In Aber Wrac'h, *Amoco Cadiz* oil still persisted 39 months after the spill in muddy subsurface sediments of the lower intertidal zone (stations 1-8, except for three at the mouth of the Aber) in June 1981, whereas nearby upper intertidal sediments (stations 51 and 61) were virtually free of oil (Fig. 19.6) (28). High concentrations (up to 11,000 μg/mL) of highly weathered oil were also present in mid-1981 in surficial sediments of the Ile Grande salt marsh. Three years after the spill, however, oil distribution was highly patchy, depending principally on the original extent of oiling, the cleanup activities in the area, and the variability of natural weathering processes (28).

Figure 19.6 illustrates the persistence over a 3-year period of *Amoco Cadiz* oil in subtidal muddy sediment of Aber Wrac'h. The oil that remained after 3 years was highly weathered. Lower molecular weight alkanes and aromatics had disappeared and n-alkanes were depleted relative to branched alkanes such as isoprenoids. After 2 years, pentacyclic triterpanes, alkylated dibenzothiophenes and phenanthrene compounds, which were not prominent earlier in the weathering process, became the major molecular markers in the saturated and aromatic fractions. An unresolved complex mixture also became a prominent feature of chromatograms for both fractions (28).

In low-energy sedimentary environments, weathered oil may persist in appreciable quantities for decades. Sediments from the West Falmouth area impacted by the 1969 Florida spill contained 1 to 3 mg aromatics per gram dry weight at least through July 1976, compared to 0.04 mg or less from control stations (J. N. Butler, personal communication-NAS background paper). High sediment concentrations persisted years after the *Amoco Cadiz* spill in the Abers and the Ile Grande marsh, but in subtidal sediments elsewhere, concentrations had generally diminished nearly to background levels (28).

Figure 19.8 illustrates a hypothetical mass balance for a typical oil spill in the marine environment (45).

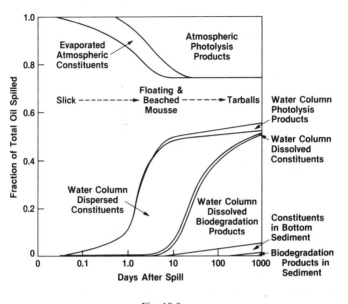

Fig. 19.8
Hypothetical mass balance of oil through time from
a "typical" marine oil spill (45).

This budget suggests that after 3 years, most of the oil will have undergone at least some photolysis and/or biodegradation, and that unmodified petroleum constituents will persist either as floating or beached tar or in sediments. The detailed evolution of spilled oil obviously depends upon the specific circumstances associated with each individual spill, but it is instructive to compare Figure 19.8 with the estimates made by Gundlach et al. (28) for the disposition of *Amoco Cadiz* oil after 30 d. Evaporation accounted for the greatest fraction of the *Amoco Cadiz* oil (30%), followed by onshore deposition (28%), dispersed and degraded in water column (13.5%, and deposited in subtidal sediments (8%). The balance of 20.5% was assumed to have left the scene as surface slicks and tar balls. A significant portion of the oil that came ashore initially and was not cleaned up was probably dispersed subsequently into shallow nearshore sediments or the water column where it underwent biodegradation. Over the long term, biodegradation of dissolved, dispersed, and sedimented petroleum is a major process in the self-cleansing of the marine environment.

REFERENCES

1. **National Academy of Sciences.** 1985. Oil in the Sea. Inputs, Fates, and Effects. U.S. National Academy of Sciences. Washington, DC.

2. **Wolfe, D. A.** 1985. Fossil fuels: transportation and marine pollution. In I. W. Duedall, D. R. Kester, P. K. Park, and B. H. Ketchum, eds., *Wastes in the Ocean*, Vol. IV—Energy Wastes in the Ocean. John Wiley & Sons, New York. pp. 45-93.

3. **Boehm, P. D.** Transport and transformation processes regarding hydrocarbon and metal pollutants in offshore sedimentary environments. In Boesch, D. F. and N. Rabalais, eds., *The Long-Term Effects of Offshore Oil and Gas Development: An Assessment and a Research Strategy.* Report to the Interagency Committee on Ocean Pollution Research, Development, and Monitoring. NOAA National Marine Pollution Program Office, Rockville, MD. Submitted by Louisiana Universities Marine Consortium, Chauvin, LA (in press).

4. **Kolpack, R. L.** 1971. *Biological and Oceanographical Survey of the Santa Barbara Channel Oil Spill 1969-1970,* Vol. 11—Physical, Chemical and Geological Studies. University of Southern California, Allan Hancock Foundation, Los Angeles, CA.

5. **Kolpack, R. L., J. S. Mattson, H. B. Mark, Jr. and T. C. Yu.** 1971. Hydrocarbon content of Santa Barbara Channel sediments. In R. L. Kolpack, ed., *Biological and Oceanographical Survey of the Santa Barbara Channel Oil Spill 1969-1970,* Vol. II—Physical, Chemical and Geological Studies. University of Southern California, Allan Hancock Foundation, Los Angeles, CA, pp. 276-295.

6. **Drake, D. E., P. Fleisher and R. L. Kolpack.** 1971. Transport and deposition of flood sediment, Santa Barbara Channel, California. In R. L. Kolpack, ed., *Biological and Oceanographical Survey of the Santa Barabara Channel Oil Spill 1969-1970,* Vol. II—Physical, Chemical and Geological Studies. University of Southern California, Allan Hancock Foundation, Los Angeles, CA, pp. 181-217.

7. **Patton, J. S., M. W. Rigler, P. D. Boehm and D. L. Fiest.** 1981. IXTOC-I oil spill: Flaking of surface mousse in the Gulf of Mexico. *Nature* 290:235-238.

8. **Shaw, D. G.** 1977. Hydrocarbons in the water column. In D. A. Wolfe, ed., *Fate and Effects of Petroleum Hydrocarbons in Marine Organisms and Ecosystems.* Pergamon Press, Elmsford, NY, pp. 8-18.

9. **Parks, G. A.** 1975. Adsorption in the marine environment. In J. P. Riley and G. Skirrow, eds., *Chemical Oceanography,* Vol. I. Academic Press, NY, pp. 241-308.

10. **Gearing, J. N., P. J. Gearing, T. Wade, J. G. Quinn, H. B. McCarty, J. Farrington and R. F. Lee.** 1979. The rates of transport and fates of petroleum hydrocarbons in a controlled marine ecosystem and a note on analytical variability. Proceedings of 1979 Oil Spill Conference. American Petroleum Institute, Washington, DC, pp. 555-564.

11. **Winters, J. K.** 1978. Fate of petroleum derived aromatic compounds in sea-water held in outdoor tanks. Chapter 12 in *South Texas Outer Continental Shelf Study.* Draft Report submitted to Bureau of Land Management.

12. **de Lappe, B. W., R. W. Risebrough, J. C. Shropshire, W. R. Sisteck, E. F. Letterman, D. R. de Lappe and J. R. Payne.** 1979. The partitioning of petroleum related compounds between the mussel *Mytilus californianus* and seawater in the Southern California Bight. Cited by R. E. Jordan and J. R. Payne. 1980. *Fate and Weathering of Petroleum Spills in the Marine Environment.* Ann Arbor Science Publishers, Ann Arbor, MI.

13. **Meyers, P. A. and J. G. Quinn.** 1973. Association of hydrocarbons and mineral particles in saline solutions. *Nature* 244:23-24.

14. **Meyers, P. A. and T. A. Oas.** 1978. Comparison of associations of different hydrocarbons with clay particles in simulated sea water. *Environ. Sci. Technol.* 12:934-937.

15. **Karickhoff, S. W., D. S. Brown and T. A. Scott.** 1979. Sorption of hydrophobic pollutants on natural sediments. *Water Res.* 13:241-248.

16. **Means, J. C., S. G. Wood, J. J. Hassett and W. L. Banwart.** 1980. Sorption of polynuclear aromatic hydrocarbons by sediments and soils. *Environ. Sci. Technol.* 14:1524-1528.

17. **Karickhoff, S. W.** 1981. Semi-empirical estimation of sorption of hydrophobic pollutants on natural sediments and soils. *Chemosphere* 10:833-846.

18. **Boehm, P. D. and J. G. Quinn.** 1973. Solubilization of hydrocarbons by the dissolved organic matter in seawater. *Geochim. Cosmochim. Acta* 37:2459-2477.

19. **Boehm, P. D. and J. G. Quinn.** 1974. The solubility behavior of the No. 2 fuel oil in sea water. *Mar. Pollut. Bull.* 5:101-105.

20. **Leversee, G. J., P. F. Landrum, J. P. Giesy and T. Fannin.** 1983. Humic acids reduce bioaccumulation of some polycyclic aromatic hydrocarbons. *Canadian Journal of Fisheries and Aquatic Science* 41(Suppl. 2):63-69.

21. **Boehm, P. D.** 1980. Evidence for the decoupling of dissolved, particulate and surface microlayer hydrocarbons in northwestern Atlantic continental shelf water. *Mar. Chem.* 9:255-281.

22. **Payne, J. R., G. S. Smith, P. J. Mankiewicz, R. F. Shokes, N. W. Flynn, V. Moreno and J. Altamirano.** 1980. Horizontal and vertical transport of dissolved and particulate-bound higher-molecular-weight hydrocarbons from the IXTOC-I blowout. Proceedings of a Symposium, Preliminary Results from the September 1979 *Researcher/Pierce* IXTOC-I Cruise. NOAA, Office of Marine Pollution Assessment Bay Biscayne, FL, pp. 119-167.

23. **Riley, R. C., E. A. Crecelius, D. C. Mann, K. H. Abel, B. L. Thomas and R. M. Bean.** 1980. Quantitation of pollutants in suspended matter and water from Puget Sound. NOAA Technical Memorandum ERL MESA-49, Boulder, CO.

24. **Payne, J. R., B. E. Kirstein, G. D. McNabb, Jr., J. L. Mabach, R. Redding, R. E. Jordan, W. Hom, C. De Oliveira, G. S. Smith, D. M. Baxter and R. Gaegel.** 1984. Multivariate analysis of petroleum weathering in the marine environment subarctic, Volume I— Technical Results. NOAA, Outer Continental Shelf Environmental Assessment Program, Contract NA80RACOOO18, Washington, DC.

25. **Boehm, P. D., D. L. Fiest, D. Mackay and S. Paterson.** 1982. Physical chemical weathering of petroleum hydrocarbons from the IXTOC I blowout. Chemical measurements and a weathering model. *Environ. Sci. Technol.* 16:498-505.

26. **Groves, M. J.** 1978. Spontaneous emulsification. *Chem. Ind.* 12:417.

27. **Calder, J. A. and P. D. Boehm.** 1981. The chemistry of *Amoco Cadiz* oil in Aber Wrac'h. Proceedings of Symposium on Fate and Effects of the *Amoco Cadiz* Oil Spill. Centre National pour l'Exploitation des Oceans, Paris, pp. 149-158.

28. **Gundlach, E. R., P. D. Boehm, M. Marchand, R. M. Atlas, D. M. Ward and D. A. Wolfe.** 1983. The fate of *Amoco Cadiz* oil. *Science* 221:122-129.

29. **Grahl-Neilson, O.** 1978. The EKOFISK-BRAVO blowout: Petroleum hydrocarbons in the sea. In Proceedings of the Conference on Assessment of Ecological Impacts of Oil Spills. American Institute of Biological Sciences, Washington, DC, pp. 476-487.

30. **Grose, P. J. and J. S. Mattson.** 1977. The *Argo Merchant* oil spill: A preliminary scientific report. NOAA, Boulder, CO.

31. **Levy, E. M.** 1971. The presence of petroleum residues off the east coast of Nova Scotia, in the Gulf of St. Lawrence, and the St. Lawrence River. *Water Res.* 5:723-33.

32. **Boehm, P. D. and D. L. Fiest.** 1982. Subsurface distributions of petroleum from an offshore well blowout: The IXTOC-I Blowout, Bay of Campeche. *Environ. Sci. Technol.* 16:67-74.

33. **Clark, R. C., Jr., and W. D. MacLeod, Jr.** 1977. Inputs, transport mechanisms, and observed concentrations of petroleum in the marine environment. In D. C. Malins, ed., *Effects of Petroleum on Arctic and Subarctic Environments and Organisms,* Vol. I— Nature and Fate of Petroleum. Academic Press, New York, pp. 91-223.

34. **Morris, B. F. and J. N. Butler.** 1973. Petroleum residues in the Sargasso Sea and on Bermuda beaches. Proceedings of 1973 Joint Conference on Prevention and Control of Oil Spills. American Petroleum Institute. Washington, DC, pp. 521-529.

35. **Conover, R. J.** 1971. Some relations between zooplankton and bunker C oil in Chedabucto Bay followed the wreck of the tanker *Arrow. J. Fish. Res. Board Can.* 28:1327-1330.

36. **Parker, C. A.** 1970. The ultimate fate of crude oil at sea; uptake of oil by zooplankton. Cited by R. E. Jordan and J. R. Payne. 1980. *Fate and Weathering of Petroleum Spills in the Marine Environment.* Ann Arbor Science Publishers, Ann Arbor, MI.

37. **Lannergren, C.** 1978. Net and nanoplankton: Effects of an oil spill in the North Sea. *Bot. Mar.* 21:353-356.

38. **Johansson, S.** 1980. Impact of oil on the pelagic ecosystem. In J. J. Kineman, R. Elmgren, and S. Hansson, eds., *The Tsesis Oil Spill.* NOAA, Boulder, CO, pp. 61-80.

39. **Parker, C. A., M. Freegarde, and C. G. Hatchard.** 1971. The effect of some chemical and biological factors on the degradation of crude oil at sea. Cited by R. E. Jordan and J. R. Payne. 1980. *Fate and Weathering of Petroleum Spills in the Marine Environment.* Ann Arbor Science Publishers, Ann Arbor, MI.

40. **Johansson, S., U. Larsson and P. D. Boehm.** 1980. The "Tsesis" oil spill: Impact on the pelagic ecosystem. *Mar. Pollut. Bull.* **11**:284-293.

41. **Gundlach, E. R. and M. O. Hayes.** 1978. Vulnerability of coastal environments to oil spill impacts. *Marine Technological Society Journal* **12**:18-27.

42. **d'Ozouville, L., S. Berne, E. R. Gundlach and M. O. Hayes.** 1981. Evolution de la pollution du littoral Breton par les hydrocarbures de l'Amoco Cadiz entre Mars 1978 et Novembre 1979. Proceedings of Symposium on Fates and Effects of the *Amoco Cadiz* oil Spill. Centre National pour l'Exploitation des Oceans, Paris. pp. 55-78.

43. **Long, B. F. N., J. H. Vandermeulen and T. P. Ahern.** 1981. The evolution of stranded oil within sandy beaches. Proceedings of 1981 Oil Spill Conference. American Petroleum Institute, Washington, DC, pp. 519-524.

44. **Long, B. F., J. H. Vandermeulen and D. E. Buckley.** 1981. Les processus de la migration du pétrole échoue dans les estrans sableux: Contamination des nappes phreatiques. Proceedings of Symposium on Fates and Effects of the *Amoco Cadiz* Oil Spill. Centre National pour l'Exploitation des Oceans, Paris, pp. 79-94.

45. **Mackay, D.** 1981. Fate and behaviour of oil spills. In J. B. Sprague, J. H. Vandermeulen and P. G. Wells, eds., *Oil Dispersants in Canadian Seas—Research Appraisal and Recommendations.* Environment Canada, Toronto, pp. 7-27.

Chapter 20

Overview of the Influence of Dredging and Dredged Material Disposal on the Fate and Effects of Sediment-Associated Chemicals

Richard K. Peddicord

INTRODUCTION

Approximately 325 million m^3 of sediment are dredged annually for navigation purposes in the United States from 40,000 km of waterways and approximately 500 commercial harbors. An appreciation of the magnitude of this volume may be obtained by attempting to visualize the Super Dome stadium in New Orleans. The Super Dome is a 27-story-high inverted "gravy boat" with a 5.3 ha base, and encloses a volume of about 5.3 million m^3. If we call this volume one Super Dome Unit (SDU), then the volume of sediment dredged annually is more than 60 SDUs. Another perspective is gained by considering that the Mississippi River annually discharges approximately 272 million m^3 of sediment to the ocean. About 46 million m^3 of dredged material are disposed annually in the ocean, and the remaining 279 million m^3 are placed in upland sites, inland waters, and estuaries.

All dredging and disposal in waters of the United States is regulated by the Corps of Engineers on a predictive, project-by-project basis. Even though many projects must be dredged annually, each one is evaluated every time it is proposed for dredging. Many projects involve only small volumes of sediment and are regulated under various general permits. In contrast, many projects involve thousands to several million cubic meters in a single operation, and the potential environmental impact of each of these is individually evaluated according to the applicable regulations before the dredging is allowed. While records are not kept in such a way as to allow determination of the total number of dredging permit applications received or permits issued annually, it is clear that at least several thousand dredging projects receive some degree of individual predictive environmental evaluation each year by the Corps of Engineers.

Projects range in size from lows of a few hundred cubic meters or less at small marinas, to highs of several million cubic meters in large navigation channels, to perhaps a few hundred million cubic meters in proposed major expansions of several large harbor systems. Over this range it is meaningless to calculate an average size for a dredging project, but many are in the 100,000 to 1,000,000-m^3 range. Some projects require dredging only once in 20 years or so, but many projects, even very large ones, must be dredged annually because of high sedimentation from spring runoff or shifting bars in major rivers or some coastal areas. Regardless of frequency, each project is evaluated every time it is dredged. The actual dredging operation for small projects may require only a few days, and often is not conducted on a round-the-clock basis. Large projects may involve several months or more of continuous dredging, interrupted only by occasional mechanical problems. The proposed major harbor expansions would require several years of essentially continuous dredging and disposal.

OPERATING CHARACTERISTICS OF
MAJOR DREDGE TYPES

Mechanical Dredges

The grab, bucket, or clamshell dredge consists of a bucket or clamshell operated from a crane or derrick mounted on a barge. It is used extensively for removing relatively small volumes of material (i.e., a few tens or hundreds of thousands of cubic meters) particularly around docks and piers or within other restricted areas. The sediment is removed at nearly its in situ density and usually placed in barges or scows for transportation to the disposal area.

Most of the sediment dispersion by a typical clamshell operation is the result of sediment resuspension occurring when the bucket impacts on and is pulled off the bottom. Also, because most buckets are not covered, the surface material in the bucket and the material adhering to the outside of the bucket are exposed to the water as the bucket is pulled up through the water column. When the bucket breaks the water surface, turbid water may spill out of the bucket or may leak through openings between the jaws. In addition to inadvertent spillage of material during the barge loading operation, turbid water in the barges is often intentionally overflowed (i.e., displaced by higher density material) to increase the barge's effective load. Sediment is also dispersed in the water column at the disposal site as the material descends to the bottom.

Hydraulic Pipeline Dredges

The hydraulic pipeline dredge is the most commonly used dredge in the United States. With this type of dredge, a rotating cutter at the end of a ladder usually excavates the bottom sediment and guides it into the suction pipe. The excavated material is picked up and pumped to a designated disposal area through a pipeline as a slurry with a typical solids content of 10 to 20% by eight.

Most of the sediment resuspension by a hydraulic pipeline dredging operation (exclusive of disposal) is usually found in the vicinity of the cutter. The levels of turbidity are directly related to the type and quantity of material cut but not picked up by the suction. At open-water disposal sites, a suspended sediment plume of variable size and concentration extends away from the site during the discharge operation.

Hopper Dredges

In those areas characterized by heavy ship traffic or rough water, a hopper dredge will probably be used. Hopper dredges are self-propelled seagoing ships with the molded hulls and lines of ocean vessels. During a hopper dredging operation, as the dredge moves forward, the bottom sediment is hydraulically lifted from the channel bottom through a draghead, up the dragarm (i.e., trailing suction pipe), and temporarily stored in hopper bins in the ship's hull. Most modern hopper dredges have one or two dragarms mounted on each side of the dredge and have storage capacities ranging from several hundred to over 9,000 m³. The hoppers are either emptied by dumping the dredged material through doors in the bottom of the ship's hull or by direct pumpout through a pipeline. Resuspension of fine-grained maintenance dredged material during hopper dredge operations is caused by the dragheads as they are pulled through the sediment, turbulence generated by the vessel and its prop wash, overflow of turbid water during hopper filling operations, and dispersion of dredged material during open-water disposal.

DISPOSAL OPTIONS AND ASSOCIATED
GEOCHEMICAL CONDITIONS

Aquatic Disposal

Aquatic or open-water disposal is probably the most familiar dredged-material disposal option to most people, although probably no more than half the material dredged is disposed in this manner. In rivers, aquatic disposal usually occurs in long, narrow disposal sites adjacent to the channel being dredged. These sites may be permanently designated, but are usually created wherever and whenever dredging is required. In large bays, lakes, and estuaries, dredged material from several different dredging sites may be transported to a single disposal site, which may be used on a more or less continuous basis. Specification of aquatic disposal sites in inland waters and estuaries must comply with the dredged-material disposal guidelines of Section 404 of the Clean Water Act. Dredged-material disposal sites in the ocean are formally designated at fixed locations pursuant to the applicable criteria of the Marine Protection, Research and Sanctuaries Act. There are approximately 130 such ocean disposal sites for dredged material in United States coastal waters. These are mostly small, shallow sites nearshore. For example, 48% of them are less than 0.5 square nautical miles, and only 8% are greater than 4.0 square nautical miles. Approximately 54% are within 3 nautical miles of shore, and only 5% are greater than 12 nautical miles offshore. The mean depth at 70% of the sites is less than 20 meters, and only 14% of the sites are deeper than 100 meters (1).

The physicochemical characteristics of dredged material deposited in non-dispersive aquatic disposal sites are virtually identical to undisturbed sediment in place in waterways. It remains saturated with water, and since in the dredging process sediment is slurried with only 4-6 additional volumes of water usually containing less than 10-12 ppm dissolved oxygen while the oxygen demand of most fine-grained sediments is several times this per hour; such dredged materials remain anoxic and near neutral in pH at nondispersive aquatic disposal sites (2).

Upland Disposal

Upland disposal usually involves the containment of dredged material behind dikes on land or above the water level so that the surface of the dredged material dries out. These sites may be built on land or in the water with emergent dikes, and they range from a few to several thousand hectares in size. Most are used over a period of years until they are filled. As the dredged material dries, the surface forms a crust that thickens over time until the top meter or more of material resembles normal soil in moisture content. As the material dries, oxygen penetrates it, and reduced compounds become oxidized. This can result in substantial decreases in pH in some sediments, especially those high in sulfur and low in basic buffering capacity. These dry, oxic, acidic soil conditions constitute the most drastic change in physico-chemical conditions for dredged material, and the release and biological availability of contaminants can be very much greater than from the same sediment if it were in a water-saturated, anoxic, neutral condition. Substantial increases in release and bioavailability of contaminants, especially metals, have been demonstrated under upland disposal conditions for freshwater, estuarine, and marine sediment (3, 4). This can be particularly important since upland disposal sites can be adjacent to productive upland or aquatic habitat, often are recolonized by a variety of plants and animals, and become productive habitat themselves, or they may overlie groundwater aquifers. Upland disposal has sometimes been viewed as a universal solution to problems with contaminated dredged material because it removes the material from the aquatic ecosystem and contains it within dikes. However, the very fact that it is removed from the water and consequently can undergo radical geochemical changes makes upland disposal an environmentally unsound option with those materials whose contaminant bioavailability would be enhanced under these conditions, unless special site selection and management actions are taken.

Other Disposal Options

Dredged material also has a variety of productive or beneficial uses when its chemical and engineering properties are suited to the intended purpose and the site conditions. These uses include beach nourishment, strip mine reclamation, enhancement of marginal farmlands, development of fastland for industrial, municipal, or recreational purposes, and creation of upland or wetland habitat. Many of these uses require dredged material relatively low in contaminants, although not necessarily contaminant-free. Any use in which the dredged material dries will result in geochemical conditions similar to upland disposal, and any use in which the material remains water-saturated will result in geochemical conditions similar to aquatic disposal. With this in mind, the suitability of a particular dredged material for a specific project can be determined before the project is initiated.

CHARACTERISTICS OF DREDGED MATERIAL

Dredged material is sediment that has accumulated in places where it is not wanted and has all the characteristics of geologically similar local sediments that are not interfering with navigation. It may be sand with essentially no silt, clay, or organic matter, and thus essentially no capacity to sorb contaminants. This is the case in the outer reaches of most coastal channels and in quickly shoaling areas of major rivers. At the other extreme is the very fine-grained sediment consisting mostly of silts and clays characteristic of most inner harbor areas. The in situ water content of these sediments is approximately 50% and may be even higher in sediments accumulated in channels where they are frequently resuspended by ship turbulence. Typical fine-grained sediments are composed of 92% to 98% or more mineral solids and contain 2% to 8% organic matter, with a few containing approximately 20% organic matter.

Chemicals may be associated with sediments in a number of ways. All the natural elements, including toxic heavy metals, may be associated with natural sediments as (1) part of the mineralogical structures of the sediment particles, (2) as complexes with reducible iron and manganese coatings on mineral particles, (3) associated with organic material or sulfides in the sediment, (4) ionically bound to the mineral particles, and (5) dissolved in the interstitial water (5). The general pattern is that the environmental availability of metals associated with sediments increases in the order listed, but the amount of metal incorporated in these ways decreases in the same order.

Sediments also act as a sink for both metals and organic chemicals introduced to waterways directly or indirectly by human activities. Anthropogenic metals may become associated with sediments in all the ways mentioned in the preceding paragraph except the mineralogical structure of the particles, and it is practically impossible through chemical analysis to distinguish natural from anthropogenic metals in a sediment.

Nonpolar organic compounds also associate with sediment in several compartments, but by different mechanisms. Natural nonpolar compounds tend to associate with the dissolved or particulate organic carbon present in the interstitial water or sediment (6, 7). Man-made organic chemicals behave similarly. Although the natural concentration of such chemicals in sediment is zero, a number of persistent synthetic organic compounds have almost ubiquitous distribution in the ecosphere and are present at low levels even in sediments in pristine areas far removed from any but the most indirect sources of input. In contrast, harbor sediments are much more directly exposed to a variety of anthropogenic inputs of both natural and synthetic organic compounds and may contain quite high concentrations. This is true of sediments in and out of navigation channels in such areas.

Because dredged material is sediment, technically sound procedures for evaluating in situ contaminated sediment should be applicable to dredged material disposed in the aquatic environment, where its geochemical characteristics remain unchanged. Dredged-material disposal under upland conditions must be evaluated using techniques that take into account the different geochemical conditions. Modifications of traditional agronomic soil evaluative techniques have potential applications for this purpose.

INFLUENCE OF DREDGING AND DISPOSAL ON SEDIMENT/CONTAMINANT INTERACTIONS

Resuspension and Water Column Interaction

Only with mechanical dredges (and hopper operations with overflow) is there substantial resuspension of sediment at the dredging site, and its chemical behavior is similar to that of the suspensions at aquatic disposal sites caused by all types of dredges. Although the turbidity plumes resulting from aquatic disposal operations may be highly visible, they usually contain suspended solids concentrations in the range of tens to hundreds of milligrams solids per liter, extend less than 1,000 to 2,000 m from the discharge point, and disappear shortly after cessation of disposal. In all cases, the majority of the dredged material goes directly to the bottom, and only 1% to 3% of the total material remains in suspension (8-10).

The major geochemical change affecting contaminants on the suspended particles is the presence of oxygen. High concentrations of soluble reduced forms of iron and manganese are present in the anoxic deposited sediment. When these are resuspended and oxidized, they form precipitates and re-settle to the bottom. Other metals are co-precipitated with Fe and Mn, so that net release of soluble metals from the resuspended dredged material is low or absent. The pH is not altered substantially by oxygenation during resuspension; such changes occur only after several months of oxic conditions in upland disposal sites. There is often some release of ammonia. The rate of desorption of organics is increased because of the great sediment/water interface area, but since most organics of environmental concern in sediments are of low solubility, the dissolved concentrations have not been reported to reach high levels (9). Such compounds may be of more concern in relation to long-term exposure to deposited sediment.

Deposited Sediment at Aquatic Disposal Sites

Almost all the sediment released at nondispersive aquatic disposal sites is deposited on the bottom. Although it has been physically disturbed by the dredging and disposal process, its physicochemical condition is unchanged, and it remains geochemically very similar to the undisturbed sediment around the dredging site. Most metals are largely bound in the mineral structure and reducible coatings on the particles or ionically bound to charged silicate clay surfaces, where they are relatively unavailable for environmental interactions (11). Most organic contaminants are associated with the organic carbon in the sediment. Complete exchange between organic matter and the interstitial water is relatively slow (6, 7), as is the exchange of interstitial water with the overlying water, and most contaminants reach the water column via this route at a slow rate. Even so, exposure to contaminated sediment can result in increased contaminant concentrations in the tissues of organisms (12, 13) (Rodgers et al., this volume). Only the surface of the deposit interacts with the environment, and those contaminated sediments deeper than the depth of bioturbation (0.25 to 0.5 meters) play a negligible role in contaminant release. Even though the surface may be oxidized, pH changes are usually negligible as a result of the buffering capacity of the surface and underlying sediments, as well as the constant mixing, burial, and re-exposure of the top few centimeters of sediment by physical or biological action.

Upland Disposal Sites

Because of the great changes in oxidation and perhaps pH resulting from drying, dredged material in upland disposal sites has a different environmental behavior than the same material would have at an aquatic disposal site. The altered geochemistry can result in elevated contaminants in surface water runoff (14) and leachate (15) and in increased contaminant bioaccumulation in plant tissues (14, 16). Therefore, confined upland disposal cannot be regarded as a universal solution to problems of contaminated dredged material, although in many cases it is acceptable and in others can be made so by proper management techniques such as liming, liners, etc. This must be evaluated on a case-by-case basis, using techniques that take the prevailing geochemical conditions into account. Thus, techniques for evaluating in situ sediment may be applicable for predicting effects of dredged material at aquatic sites but may be wholly inappropriate for predicting impacts of the same material in an upland disposal site.

ENVIRONMENTAL PERSPECTIVE ON DREDGING AND DREDGED MATERIAL

Dredging removes sediments from channels that are chemically much like sediments in surrounding nondredged areas. Although these sediments may be highly contaminated with a variety of chemicals, the contaminants are not introduced or changed by the dredge, but are a result of the sediment acting as a sink for contaminants introduced to the water from other sources. Nor does the dredging and disposal process at aquatic sites alter the sediment geochemistry substantially. Therefore, aquatic disposal does not cause great changes in the rate and magnitude of contaminant release to the environment from sediment that has been dredged.

Dredging and disposal does relocate sediment, which sometimes involves removing contaminated sediment from a highly contaminated harbor and depositing it in a relatively uncontaminated area. In such cases, the unaltered release of contaminants from the dredged material may be of considerably greater environmental concern at the disposal site than it was at the dredging site. Predictions of environmental impact of dredging and disposal must therefore consider the entire operation in relation to its environmental setting. The degree of contamination in the sediment, water, and organisms at the proposed dredging site must be considered in assessing the influence of disturbing the sediment in the channel and the effects, if any, of removing the sediment from the channel and relocating it to another site. Sediment, water, and biota contamination in the disposal site vicinity must also be considered in assessing the potential for adverse effects from the disposal operation. The influence of the concentration, duration, and frequency of turbidity caused by dredging must be evaluated in relation to the same characteristics of turbidity caused by ship traffic in the shallower undredged channels, wind and storm resuspension over the surrounding undredged area, other activities such as shrimping (17), and nonsediment- associated contaminant sources. Effects of suspended and deposited dredged material at the disposal site must also be evaluated in light of all except the first of these considerations. It is also necessary to consider the temporal and spatial scale of the dredging and disposal operations in relation to the size of the water body in which they take place. The duration of the operation and the frequency with which the operation must be repeated are also important considerations.

CONCLUSIONS

- Dredged material is sediment and behaves environmentally like sediment, unless it is disposed under conditions that alter its geochemistry.

- Evaluative procedures technically appropriate to in-place sediment may be used for dredged material deposited at aquatic disposal sites.

- Dredged material in nonaquatic disposal sites where its geochemistry is altered must be evaluated by procedures that take those alterations into account.

- The potential environmental impacts of dredged-material disposal can be accurately evaluated using procedures that consider its geochemistry and the physicochemical and biological conditions at the disposal site, if results are interpreted in light of considerations such as those mentioned in the section on environmental perspectives on dredging and dredged material.

REFERENCES

1. **Pequegnat, W. E., L. H. Pequegnat, B. M. James, E. A. Kennedy, R. R. Fay and A. D. Fredericks.** 1981. Procedural guide for designation surveys of ocean dredged material disposal sites. Technical Report EL81-1. U.S. Army Engineer Waterways Experiment Station, Vicksburg, MS. NTIS Access No. AD-AO96-955.

2. **Gambrell, R. P., R. A. Khalid and W. H. Patrick, Jr.** 1978. Disposal alternatives for contaminated dredged material as a management tool to minimize adverse environmental effects-Synthesis report. Technical Report DS-78-8. U.S. Army Engineer Waterways Experiment Station, Vicksburg, MS. NTIS Access No. AD-AO73-158.

3. **Lee, C. R., B. L. Folsom, Jr. and R. M. Engler.** 1982. Availability and plant uptake of heavy metals from contaminated dredged material place in flooded and upland disposal environments. *Environment International* 7:65-71.

4. **Folsom, B. L., Jr., C. R. Lee and D. J. Bates.** 1981. Influence of disposal environment on availability and plant uptake of heavy metals in dredged materials. Technical Report EL-81-12. U.S. Army Engineer Waterways Experiment Station, Vicksburg, MS. NTIS Access No. AD-A112-279.

5. **Brannon, J. M., R. M. Engler, J. R. Rose, P. G. Hunt and I. Smith.** 1976. Selective analytical partitioning of sediments to evaluate potential mobility of chemical constituents during dredging and disposal operations. Technical Report D-76-7. U.S. Army Engineer Waterways Experiment Station, Vicksburg, MS. NTIS Access No. AD-AO35-247.

6. **Karickhoff, S. W. and K. R. Morris.** Pollutant sorption: Relationship to bioavailability. This volume, pp. 75-82.

7. **Podoll, R. T. and W. R. Mabey.** Factors to consider in conducting laboratory sorption/desorption tests. This volume, pp. 99-108.

8. **Barnard, W. D.** 1978. Prediction and control of dredged material dispersion around dredging and open-water pipeline disposal operations—Synthesis report. Technical Report DS-78-13. U.S. Army Engineer Waterways Experiment Station, Vicksburg, MS. NTIS Access No. AD-AO59-573.

9. **Jones, R. A. and G. F. Lee.** 1978. Evaluation of the elutriate test as a method of predicting contaminant release during open-water disposal of dredged sediments and environmental impact of open-water dredged material disposal. Technical Report D-78-45. U.S. Army Engineer Waterways Experiment Station, Vicksburg, MS. NTIS Access Nos. AD-A064-014 and AD-A061-710.

10. **Schubel, J. R., H. H. Carter, R. E. Wilson, W. M. Wise, M. G. Heaton and M. G. Gross.** 1978. Field investigations of the nature, degree, and extent of turbidity generated by open-water pipeline disposal operations. Technical Report D-78-30. U.S. Army Engineer Waterways Experiment Station, Vicksburg, MS. NTIS Access No. AD-A058-507.

11. **Brannon, J. M., R. H. Plumb, Jr. and I. Smith, Jr.** 1980. Long-term release of heavy metals from sediments. In R. A. Baker, ed., *Contaminants and Sediments*, Vol. 2—Analysis, Chemistry, Biology. Ann Arbor Science Publishers, Ann Arbor, MI, pp. 221-266.

12. **Adams, W. J.** Bioavailability and safety assessment of lipophilic organics sorbed to sediments. This volume, pp. 219-244.

13. **Rodgers, J. H., Jr., K. L. Dickson, F. Y. Saleh and C. A. Staples.** Bioavailability of sorbed chemicals to aquatic organisms—some theory, evidence, and research needs. This volume, pp. 245-266.

14. **Lee, C. R. and J. G. Skogerboe.** 1983. Predictions of surface runoff water quality from an upland dredged material disposal site. Proceedings of the International Conference on Heavy Metals in the Environment, Heidelberg, GE, Sept. 1983, pp. 932-935.

15. **Yu, K. Y., K. Y. Chen, R. D. Morrison and J. L. Mang.** 1978. Physical and chemical characterization of dredged material sediments and leachates in confined land disposal areas. Technical Report D-78-43. U.S. Army Engineers Waterways Experiment Station, Vicksburg, MS. NTIS Access No. AD-A061-846.

16. **Folsom, B. L., Jr. and C. R. Lee.** 1983. Contaminant uptake by *Spartina alterniflora* from an upland dredged material site—Application of a saltwater plant bioassay. Proceedings of the International Conference on Heavy Metals in the Environment, Heidelberg, GE, Sept. 1983, pp. 646-648.

17. **Schubel, J. R., H. H. Carter and W. M. Wise.** 1979. Shrimping as a source of suspended sediment in Corpus Christi Bay (Texas). *Estuaries* 2:201-203.

Chapter 21

Polynuclear Aromatic Hydrocarbons in the Elizabeth River, Virginia

R. J. Huggett, M. E. Bender and M. A. Unger

INTRODUCTION

The Elizabeth River is likely the most polluted estuary in Virginia. It is bordered by numerous civilian and military facilities. It has been a center of modern human activities for approximately three centuries. Chemical pollutants, both inorganic and organic, from numerous sources have collected in the sediments, and in the case of polynuclear aromatic hydrocarbons (PAHs), have reached dangerous and toxic levels.

Chemical data show that concentrations of PAHs exceed thousands of ppm in the bottom sediments from some areas of the river. It appears to have the highest concentrations of PAHs of any estuary in the world. Biological data derived from the collection and analysis of organisms from the river reveal that the natural community has been impacted and that the impacts are greatest in the areas where the PAHs are in highest concentrations.

Recently both state and private entities began contemplating the development of more and better port facilities in the Hampton Roads area. The Elizabeth River is included. The basic question then arises: Will deepening of the Elizabeth River, either during dredging or afterwards, create a situation whereby PAHs will impact the biota outside the river? Nearby, in the James River, seed oysters are produced that supply the majority of stock for the private oyster industry in Virginia. Just downstream in Chesapeake Bay, the majority of the bay's blue crabs spawn. All fish entering the bay spend some period of time near its mouth, and therefore near the Elizabeth River, to equilibrate to the bay's lower salinity. Clearly, this area is quite sensitive from an ecological standpoint, and potential threats to its inhabitants must be carefully evaluated.

A secondary but also important reason to study the PAHs in the Elizabeth River stems from the fact that these compounds are the major organic pollutants in the Chesapeake Bay proper [1]. As expected the levels are much lower than in the Elizabeth River but with the increasing human and industrial densities around the Bay, the concentrations are likely to increase. By studying the Elizabeth, information may be gained which will be applicable to better management of the Bay itself. In addition, knowledge gained will be of use elsewhere since PAHs are such widespread contaminants.

This paper is not intended to address the questions of biological impacts from dredging activities in the river, but rather describes some of the observations, both chemical and biological, which indicate that PAHs cause biological damage in estuarine systems. The results discussed deal with (1) bioavailability of PAHs in the system; (2) the distribution of contamination within the river as indicated by sediment loads; and (3) abnormalities observed in fish resident in the river. Other field studies are being conducted in the system, to determine PAH effects on benthic microbial and animal populations and these will be the subject of future papers.

METHODS

Oysters for transplanting were collected in September 1983 by dredge from two rocks located ~35 km upstream from the mouth of the Rappahannock River. Salinities at the time of collection ranged between 15-17%. The day after collection, the oysters were transplanted to five stations located 7, 12, 17, 19 and 24 km upstream from the mouth of the Elizabeth River (Fig. 21.1). At each station, seventy-six oysters were placed in steel mesh trays that were suspended from a stake ~0.3 m above the bottom. Salinities at the transplant locations ranged from 21% at the most downstream location to 12% at the upstream station. Resident oysters were sampled at the Hospital Point station on the day the transplants were initiated. At intervals of 7, 14, 28, 42 and 63 d, twelve oysters were removed from each sampling location for PAH analysis. Prior to extraction, the transplanted oysters were depurated for 24 h in the York River at Gloucester Point. This brief period of depuration was conducted to reduce the contribution of PAHs sorbed on solids in the digestive system of the oysters. After depuration, the oysters were cleaned by vigorously scrubbing each shell under flowing tap water with a brush and then frozen at $-20°C$. The frozen oysters were allowed to thaw slightly and shucked by opening through the bill. The tissues and fluids were then homogenized at high speed in a Virtis tissue blender and freeze dried. The dried samples were ground by mortar and pestle and a 10-g sample of pulverized tissue was spiked with internal standard (2,2′ binaphthyl) and soxhlet extracted for 24 h with ~350 ml of methylene chloride.

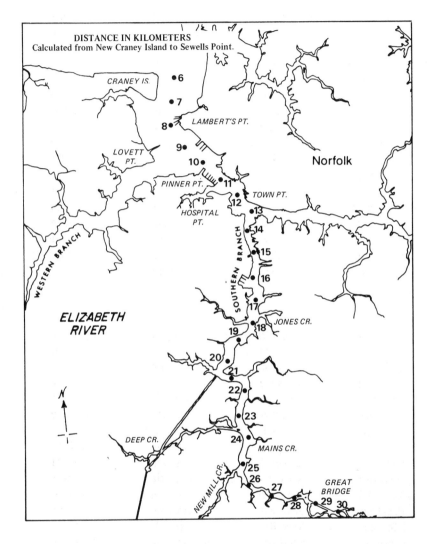

Fig. 21.1
Reference view of Elizabeth River with distance
from the river mouth shown in kilometers.

Sediments were collected by gravity cores and with a 0.1-m² Smith-MacIntyre grab. Samples were freeze dried or desiccated with a 9:1 mixture of anhydrous sodium sulfate and precipitated silica. The dried samples were ground with a mortar and pestle, spiked with internal standard, and soxhlet extracted for 24 h with methylene chloride.

Fish were sampled by bottom trawling with a 9-m (30 ft) lined semi-balloon trawl. At each station, two 5-min tows were made, one with and one against the tide. Fish collected in each tow were identified, examined for abnormalities, counted and weighed by species. Depending on abundance, 25-50 fish of a species were measured for total length in millimeters. Surveys were conducted during October, November and December 1983.

High concentrations of biogenic compounds are often found in environmental samples, necessitating a "clean-up" step to remove as many interferences as possible prior to gas chromatographic analysis. Extracts of oysters and sediments were reduced in volume with a rotary evaporator and "cleaned" by gel permeation chromatography on Biobeads S-X8 size-exclusion resin using methylene chloride as the elution solvent. Most biogenic molecules, which are generally larger than simple hydrocarbons, are un-retained by the resin and eluted before the molecules of interest [1].

Two fractions called G1 + G2 and G3 were collected. Aromatic hydro-carbons and many polar anthropogenic substances are eluted in the G3 fraction, which was then separated into two subfractions, G3.1 + G3.2, of increasing polarity using high pressure liquid chromatography. HPLC frac-tionation was carried out on a semi-preparative normal phase column with a cyano-amino phase bond to silica. The first, non-polar, subfraction was eluted with hexane, after which methylene chloride was programmed into the solvent mixture. Twenty-five percent methylene chloride in hexane was used to elute the aromatic fraction. Compound classes eluted in each fraction are given below:

G3.1   aliphatic (stored)

G3.2   polynuclear aromatic hydrocarbons (PAHs)
        polychlorinated biphenyls (PCBs)
        DDT
        DDD
        DDE
        mononitro-PAHS

The G3.2 fraction was analyzed by capillary column gas chromatography using flame ionization detection. A Varian 3700 gas chromatography temperature programmed from 75°C to 300°C at 6°/min was used for all analyses. Persilated glass capillary columns coated with 0.2 μm of SE52 were prepared in this laboratory according to the method of Grob and Grob [2]. Columns were approximately 25 m $\times$ 0.32 mm id and used Helium carrier gas at a linear flow of 27 cm/s. Data were collected and stored on a Hewlett Packard 3354B laboratory data system. Peak identification on the G3.2 fraction was done using the aromatic retention index of Bieri et al. [1]. Selected marker peaks from each chromatogram were identified by visual comparison with standard runs the same day. Using these markers, computer programs written in this laboratory used the stored data to assign each peak an aromatic retention index (ARI). The ARI is calculated by the formula:

$$ARI = \frac{T_x - T_{mp}}{T_{mf} - T_{mp}} \times 100 + ARI_{mp}$$

where T_x = retention time of peak x,

T_{mp} = retention time of the last marker preceding peak x,

T_{mf} = retention time of the next marker following peak x,

ARI = ARI defined for the last marker preceding x (ARI of the markers are defined as 000 = naphthalene; 100 = biphenyl; 200 = phenanthrene; 300 = pyrene; 400 = chrysene/triphenylene; 500 = perylene; 600 = benzo(ghi)perylene).

Using the calculated ARI, computer programs then identified peaks whose ARIs are known from previously injected standards and mass spectral identifications. Quantitation of these chromatograms was done using the internal standards added prior to extraction (2,2' binaphthyl). This method corrects automatically for extraction efficiency variations and losses of material during the analytical procedure. Recoveries from spiked samples usually range between 50 to 80%; duplicate extractions of the same sample typically yield results with 10 to 20%.

RESULTS

Sediments

Before presenting data on PAH concentration in the river's sediments, a brief statement as to their likely origin is in order. The Elizabeth River is bordered by three highly populated cities: Norfolk, Portsmouth and Chesapeake. Being a relatively deep and sheltered port, it has a major naval base, private shipyards, numerous coal and cargo handling piers, sewage treatment plants, and recreational marinas along its borders. Therefore, the Elizabeth River sediments could be expected to contain higher than natural background levels of PAHs originating from combustion sources and oil spills. In addition, since the turn of the century, five wood treatment facilities (which used creosote as a preservative for telephone poles, pilings and railroad ties) have, at one time or another, operated along the river [3]. Since creosote contains numerous PAHs at high concentrations, it is obviously a potential source for the compounds in sediment. In fact, grab or core samples from some parts of the river contain what appears to be (both visually and chemically, globular creosote inclusions, sometimes measuring several centimeters in diameter.

All but one of the wood treatment facilities have ceased operation, but creosote still seeps in to the river from contaminated soils at abandoned plant sites. Lu (3) analyzed PAHs in a 1-m core collected from the river near one of the abandoned facilities. She found a total aromatic hydrocarbon concentration of 400 mg/kg in the top 4.5 cm of the sediment and 13,000 mg/kg at a depth between 24.5 cm to 30.5 cm. At several depths in the sediment, she found unusually high concentrations of both total aromatic hydrocarbons and individual PAHs. She attributed these to documented spills from the wood treatment facility and by doing so, calculated a sedimentation rate of 2 cm/yr for this area. This agrees with a current estimate obtained by cesium-137 profiles. These observations suggest that past inputs of creosote to the river were much higher than those presently entering.

Bieri et al. (4) collected surface sediment from the river and analyzed them for PAHs. They found a gradient of increasing concentrations from the mouth to approximately 19 km upstream, the most upstream sample collected. In the study, over 300 aromatic compounds were identified by glass capillary gas chromatography-mass spectrometry. Most were PAHs. The concentrations of selected PAHs, which were determined in the top 2 cm of sediment, are presented in Table 21.1. Since then, other samples have been collected and analyzed to give a more complete picture of the distribution. The concentration of benzo(a)pyrene in the top 2 cm of bottom sediments are given in Figure 21.2. Progressively high concentrations in upstream areas confirm the trend noted by Bieri et al. (4). In addition where transects have been completed, the resulting data show that considerable variability in concentrations exist in any one segment of the river. This may be due to either different depositional patterns from one site in the river to another or perhaps past dredging activities that could have removed contaminated sediments.

Table 21.1   Concentrations of Selected PAHs in Surface Sediments from the Elizabeth River (ng/g dry weight)

Sample[a] code	Phe	Fla	Pyr	BbF	BaA	Chr	BFls	BaP	BeP	IPy	Bghi
06	86	290	250	89	93	160	210	90	84	21	28
07	110	300	260	82	79	140	150	57	53	n.d.	n.d.
08	180	460	410	160	190	340	590	280	260	100	76
09	200	840	880	470	320	470	1,100	420	480	360	350
10	130	320	440	160	150	290	870	380	380	260	230
11	410	860	960	290	350	590	1,100	480	520	200	110
12	580	1,500	1,400	530	560	970	1,700	740	740	280	170
13	670	1,900	1,800	720	880	1,400	2,600	1,200	1,200	550	500
14	750	2,200	2,000	900	840	1,400	2,700	1,200	1,200	660	560
15	760	2,200	2,800	1,200	1,000	1,700	4,100	1,300	1,700	940	620
16	2,300	5,500	4,600	2,000	1,900	3,200	4,900	2,000	2,100	730	340
17	950	3,800	3,600	1,500	1,500	3,500	6,300	2,800	2,600	1,400	750
18	710	2,600	2,000	760	940	1,700	2,400	970	1,000	200	84
19	25,000	42,000	28,000	12,000	11,000	19,000	17,000	6,300	8,700	2,100	1,600

[a] Numbers correspond to kilometers upstream from the mouth of the estuary.

Abbreviations: Phe = phenanthrene; Fla = fluoranthene; Phr = pyrene; BbF = benzo(b)fluorene; BaA = benz(a)enthracene; Chr = chrysene; BFls = benzofluoranthenes (j,b,k); BeP = benzo(e)pyrene; BaP = benzo(a)pyrene; IPy = indenopyrene; Bghi = benzo(ghi)perylene

FSBC—L*

Fish

Fish collected along the river show responses correlated to the PAH contamination levels of the sediments through changes in abundance and by increasing frequency of several gross abnormalities. Averages from the three sampling periods for biomass, total number of individuals, and abundance of selected species are presented in Table 21.2.

Table 21.2   Species Distribution of Fish Observed from Samples of the Elizabeth River Taken at Various Distances from The River Mouth. Data are means of three samples taken at October, November and December 1983.

| | Kilometers from the mouth | | | | | | | | | | |
	6.5	8.5	10.5	12.5	15	17	19	21.5	23.5	25.5	28
Biomass (kg)	34	26	21	13	9	7	4	11	8	3	4
No. of Ind.	1,335	1,200	1,125	705	380	320	245	495	430	795	325
No. Hogchoker[a]	50	140	145	240	50	70	25	165	210	40	60
No. Spot[b]	840	800	610	205	240	160	90	230	175	100	100
No. Gray Trout[c]	190	135	170	240	65	55	40	80	70	15	25
No. Croaker[d]	150	50	90	90	50	55	40	70	105	40	55

[a] *Trinectes maculatus*;
[b] *Leiostomus xanthurus*;
[c] *Cynoscion regalis*;
[d] *Micropongonias undulatus*.

Contamination of sediment by PAHs is most severe between 19 to 21 km upstream from the river mouth, as indicated from Figure 21.2. Depressions in biomass, total numbers of individuals, and abundance of selected species are indicated in this region of the river (Table 21.2). Even more dramatic is the increasing frequency with which abnormalities occur, as one progresses upstream into the more heavily contaminated regions (Table 21.3). Figure 21.3 graphically depicts these results for cataracts in croaker (*Micropogon undulatus*) and gray trout (*Cynoscion regalis*). For individual species, the frequency of cataracts increases with size. Figure 21.4 shows the percent occurrence of cataracts for three different size classes of gray trout, fish below 120-mm total length showed no incidence of cataracts. A similar increase in the occurrence of cataracts with increasing size was noted for croaker.

Chemical analysis for PAHs was performed on the flesh of selected fishes from the October sampling period and only trace (< 1 ng/L) levels of PAHs were detected. Analysis of other organs (e.g., liver) and for metabolites is planned.

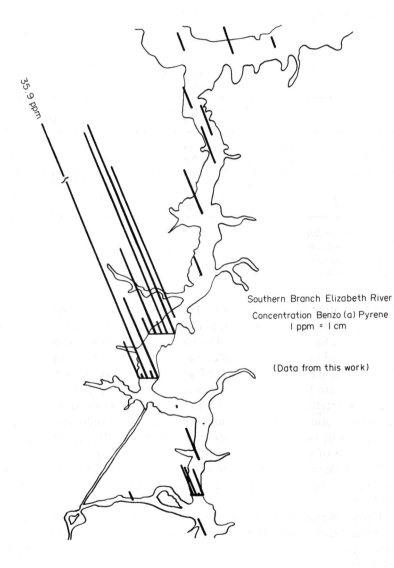

Fig. 21.2
Surface sediment concentrations (mg/kg) of benzo(a)pyrene
along the Elizabeth River.

Table 21.3   Percentage of Fish Showing Gross Abnormalities from
                  Exposure to Contaminants in the Elizabeth River.
                  Data are means of three samples taken
                  October, November and December 1983.

				Kilometers from the mouth						
	6.5	8.5	10.5	12.5	15.0	17	19	21.5	23.5	25.5	28
Fin Erosion											
Hogchoker	0.7	0	0	0.4	1.4	5.5	4.3	11.2	1.9	0	0.5
Toadfish[a]	0	0	11.0	5.0	0	11.5	30.1	26.3	25.0	0	0
Cataracts											
Spot	0	0	0.1	0	3.0	0.8	9.6	6.0	0.2	0.3	0
Gray Trout	0.2	0	0	0.8	1.0	1.8	3.5	14.0	21.0	2.5	7.5
Croaker	3.3	1.4	1.5	2.2	4.5	7.9	15.8	15.9	18.1	2.5	5.6

[a] *Opsanus tau.*

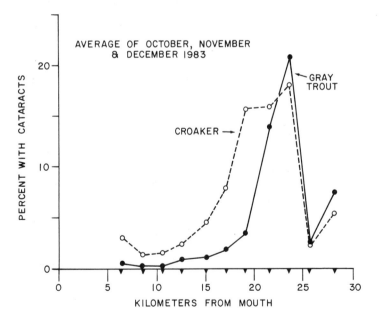

Fig. 21.3
Average occurrence of cataracts in croaker and gray trout
from stations along the Elizabeth River.

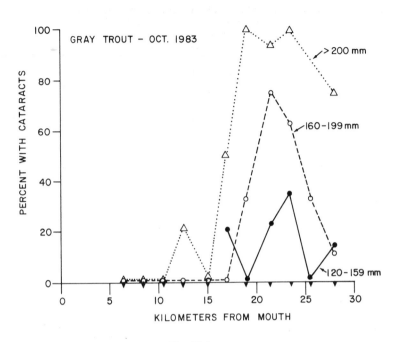

Fig. 21.4
Percent occurrence of cataracts for three size classes
of gray trout taken from the Elizabeth River.

Oysters

The bioavailability of PAHs in the river is indicated by their accumulation in transplanted oysters (Table 21.4 and Figure 21.5). Uptake was rapid, reaching 27 μg/g of total resolved aromatics within 1 week at the station located 17 km upstream.

Table 21.4   Total Resolved Aromatic Hydrocarbons in Oysters
Taken from the Elizabeth River.
Data are expressed in μg/g dry weight.

Station (km)	Exposure in weeks				
	1	2	4	6	9
7	1.9	6.1	—	6.8	13.9
12	5.5	7.4	16.2	8.8	20.5
17	27.0	31.0	57.3	31.8	60.2
19	19.3	25.7	25.8	22.5	36.3
24	3.2	7.8	11.7	7.0	16.5

Figure 21.5 shows the accumulation of total resolved aromatics at each station over the 63-d exposure period.

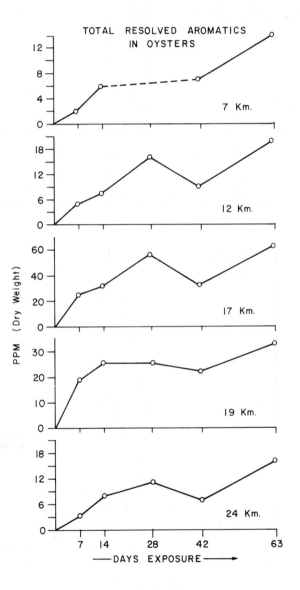

Fig. 21.5
PAH uptake by oysters along the Elizabeth River
(μg/g-dry weight).

Uptake at most of the stations appeared to plateau between 2 to 4 weeks. Concentrations decreased at most stations after exposures of 6 weeks and then increased again. Analysis of the data for individual compounds shows considerable fluctuation in levels with time, particularly at some locations. These two observations indicate that some variability is present in the source of PAHs for the oysters in the system. PAH residues as a function of distance in the river are shown in Figure 21.6.

Table 21.5   Unresolved Complex Mixture in Oysters
Taken from the Elizabeth River.
Data are expressed in μg/g dry weight.

| | Exposure in weeks | | | | |
Station (km)	1	2	4	6	9
7	10.1	41.4	—	25.8	91.7
12	21.5	35.3	95.6	40.2	118.9
17	51.3	127.0	223.5	104.0	150.5
19	37.4	103.0	144.5	51.8	159.7
24	6.7	26.5	44.5	16.6	59.4

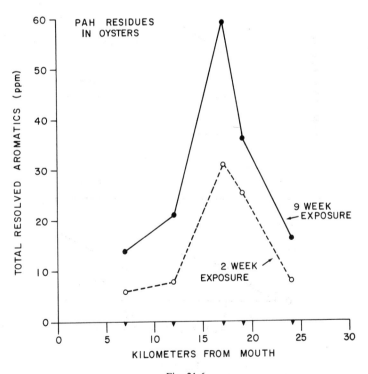

Fig. 21.6
Total resolved aromatics in oysters a function of distance
from the mouth of the Elizabeth River (μg/g-dry weight).

The highest residues are found in oysters located 17 km upstream from the river mouth, whereas the highest sediment burdens are found 2 to 3 km further upstream. However, the 17-km site is located near the only operational wood treatment on the river, and runoff from the plant site may be reponsible for the higher residues observed.

The most consistently abundant compounds identified were benzofluoranthene, benzo(a&b)fluorene, benzo(a)pyrene, fluoranthene, pyrene, benz(a)-anthracene and chrysene/triphenylene. Concentrations of benzo(a)pyrene ranged from 2 μg/g at the 17-km site to 0.1 μg/g near the river mouth.

DISCUSSION

Laboratory experiments conducted with spot (*Leiostomus xanthurus*) and contaminated sediments from the Elizabeth River have shown that relatively brief exposure (8 d) to contaminated sediment can induce many pathological changes in fish. Hargis et al. (5) observed: (1) integumental lesions within 8 d after exposure began leading to severe fin and gill erosion; (2) hematocrits were significantly reduced; (3) pancreatic and liver alterations were observed in some specimens; and (4) opacity of the eyes was observed in fishes exposed to contaminated effluent from the sediment exposure tank. Weeks and Warinner (6) found that the phagocytic efficiency of macrophages from spot and hogchokers (*Trinectes maculatus*) taken from the Elizabeth River was markedly reduced when compared to fish from control locations. The macrophage phagocytic activity of Elizabeth River fish returned to normal after fish were held in clean water for several weeks, indicating that the decreased activity was related to exposure to Elizabeth River pollutants.

The studies discussed above support our field observations that fishes in the Elizabeth River are severely stressed because of contamination of river water by pollutant discharges into the river and PAH contaminated sediments. Observations from the field studies that occurrence of cataracts is related to the size of the individuals are at present unexplained. However, if we assume that size is an indication of length of residence and hence exposure duration, this simple hypothesis could be presented as a possible explanation. In addition, physiological changes related to the uptake and/or metabolism of PAHs may be occurring as the fish grow thus making them more susceptible to the formation of cataracts.

Additional studies are needed to define the sediment contamination levels necessary to cause effects. As mentioned in the introduction field studies are in progress that address this question for benthic animals. Longer-term exposures with fishes to graded contamination levels in sediments will be required to refine estimates of chronic toxicity levels. In addition, questions remain as to the sources of PAHs accumulated by oysters. As well, the biological availability of the PAHs in the Elizabeth River may be considerably different than PAHs only sorbed to particulate matter.

Acknowledgments—The help of our colleagues must be acknowledged: H. Slone, J. Greene and D. Hunt for assistance in chemical analysis; and J. Colvocoresses, W. J. Hargis, Jr. and J. E. Warinner for help in collecting and examining the fishes. Financial assistance for this program was provided by the Commonwealth of Virginia, the Coastal Energy Impact Program through the State Council on the Environment, and the U.S. Environmental Protection Agency.

REFERENCES

1. **Bieri, R. H., P. deFur, R. J. Huggett, W. MacIntyre, P. Shou, C. L. Smith and C. W. Su.** 1981. Organic compounds in surface and sediments and oyster tissues from the Chesapeake Bay. Report on file with Chesapeake Bay Program, U.S. EPA Environmental Research Lab, Annapolis, MD.

2. **Grob, K. and G. Grob.** 1982. Preparation of inert glass capillary columns for gas chromatography—a revised comprehensive description. *J. Chromo.* **244**:197-208.

3. **Lu, M. Z.** 1982. Organic compound levels in a sediment core from the Elizabeth River of Virginia. M. S. Thesis. College of William and Mary, Williamsburg, VA. 157 pp.

4. **Bieri, R. H., C. Hein, R. J. Huggett, P. Shou, H. Slone, C. Smith and C. W. Su.** 1982. Toxic organic compounds in surface sediments from the Elizabeth and Patapsco rivers and estuaries. Report on file with Chesapeake Bay Program, U.S. EPA Environmental Research Lab, Annapolis, MD.

5. **Hargis, W. J. Jr., M. H. Roberts, Jr. and D. E. Zwerner.** 1984. Effects of contaminated sediments and sediment-exposed effluent water on an estuarine fish: Acute toxicity. *Marine Environ. Res.* **14**:337-354.

6. **Weeks, B. A. and J. E. Wariner.** 1984. Effects of toxic chemicals on macrophage phagocytosis in two estuarine fishes. *Marine Environ. Res.* **14**:327-335.

Chapter 22
Synopsis of Discussion Session 4:

Sediment-Associated Chemicals in The Aquatic Environment: Case Studies

K. Malueg (*Chairman*), R. J. Breteler, R. J. Huggett, R. K. Peddicord, D. Wolfe and D. Woltering

INTRODUCTION

In several of the past Pellston conferences, workshop participants had the opportunity to develop the hazard evaluation approach for assessing and predicting potential effects of chemical discharges to aquatic systems. Through these successive workshops evolved the now accepted tiered approach that provides for development of chemical fate and biological effects data of aquatic contaminants, thus allowing estimates of risks to be made after each tier of testing. Tests proceed from laboratory acute toxicity tests through short- and long-term chronic tests, bioaccumulation tests, microcosm and mesocosm studies, to field studies under actual use conditions. The initial steps are dependent upon tenuous laboratory to field extrapolations, recognizing that laboratory and microcosm studies and mathematic modeling, no matter how detailed, cannot ultimately determine the magnitude of hazard. In the final analysis, studies of natural field conditions are viewed as the ultimate step in ascertaining the environmental safety of chemicals and are essential to validate laboratory and modeling efforts.

This workshop, the sixth Pellston conference, focused on the role of sediments in determining the fate and effects of aquatic contaminants. Following the format adopted in earlier workshop proceedings, several case studies were presented that served to illustrate our current understanding of the interactions of physical, chemical, and biological processes that ultimately determine short- and long-term impacts of sediment-associated contaminants in aquatic systems.

A case study may be viewed as a detailed examination and evaluation of the pollution status in an actual field situation in order to understand pollutant transport and bioavailability processes and to make long-term risk assessments. Case studies are required to verify the validity and applicability of laboratory results to field situations, and to determine the adequacy of assessment models and regulatory schemes for protecting the environment. A case study should be closely coupled to related laboratory experiments designed to provide quantitative data under carefully controlled conditions.

Based on results of laboratory experiments and tests, appropriate models should be constructed to predict or describe expected behavior of contaminants under field conditions. The case study should then be designed to verify the adequacy of any assumptions used in the model, or to test hypotheses that are implicit in the model. The key elements of a case study depend totally on the specific model or hypothesis under examination. A case study must include detailed, quality-controlled chemical analyses and a carefully designed sampling effort, preferably covering various seasons and extreme weather conditions (i.e., influence of environmental variability). Thus, the case study is an integral part of an interactive process that includes laboratory studies, modeling, and field studies. The initial modeling effort should be followed by field validation through at least one and preferably more additional sample collection and analysis efforts until confidence is achieved in the developing model.

The session participants agreed that the usefulness of a case study depends on the degree of effort spent prior to initiation of the study. A set of attainable objectives must be clearly defined and a study plan designed to meet these objectives. Failure to do so often results in substantial loss of information and inability to properly model and interpret the chemical and biological processes. However, a carefully designed case study allows us to compare predicted effects against real world events. Case studies are essential for gauging our ability to accurately assess the impact of environmental insults on ecosystems and human health.

Within the objective of this session, the participants critically examined several case studies presented during the symposium. In addition, supplemental case studies were discussed where appropriate to provide information to support or refute questions identified as important by the session work groups.

SUMMARY OF PRESENTATIONS

Three papers specifically dealing with case studies were presented at the workshop. Summaries of these papers are provided here as part of the synopsis of this session; for details, the reader is referred to each author's paper.

Oil Spills

Using observations and examples from pertinent oil spills and from laboratory evidence, Wolfe [1] discussed interactions and transformations between petroleum and either suspended particulate matter or sediments in the marine environment. Numerous interactions may occur, depending on the specific location and the circumstances surrounding a spill. Suspended particulate material may become entrained in floating oil or mousse; conversely, dissolved or finely dispersed oil may be sorbed onto suspended particulate material. Weathered oil particles may be generated through the processes of emulsification and photo-oxidation. Oil may also be ingested by zooplankton and become incorporated into fecal pellets. Oil slicks that come ashore are entrained to varying degrees in the intertidal beach sediment, depending on the specific characteristics of the beach. Once incorporated into intertidal or subtidal sediments, the oil may be re-introduced into the water column through erosion, resuspension, or dissolution.

In general, the interactions of oil with suspended particulates under spill circumstances are still poorly understood, probably because of the transient character of most spills and highly variable conditions under which spills occur in the marine environment. In addition, oil itself is a very complex chemical mixture, containing hundreds of compounds with different physical properties. This mixture undergoes changes in composition, both with evaporation and dissolution of the more volatile and soluble compounds, and as biodegradation and photo-oxidation proceed. These physical and chemical transformations strongly affect the character and magnitudes of interactions between spilled oil and suspended particulates.

By contrast, oil interactions with intertidal and subtidal sediments are better understood because oil persists in sedimentary regimes and is not as likely to undergo rapid transitions. Oil penetration into sediments depends on the coarseness of the sediment, the extent of physical mixing and, probably, the organic content of the sediments. Penetration increases as the grain size increases from sand to cobble to rocky beaches, and with this penetration, the ease of cleanup decreases. In sedimentary environments, such as sheltered tidal flats and salt marshes, the sediments are characterized by fine silts with high organic content. Here oil may be thoroughly incorporated into the sediments and persist for very long periods. The potential persistence of spilled oil thus can be estimated on the basis of coastal geomorphology and sediment characteristics. These factors are important both in spill contingency planning and in mitigative responses when a spill occurs.

PAHs in Sediments

Huggett (2) presented data that show that the biota of the Elizabeth River, an industrialized estuary of the Chesapeake Bay in Virginia, are being affected by sediments contaminated with polynuclear aromatic hydrocarbons (PAHs).

A brief description was given of the chemical sampling and analytical methodologies, involving soxhlet extraction of the sediments followed by gel permeation chromatography, high performance liquid chromatography, glass capillary gas chromatography, and gas chromatography mass spectrometry. Chemical analyses of bottom sediments were then presented. The concentrations of PAHs in the sediments increased from the mouth of the river to areas upstream that were once or are still the sites of wood treatment facilities that utilized creosote. Photographic slides of liquid inclusions of creosote in the sediments were shown. These inclusions, buried as deep as 60 cm in the sediment, indicate that little chemical weathering or biological degradation of the PAH-laden material has occurred.

Biological data from laboratory bioassays involving the contaminated sediments and from field samplings were discussed. Similar effects were noted in these laboratory and field studies. Effects included fin erosion, skin lesions, eye cataracts, and gill damage to fish. Field observations showed maximum occurrences of these abnormalities where the sediment PAHs are at highest concentrations. The data also show that the biomass of finfish collected and the number of individuals were at a minimum where the sediment PAHs were highest.

The author also reported that biological availability of the sediment PAHs had been demonstrated by transplanting oysters from a clean environment to the Elizabeth River and measuring uptake. Concentrations of resolved aromatic hydrocarbons reached levels of 60 μg/l (dry weight) after 9 weeks. Concentrations of the unresolved complex mixtures reached 150 μg/l (dry weight) after the same time period.

Huggett then hypothesized that in severely polluted aquatic environments, organic contaminants in sediments may actually be in liquid rather than sorbed phases. This in turn would likely alter the partitioning characteristics and biological availabilities of the contaminants.

Dredged Material Disposal

Peddicord (3) provided an overview of dredging and dredged material disposal as they relate to ecological and human health concerns. Several hundred million cubic meters of sediment are dredged annually in the United States for navigation purposes, requiring predictive evaluation of potential environmental impacts.

Dredging is accomplished by mechanical or hydraulic dredges, and the excavated material is placed in aquatic or upland disposal sites. Aquatic disposal sites are usually small and relatively close to the dredging sites. Upland sites confine the dredged material behind dikes and may be as large as a few thousand acres. Peddicord described dredged material as sediment very similar to adjacent undisturbed sediment. It may range from uncontaminated to highly contaminated with a wide variety of chemicals, just as any other sediment.

The dredging and disposal process does not introduce contaminants into sediments, although it relocates sediments and any associated chemicals. If the material is disposed at an aquatic site, the sediment/contaminant interactions remain essentially unaltered. However, if upland disposal is utilized, the physico-chemical environment usually changes, and dramatic changes in sediment/contaminant interactions can occur as the material dries and oxidizes. Therefore, Peddicord identified the physico-chemical conditions at the disposal site as a major factor controlling contaminant behavior in dredged material. Since dredged material at aquatic disposal sites behaves like undredged sediment, technically sound sediment evaluation techniques may be applied to dredged material at aquatic sites. Evaluation of upland disposal sites requires the use of techniques that consider the physico-chemical conditions of terrestrial environments.

APPLICATIONS AND RESULTS OF CASE STUDIES

As mentioned in the introduction to this chapter, any model of the fate and effects of a pollutant in aquatic systems should be verified by field observations if possible. These observations serve not only to give robustness to the model predictions, but also to give direction for future laboratory experimentation. To be effective, case studies should be carefully focused on a priori hypotheses. Based on the results of these studies, hypotheses of the roles of sediments in the fate and effect of pollutants can be validated and subjected to further laboratory and field testing. During the workshop, a number of issues were raised repeatedly as basic concerns requiring field confirmation. We have listed below several of the fundamental questions that have received attention in field studies; we have also identified some of the pertinent observations that support or refute each question. The list is not all-inclusive and is intended only to illustrate the applicability of field observations for hypothesis testing and model verification.

1.   *Can suspended sediments play a major role in the transport of pollutants from the water column to the bottom and/or from one location in a system to another?*

Kepone in the James River, Virginia, has been shown to be associated with particulate matter. The major mode of transport of kepone in this river is via the movement of these suspended sediments. Because of the deposition of these materials in the zones of maximum turbidity of the river, highest bottom sediment concentrations of kepone were found approximately 50 km downstream from the source (4).

Data from oil spills show that petroleum hydrocarbons partition onto suspended particulate material, with partition coefficients approximating those measured in laboratory experiments (1, 5, 6). Sediment traps deployed in the vicinity of the TSESIS oil spill 1 week after the spill captured 30-60 mg of oil/m^2 day. Sedimentation of oil adsorbed on suspended particulates may have been augmented by settling of fecal pellets (7). At the *Amoco Cadiz* oil spill, a major transport pathway for oil into subtidal sediments was the seaward transport of oil sediments through successive resuspension from the beaches (8).

Long-term leaching of metallic and phenyl mercury from a point source upstream of Berry's Creek (a tributary of the Hackensack River, New Jersey) resulted in a distinct gradient of mercury concentrations in creek and river sediments from greater than 100 μg/g down to smaller amounts some 20 km downstream (9). It was demonstrated that mercury was predominantly associated with suspended organic particulates and subsequently was deposited in bed sediments. A similar situation has been documented by Breteler et al. (10) in the Niagara River, which for many years received large inputs of predominantly elemental mercury from two chloralkali plants. In this case study, mercury was absent in the high-flow, rocky river bed but was found in sediments of Lake Ontario in high depositional areas. The mercury discharge pattern could be clearly reconstructed by radiodating techniques.

2. *Are "pollutants" sorbed to bottom sediments biologically available?*

While it is impossible to state emphatically that all sediment-sorbed pollutants are biologically available, there are numerous observations from case studies which show that many are. It is unknown whether the pollutant desorbs onto the overlying or pore waters before entering the plant or animal or whether the biota accumulate the substances directly via ingestion. In any case, sediment-associated pollutants can be biologically available, although some may be so slightly available as to be of little concern (2, 4, 11-13).

Concentrations of petroleum hydrocarbons from the *Amoco Cadiz* remained elevated in intertidal fauna, including cultured oysters, 2 years after the hydrocarbon concentrations in the water column had declined to ambient levels. Hydrocarbon concentrations in the oysters paralleled the decreasing amounts in intertidal sediments 1 to 3 years after the spill (8, 14, 15). Similarly, PCBs are highly concentrated in the sediments of the Hudson River (16) and of New Bedford Harbor, Massachusetts (17). In both these cases, commercial fishing has been closed as a result of the PCB contaminations. Again, the specific mechanisms for bioavailability of the PCBs from the sediments to the organisms are not known. Near the sewage outfall in the

Southern California Bight, benthic organisms retained elevated concentrations of DDT residues in 1974, despite a substantial reduction of DDT input to the system after 1971. This contamination was attributed to the high DDT levels in the sediments [18].

In an extensive field study, Bryan [19] compared concentrations of As, Cd, Co, Cu, Mn, Pb, and Zn in a variety of estuarine organisms to concentrations of these metals in sediments of the Fal Estuary (Great Britain). This author found statistically significant correlations between sediment concentrations of As, Cu, and Zn and those in tissues of some species. Breteler et al. [9] and Giblin et al. [20] conducted a series of field studies at Great Sippewissett Marsh (Cape Cod, Massachusetts) contaminated with metal-containing sewage sludge. The long-term fate of seven heavy metals, their retention in the marsh sediments, and their availability to several salt marsh organisms were examined. These studies revealed that bioavailability of Cr, Fe, Hg, Pb, and Zn was low to absent, while tissue concentrations of Cu and Cd were significantly higher in animals sampled from the contaminated areas as compared to control marsh areas.

3. *Are interstitial waters involved in the biological availability of contaminants in sediments?*

Petroleum constituents are highly mobile in interstitial waters. In heavily oiled beach sediments, interstitial waters contain much higher concentrations of aromatic hydrocarbons than are present in the overlying water column, and beach sediments are cleansed by horizontal and vertical movements of water through the sediments. The importance of the mobility to bioaccumulation has not been demonstrated [1, 21]. Adams [12] reported on laboratory studies that demonstrated interstitial water was a key exposure route for some chlorinated organics in freshwater systems.

4. *Can pollutants in sediments impact the indigenous biota or impair their usefulness to man?*

Data presented at this conference on PAHs [2], kepone in the James River [11], and PCBs in the Hudson River [22] show that indeed sediment-associated contaminants can impact indigenous organisms and impede their usefulness to man. High levels of mercury in salt marsh sediments of Berry's Creek, Hackensack Meadowlands, resulted in diminished natural populations of fiddler crabs [9], while blue crabs and muskrats contain elevated levels of this metal, such that their consumption has been disadvised. In New Bedford Harbor, Massachusetts, all commercial and recreational fisheries have been banned by the State Department of Public Health because of the presence of high concentrations of PCBs and other heavy metals.

The incidence of fin erosion, skin tumors, and liver pathologies in fish observed near several municipal sewage outfalls such as Boston Harbor and several embayments within central and southern Puget Sound were linked by Malins (23) to the degree of sediment contamination. Specific causal mechanisms have not yet been defined for these observed biological effects.

5. *Are physical characteristics of sediments (e.g., grain size) important in sediment pollution concentrations?*

For sorption reactions that involve surfaces, sediments of finer grain sizes will normally contain higher concentrations of contaminants than coarser-grained ones with everything else being equal (24). In the New York Bight in the general vicinity of the sewage sludge dumpsite, sediment concentrations of metals and organic contaminants are strongly related to grain size and organic content. The areas of highest contaminants also exhibit a depauperate benthic fauna (25, 26).

KEY CONCLUSIONS RELATED TO CASE STUDIES
OF SEDIMENT SYSTEMS

- Thorough investigations of chemical impact on aquatic ecosystems have included the sediment compartment.

- Case studies are essential for understanding the relationships between sediments and the fate and effects of particular chemicals in aquatic ecosystems.

- Design of a case study requires careful integration of physical/chemical characterization, mathematical models, laboratory and field investigations.

- The specific investigations necessary to meet stated needs will be different for different case studies. Objectives must be carefully defined, applicable approaches/methods selected, and supporting data assembled in advance of developing a case study.

- A proper case study incorporates relevant sediment geochemistry; hydrodynamic, physico-chemical and biological conditions; and temporal and spatial elements of the system of interest.

- Results from case studies have been successfully integrated into existing models and have formed the basis for improved laboratory experimentation.

- Contaminated sediments in the field very often contain complex chemical mixtures. As it is difficult to differentiate toxic components in a field study, biological effects must often be expressed in terms of correlations rather than as demonstrated causal relationships.

- Case studies of aquatic systems have been underutilized both in testing of laboratory-derived hypotheses of chemical impact and in assessing the adequacy of regulations for environmental protection.

RECOMMENDATIONS

- Case studies should incorporate an interdisciplinary approach whenever possible.
- Case studies of contaminated sediment should be used to assess both ecological consequences (i.e., population and community-level effects) and human health concerns resulting from bioaccumulation in edible species.
- Case studies should be used to identify and to verify key environmental variables that control/influence fate and effects of chemicals (or chemical classes) sorbed to sediments (or sediment types). This information will help direct further laboratory studies and predictive modeling efforts.
- Field verification should be an essential component of a sediment hazard assessment scheme as it is for the conventional water column hazard assessment.
- Field investigations should differentiate between dissolved and particulate phases of chemical compounds in the water.
- Case studies should include the hydrodynamics of the study site because of its key role in sediment/water exchange reactions.
- One should capitalize upon spills and accidents because such field situations offer an opportunity to bridge the gap between laboratory-based predictions and observed effects on sediment systems.
- Carefully designed and controlled manipulations of aquatic ecosystems should be considered to assess the role of sediment on the potential impact of chemicals.
- Physical impacts (e.g., habitat alteration) of waste discharge and dredging operations should not be overlooked in case studies of sediment.
- When possible, case studies should include some assessment of recovery potential from the perspective of eventual "clean" sediment and from long-term response of the biota once chemical input has ended.

REFERENCES

1. **Wolfe, D. A.** Interactions of spilled oil with suspended materials and sediments in aquatic systems. This volume, pp. 299-316.
2. **Huggett, R. J., M. E. Bender and M. A. Unger.** Polynuclear aromatic hydrocarbons in the Elizabeth River, Virginia. This volume, pp. 327-341.
3. **Peddicord, R. K.** Overview of the influence of dredging and dredged material disposal on the fate and effects of sediment-associated chemicals. This volume, pp. 317-326.
4. **Huggett, R. J., M. M. Nichols and M. E. Bender.** 1980. Kepone contamination of the James River Estuary. In R. A. Baker, ed., *Contaminants and Sediments*, Volume 1. Ann Arbor Science Publishers, Ann Arbor, MI, pp. 33-52.

5. **Boehm, P. D.** 1980. Evidence for the decoupling of dissolved, particulate, and surface microlayer hydrocarbons in northwestern Atlantic continental shelf waters. *Marine Chemistry* **9**:255-281.

6. **Riley, R. C., E. A. Crecelius, D. C. Mann, K. H. Abel, B. L. Thomas and R. M. Bean.** 1980. Quantitation of pollutants in suspended matter and water from Puget Sound. NOAA Technical Memorandum ERL MESA-49, National Oceanic and Atmospheric Administration, Boulder, CO.

7. **Johansson, S.** 1980. Impact of oil on the pelagic ecosystem. In J. J. Kineman, R. Elmgren and S. Hansson, eds., *The TSESIS Oil Spill*. National Oceanic and Atmospheric Administration, Boulder, CO, pp. 61-80.

8. **Gundlach, E. R., P. D. Boehm, M. Marchand, R. M. Atlas, D. M. Ward and D. A. Wolfe.** 1983. The fate of *Amoco Cadiz* oil. *Science* **221**:122-129.

9. **Breteler, R. J., I. Valiela and J. M. Teal.** 1981. Bioavailability of mercury in several northeastern U.S. *Spartina* ecosystems. *Estuarine, Coastal, and Shelf Science* **12**:155-166.

10. **Breteler, R. J., V. T. Bowen, D. L. Schneider and R. Henderson.** 1984. Sedimentological reconstruction of the recent pattern of mercury pollution in the Niagara River. *Environ. Sci. Technol.* **18**:404-409.

11. **Bender, M. E. and R. J. Huggett.** 1984. Fate and Effects of Kepone in the James River Estuary. In E. Hodgson, ed., *Reviews in Environmental Toxicology*. Elsevier Science Publishers, New York, NY, pp. 5-50.

12. **Adams, W. J.** Bioavailability and safety assessment of lipophilic organics sorbed to sediments. This volume, pp. 219-244.

13. **Swartz, R. C.** Toxicological methods for determining the effects of contaminated sediment on marine organisms. This volume, pp. 183-198.

14. **Neff, J. M. and W. E. Haensly.** 1982. Long-term impact of the *Amoco Cadiz* crude oil spill on oysters *Crassostrea gigas* and plaice *Pleuronectes platessa* from Aber Benoit and Aber Wrac'h, Britanny, France. I. Oyster histopathology. II. Petroleum contamination and biochemical indices of stress in oysters and plaice. In *Ecological Study of the Amoco Cadiz Oil Spill*. Report of the NOAA-CNEXO Joint Scientific Commission. U.S. Department of Commerce, Washington, D.C., pp. 269-327.

15. **Seip, K. L.** 1984. The *Amoco Cadiz* oil spill-at a glance. *Mar. Pollut. Bull.* **15**:218-220.

16. **Turk, J. T.** 1980. Applications of Hudson River Basin PCB-transport studies. In R. A. Baker, ed., *Contaminants and Sediments*. Ann Arbor Science Publishers, Ann Arbor, MI, pp. 171-183.

17. **Weaver, G.** 1984. PCB contamination in and around New Bedford, Massachusetts. *Environ. Sci. Technol.* **18**:22A-27A.

18. **Young, D. R., D. J. McDermott and T. C. Heesen.** 1976. DDT in sediments and organisms around southern California outfalls. *J. Water Pollut. Control* **48**:1919-1928.

19. **Bryan, G. W.** 1983. Heavy metals in the Fal Estuary, Cornwall: A study of long-term contamination by mining waste and its effects on estuarine organisms. *Mar. Biol. Assoc. U.K., Occasional Publication No. 2.*

20. **Giblin, A. E.** 1982. Uptake and remobilization of heavy metals in salt marshes. Ph.D. thesis. Boston University, Boston, MA.

21. **Long, B. F. N., J. H. Vandermeulen and T. P. Ahern.** 1981. The evolution of stranded oil within sandy beaches. Proceedings of 1981 Oil Spill Conference (Prevention, Behavior, Cleanup). American Petroleum Institute, Washington, D.C., pp. 519-524.

22. **Horn, E. G., L. J. Hetling and T. J. Tofflemire.** 1979. The problem of PCBs in the Hudson River System. In W. J. Nicholson and J. A. Moore, eds., Health Effects of Halogenated Aromatic Hydrocarbons. *Ann. N.Y. Acad. Sci.* **320**:591-609.

23. **Malins, D. C., B. B. McCain, D. W. Brown, A. K. Sparks and H. O. Hodgins.** 1980. Chemical contaminants and biological abnormalities in Central and Southern Puget Sound. NOAA Technical Memorandum OMPA-2. National Oceanic and Atmospheric Administration, Office of Marine Pollution Assessment, Boulder, CO.

24. **Huggett, R. J.** 1981. The importance of natural variabilities in the total analytical scheme. *Biomedical Mass Spectrometry* **8**:416-418.

25. **Boesch, D. F.** 1982. Ecosystem consequences of alterations of benthic community structure and function in the New York Bight region. In G. F. Mayer, ed., *Ecological Stress and the New York Bight: Science and Management.* Estuarine Research Federation, Columbia, SC, pp. 543-568.

26. **Wolfe, D. A., D. F. Boesch, A. Calabrese, J. F. Lee, C. D. Litchfield, R. J. Livingston, A. D. Michael, J. M. O'Connor, M. Pilson and L. Sick.** 1982. Effects of toxic substances on communities and ecosystems. In G. F. Mayer, ed., *Ecological Stress and the New York Bight: Science and Management.* Estuarine Research Federation, Columbia, SC, pp. 67-86.

Session 5
Regulatory Implications

Chapter 23

Establishing Sediment Criteria for Chemicals—Regulatory Perspective

Gary A. Chapman

INTRODUCTION

Chemicals introduced into the environment in potentially toxic amounts include both those whose presence is almost solely a result of synthesis by man, as well as those natural elements (and a few compounds) released by man from geological sinks at rates considerably greater than normal. The fate and bioavailability of these synthesized chemicals or released elements are determined in large part by their physical-chemical properties and the medium into which they are released. They may exist as gases, liquids, or solids and may be discharged into the air, onto soil, or into water. Many of these chemicals are broken down into smaller, usually less-toxic fragments by chemical, biochemical, or physical processes. Many halogenated organic compounds are rather persistent and remain in the environment in free or complexed forms that are, at best, only slowly decomposed. Many metals are also present in contaminated sediments at concentrations higher than in natural sediments.

The environmental fate of many of the more persistent chemicals is ultimately deposition in aquatic sediments. Deposition of aerial contaminants by wet- and dry-fall mechanisms bring airborne chemicals to earth where they fall into water or onto the land. Precipitation runoff can lead directly to nonpoint source inputs from wet deposition as well as from recruitment of soluble and particulate-bound forms of chemicals previously deposited as dry-fall or applied directly to the soil by accident or design. Once introduced into a body of water, the chemicals occur bound to particulate material (either suspended or settled), in solution, or present in organic micelles or surface slicks.

In theory, all of these forms may be bioavailable via transport across biological membranes. This uptake may occur across respiratory or other dermal tissue, in digestive organs following ingestion of particulate matter (or contaminated biota), and through phagocytosis (e.g., at the protozoan level).

Legislation requires EPA to publish water quality criteria accurately reflecting the latest scientific knowledge of effects of pollutants in any body of water. These criteria are often used as guidance in establishing regulations. The development of water quality criteria has relied heavily on laboratory bioassay data on essentially pure chemical elements or compounds. This fact is due to emphasis on the use of clean dilution water, the introduction of chemicals from stock solutions, and a desire to test worst-case conditions (usually synonymous with maximum solubilization of the chemical). Through observation of tests in which chemicals precipitated or were bound to particulate material, it was rather evident that soluble forms were the more toxic; indeed, insoluble forms are frequently considered as nontoxic and not significantly bioavailable.

However, the relative bioavailability, toxicity, or re-solubility of forms bound to particulates of various types has not been demonstrated to any satisfactory degree. Thus, we are currently without any good mechanism for evaluating the biological significance of contaminated sediments. Yet there are certainly instances where contaminated sediments may be of considerable environmental interest: (1) discharge of effluents for which control is based upon effluent bioassay criteria that may maximize short-term uptake of soluble forms and ignore long-term uptake via ingestion, (2) the increasing tendency for water quality criteria to emphasize active forms of pollutants, (3) potential concentration of pollutants in small areas of sediment deposition thereby summating long periods of long-level discharges in one location, (4) similar concentrations resulting from accidental spills, and (5) intimate mixing of pollutants in sediments as a result of "safe" water column contamination with single chemicals at various times.

Toxic chemicals in sediments may partition between the particulates and the interstitial waters, but neither may be in equilibrium with the overlying water column. Thus, monitoring of water column concentrations may represent neither the concentrations found at the sediment/water interface nor in interstitial waters. Even if only the soluble forms of chemicals were toxic (bioavailable), it may be necessary to monitor the water associated with the sediment. In addition, if particulate-bound toxics are also directly available to organisms via ingestion of particulates, then additional safeguards must be considered.

For the regulatory agency there are a number of questions that need to be answered, but two major questions predominate: (1) Do we need sediment criteria to protect the environment? and (2) Do we know enough to set and use reasonable sediment criteria? The "need" question has both scientific and political aspects and is a blend of perception and fact. If there is a perception that sediment criteria are necessary and insufficient facts to either prove or disprove the need, then there will be pressure to develop sediment criteria; i.e., there is a perceived need.

Certainly there is current anecdotal information to suggest that sediment criteria in some form are needed. For example, fish, especially bottom-dwelling varieties, are showing high incidences of dermal tumors, disease, and histopathology in areas of contaminated sediments; contaminated sediments overlayed with clean water in laboratory bioassays can produce acute lethality in a variety of aquatic organisms; high levels of chemical residues are found in natural populations of aquatic organisms inhabiting areas of contaminated sediments; and benthic communities often appear to be sparse in regions of contaminated sediments. From a scientific viewpoint, we must admit that it is difficult to separate effects of sediment-bound chemicals from the effects of soluble chemicals in the water column. Similarly, it is difficult to determine whether effects are due to current exposure conditions or to conditions that were extant at some time in the past; e.g., is the effect due to what is in the sediment now or what was in the water last month? If sediment-bound chemicals produce significant environmental effects, can they attain significant levels in sediments without violating water column criteria; i.e., if we really controlled water pollution per se, could significant sediment contamination occur?

The most viable alternatives for establishing numerical sediment criteria are modeling approaches derived from water quality criteria, but water quality criteria exist for only a few chemicals. Are the majority of sediment-contaminating chemicals those for which we have current water quality criteria or are we dealing with different chemicals for which we need considerably more toxicity information?

Water quality criteria are derived from chemical concentrations producing lethality and adverse chronic effects such as poor reproduction and growth, but much of the current concern about sediment contamination is focused on tumor induction. Will typical water quality criteria or, by extrapolation, sediment quality criteria protect against oncogenicity?

In reviewing the various proposals for establishing sediment criteria, it appears that the more expedient and viable procedures entail extrapolation from water quality criteria and are based upon predictions of chemical concentrations in tissue or water derived from sediment:water:tissue partitioning theory. These procedures are limited by our knowledge of partitioning and may be less than protective inasmuch as they presume minimal importance of direct uptake of chemicals from sediment; furthermore, they do not address oncogenicity.

Finally, if reliable sediment criteria were in place (based on bulk chemistry of bottom sediments), how would the criteria be applied? Would they be used to prioritize clean-up sites? How would they be applied to point and nonpoint sources? Do we have sufficiently sophisticated fate models to apportion responsibility for sediment contamination? What are the treatment/disposal alternatives to sediment contamination?

The preceding discussion has merely outlined the technical problems of evaluating the effect of contaminated sediments and the questions facing those who would have to regulate to avoid significant sediment contamination. Where regulatory actions are mandated, the agency can do no more, or less, than the state of the art allows.

In waters that have received significant pollution with persistent chemicals, a potential problem with contaminated sediments now exists. In a report on PCB pollution in Puget Sound, Pavlou and Weston [1] suggest that sediment contamination, and presumed resulting biological effects, have occurred in the absence of water quality criteria violations. However, at most sites of sediment contamination and observed biological impact, there is no certainty that the effects seen in the field are due to sediment contamination alone or have resulted in the absence of significant water pollution.

Current adverse effects may be residual effects of past water quality problems, may be caused by current water quality problems (including violations of water quality criteria, interaction of several chemicals present at levels below water quality criteria, or presence of toxic amounts of chemicals for which no criteria exist), or may be due to a combination of sediment contamination and water quality problems. Stating that sediment has been contaminated to harmful levels in the absence of water quality criteria violations would be difficult to establish as fact without the existence of thorough historic water quality monitoring data. However, bioassays of contaminated sediments have shown convincingly that these sediments can cause toxic effects in the absence of water column pollution from any source other than sediment. What has not been demonstrated consistently are the relative toxicity/bioavailability of sediment-bound toxic materials and toxic materials desorbed from sediment into the water column/interstitial water. Evidence of the relative bioavailability from various sediment/water compartments shows considerable variability at best and could be interpreted as contradictory.

There is good agreement that soluble forms are bioavailable and considerable evidence that ingestion of contaminated organisms can be a major route of uptake. What is not well demonstrated is the role of sediment as a contaminant source for species at various biological levels in the food chain. Nevertheless, demonstrations indicating that contaminated sediments cause toxicity in laboratory tests and are associated with adverse effects in the field have led to serious proposals for the establishment of sediment criteria.

CRITERIA APPROACHES

The generally excellent report by Pavlou and Weston [1] suggests several approaches to establishing sediment criteria. Each approach is discussed by the authors, along with a listing of many of the problems associated with each approach. The four classes of approaches are:

1. Background Levels.
2. Water Quality Criteria.
3. Equilibrium Partitioning.
4. Biological Responses.

In general, I believe that each approach, or a combination of the approaches, can lead to site-specific decisions and even tentative criteria; however, I believe that in most cases none of the approaches can currently produce an acceptable criterion. Most of the shortcomings in these approaches are related to the inadequacy of the toxicity data base for those chemicals commonly associated with sediment and especially for chemicals while they are in the complex sediment environment.

In spite of these shortcomings, much more can be made of the current data base with respect to identifying those chemicals, organisms, and sites for which sediment criteria will most likely be needed. Furthermore, models can be used to suggest tentative criteria that can be evaluated in the field and the laboratory. Results from these evaluations can be used to improve present models and criteria approaches.

A number of candidate sites for testing tentative criteria are included in a draft report to EPA's Office of Criteria and Standards by Lehnig et al. (2) in which they document areas of known sediment contamination. They mention the Great Lakes, Puget Sound, and the Hudson and James rivers as sites with contaminated sediments. These sites are characterized by the presence of highly persistent chemicals such as PCBs and kepone. These and other sites below heavily industrialized areas are probably contaminated by other environmentally persistent chemicals as well. These are areas where sediment criteria should be most applicable, yet because of the high degree of uncertainty inherent in any criterion approach, on-site validation is scientifically, and probably politically, necessary. Such validation requires in situ data and would benefit from confirmatory laboratory dose-effect data. However, the complex mixture of sediment contaminants at many sites may make cause and effect relationships extremely difficult to establish. For this reason, initial validations of numerical criteria may require sites contaminated with only a very few chemicals. The reasons such validations will almost certainly be required are contained in the following critiques of the criteria approaches suggested by Pavlou and Weston (1); most of these problems were mentioned by the original authors.

Background Levels

The Background Levels approach entails measurement of normal levels of contaminants in clean sediments, where clean sediments would be defined as clean either by their geographical isolation from anthropogenic sources or by their location at sufficient depth in sediment cores to be associated with preindustrial deposition. Although some arguments may be raised as to whether either procedure truly provides an uncontaminated sediment (e.g., because of aerial deposition or vertical migration of contaminants) these arguments are minor compared to several others that essentially negate this approach as a regulatory criterion. The first point is that to prohibit any increase over natural contaminant levels basically precludes discharge of any contaminant that could reach the sediment. Allowing some additional contamination requires an upper limit that ultimately must be an upper ceiling based on probable damage, i.e., from dose-effect type data. In other words, the Background Level approach actually requires the existence of some other type of sediment criterion, in which case a sediment criterion already exists and the natural background approach is superfluous. However, knowing natural backgrounds will provide valuable knowledge and suggest when criteria, especially for metals, tend toward unrealistically low levels. This may, in turn, help identify incorrect assumptions of cause and effect and help identify important new test parameters or site characteristics that will lead to improved criteria.

Water Quality Criteria

The Water Quality Criteria approach uses existing water column criteria and applies them to interstitial/interfacial waters. As stated by Pavlou and Weston (1), this approach ignores the potential for direct transfer of contaminants from sediment to biota. If the sediment-to-biota transfer is inconsequential, then this approach has great appeal, but because the significance of the sediment route of uptake is not well investigated it cannot be considered inconsequential. Existing data on the relative importance of direct uptake from water as opposed to ingestion are inconclusive, especially for benthic and infaunal invertebrates.

Lyes (3) reported that uptake of naphthalene from sediment by the marine worm *Arenicola marina* was at a very low rate, far below the rate from water. Petrocelli and Anderson (4) found that clams and oysters accumulated from 10 to 20% of dieldrin from dieldrin-spiked sediment, but it is possible that most of the dieldrin was first released from the sediment into the water. McLeese and Metcalfe (5) concluded that the concentration in surficial water was the primary route of eight organochlorine compounds to the shrimp *Crangon septemspinosa*.

McLeese et al. (6) studied PCB uptake in worms (*Nereis virens*) and shrimp (*Crangon septemspinosa*) and concluded that it was uncertain whether contaminant was absorbed from water or from sediment. Elder et al. (7) looked at PCB uptake by the worm *Nereis diversicolor* and concluded that the question of route of uptake remained uncertain. Nearly all of the above studies were of short duration (less than 100 h).

What seems certain is that both water and sediments are possible sources of uptake; the important question concerns their relative contribution under usual conditions. Almost as certain is that the relative contribution from the two sources depends upon the feeding and respiratory mechanisms of the various biological taxa and upon the relative contaminant concentrations in the matter ingested and in the water in which the organism dwells. Finally it is critical to know whether thermodynamic equilibrium is established between organism and environment, especially for chemicals with high bioaccumulation potential.

Calculations of uptake for fish, which have rather well-defined ventilatory and feeding mechanisms, can illustrate the type of estimates possible when these factors have been quantified. A sophisticated model of the uptake of PCBs by lake trout in Lake Michigan is an excellent example [8], but a much simpler model can provide reasonable estimates of routes of uptake.

For example, most studies of uptake efficiency for chemicals such as PCBs, DDT, and dieldrin show that both respiration and digestion can be very efficient routes of uptake, with efficiencies approaching 100% in some cases. Most importantly, the relative efficiency of uptake does not appear to differ greatly ($< 10 \times$) between the two routes. A general comparability of uptake efficiency between routes means that uptake by each route is proportional to exposure by each route. Exposure through respiration is determined by multiplying ventilation volume (L/kg/d) and water concentrations (μg/L), yielding uptake potential (μg/kg/d). Exposure through feeding is similarly a function of feeding rate (kg/kg/d) and concentration of contaminant in food (μg/kg), again yielding uptake potential (μg/kg/d). The ratio of these two uptake rate potentials expresses the relative importance of each route regardless of uptake efficiency (as long as efficiencies are essentially equal between routes).

For fish, the respiratory volume depends to some extent upon temperature, body size, dissolved oxygen concentration, and activity, but in general the volume varies between 100 and 800 L/kg/d [9]. Higher levels occur, but usually only under conditions of unusual activity or hypoxia. For a simple model, I selected a rate of 200 L/kg/d, about that observed for brook and rainbow trout (Dr. James McKim, USEPA Duluth, MN, personal communication).

Feeding levels of fish range from < 0.01 to 0.1 kg/kg/d, although over a several-year period, the lower values are probably more typical. A long-term mean value of 0.02 kg/kg/d is both appropriate and convenient. Using these two rates of exposure (200 L/kg/d respiration rate and 0.02 kg/kg/d feeding rate), it is possible to determine the relative importance of each route once contaminant concentrations are known or assumed for water and food: these need only be expressed as a ratio.

As shown in Table 23.1, the food route is insignificant until the contaminant concentration in food exceeds 1,000× that in the water; at 10,000× more contaminant in food, the routes are of equal importance; at 100,000×, the food provides over 90% of the total uptake. If efficiencies of uptake are not equal, the estimates would need to be adjusted accordingly.

Table 23.1   Relative Importance of Food and Water as the Route of Uptake for Fish as a Function of the Ratio of Contamination Between Food and Water

Contaminant in food (mg/kg):water (mg/l)	Percent uptake[a] from food
1:1	<1
10:1	<1
100:1	1
1,000:1	9
10,000:1	50
100,000:1	91
1,000,000:1	99

[a] Calculations based on respiratory volume of 200 K/kg/day, food consumption of 0.02 kg/kg/day, and uptake/depuration rates independent of route of exposure.

Using PCP concentrations for water and food organisms in Lake Michigan in his detailed model of PCB uptake by lake trout, Weininger [8] estimated that only 2 to 3% of the total PCB accumulation is from water; the remainder comes from the diet. Using Weininger's data for PCBs in water and food, the simple model provided a similar estimate for PCB uptake by fish from water and food (5% and 95%, respectively).

Similar models may be possible for sediment-dwelling organisms, enabling qualitative judgment as to the relative importance of the two routes. To answer the question whether interstitial water criteria are acceptable as sediment criteria requires only a qualitative answer to the question of the role of ingested sediment in uptake. If ingestion is a significant route, then the direct application of water quality criteria to interstitial water is not acceptable. It should be pointed out that no distinction is usually made between sediment-inhabiting bacteria, protozoa, and other microbiota and the nonliving organic components of sediment; ingestion of sediment may in effect, be ingestion of microprey organisms.

It is important to note that the respiration (volume) of sediment-dwelling organisms need not be known. An exposure value based on studies with clean sediment (or without sediment) where uptake is clearly and predominantly from the water, can provide a net uptake value (L/kg/d) based upon body burden and water concentration alone. Indeed, such a value is preferable to respiratory volume, especially for passive ventilators that lack an easily definable and quantifiable ventilatory mechanism.

If contaminant concentration in interstitial/interfacial water should be the key parameter for uptake and effect, then measurement of these concentrations represents a technical problem that probably has a feasible solution. However, currently available sampling procedures may not provide a sufficient sample volume, and sampling may chemically alter the water obtained (1).

Perhaps the most disquieting caveat to the application of existing water quality criteria for interstitial waters is the fact that water quality criteria exist for only eight organic chemicals (chlordane, dieldrin, DDT, endrin, endosulfan, heptachlor, hexachlorocyclohexane, and PCBs). These and a handful of metals are the only chemicals for which it is currently possible to derive sediment criteria from a water quality criterion. Based on their persistence and their importance to water quality, these chemicals may be the most significant toxic chemicals in sediment. In this regard, EPA should examine the frequency and distribution of contaminants associated with sediments with the goal of prioritizing the development of sediment criteria.

An example of the status of the criteria for other chemicals for which water quality criteria documents exist is provided by the criteria statement for polynuclear aromatic hydrocarbons (PAHs), a group of chemicals with potential significance to sediment contamination. The criteria for PAHs (10) states:

> The limited freshwater data base available for polynuclear aromatic hydrocarbons, mostly from short-term bioconcentration studies with two compounds, does not permit a statement concerning acute or chronic toxicity.
>
> The available data for polynuclear aromatic hydrocarbons indicate that acute toxicity to saltwater aquatic life occurs at concentrations as low as 300 μg/l and would occur at lower concentrations among species that are more sensitive than those tested. No data are available concerning the chronic toxicity of polynuclear aromatic hydrocarbons to sensitive saltwater aquatic life.

The criteria documents for many organic chemicals, or classes of organic chemicals, e.g., chlorinated phenols and chlorinated benzenes, are similarly vague and incomplete because there is a very limited amount of data on the toxicity of these compounds to aquatic organisms. As a result, even if the toxic portion of sediment contaminants was only that fraction released to interstitial/interfacial waters, the absence of toxicity data for many organic chemicals would preclude the borrowing of criteria from water.

Finally, of the multitude of chemicals of potential environmental concern, there are criteria documents for only about sixty-five to seventy priority toxic chemicals. Of the approximately fifty criteria documents for organic chemicals, only eight were based on the minimum acceptable data base; the others contain only narrative criteria such as that earlier cited for PAHs. It is sobering, therefore, to reflect that Samoloff et al. (11) found that the most toxic fraction eluted from sediments of Tobin Lake, Saskatchewan contained none of the priority toxic chemicals.

Equilibrium Partitioning

The equilibrium approaches suggested by Pavlou and Weston (1) and discussed by Pavlou (12) rely on three "quasi-equilibrium constants" that are of great theoretical and empirical significance (Fig. 23.1):

The Sediment-Water Partition Coefficient (K_p):

$$K_p = \frac{\text{Concentration in Sediment}}{\text{Concentration in Interstitial/Interfacial Water}}$$

The Bioconcentration Factor (BCF):

$$\text{BCF} = \frac{\text{Concentration in Biota}}{\text{Concentration in Water}}$$

and the Accumulation Relative to Sediment (ARS):

$$\text{ARS} = \frac{\text{Concentration in Biota}}{\text{Concentration in Sediment}}$$

There is good reason to believe that K_p values can be reasonably (within $\pm$ $10\times$) estimated from measurement of octanol-water partition coefficient (K_{ow}) or water solubility. To account for large differences in organic carbon content between sediments, it will probably be necessary to calculate K_{oc} values for the partitioning between water and the organic material in the sediment (assuming sorption of organic pollutants by inorganic material is relative insignificant). Confirmation of K_p or K_{oc} may require measurement of a contaminant in field sediment and in interstitial waters. Reasonable validation can be provided by laboratory partitioning studies. Ideally, both lab and field values for K_p should be obtained.

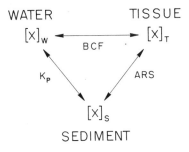

Fig. 23.1
The basic equilibria
relationships suggested
for converting among
pollutant concentrations
in water, tissue, and
sediment.

Bioconcentration Factors (BCF) have been estimated for a number of chemicals and with several species of organisms. Because of their close relationship to sediment, it is most important to determine BCF values for benthic invertebrates. Unfortunately most BCF values are for fish. For example, the water quality criteria document for PCBs (13) contains BCF values from 28 + − d exposures (as recommended by ASTM) for only three invertebrate species. In contrast, there are such values for nine species of fish. If the few BCF values for invertebrates are rather close (e.g., for PCBs they were 56,900, 108,000, and 101,000), they might be extended to other species in similar taxa with some confidence. Such BCF values still ignore the uptake via ingestion of contaminated food/sediment. For this reason, bio-accumulation factors (BAF) from all sources may be more meaningful.

All routes of uptake are integrated in the ARS value (Accumulation Relative to Sediment); however, such values (BAF or ARS) are not common-place, usually evolving from field studies or specially designed laboratory studies.

Potential applications of the partitioning constants to the development of sediment criteria were presented by Pavlou and Weston (1). Four approaches (Fig. 23.2) suggested that "rely heavily on existing water quality criteria for priority contaminants and the availability of biota threshold values above which a toxic effect may occur."

EQUILIBRIUM PARTITIONING APPROACHES
TO SEDIMENT CRITERIA

Water Quality Criteria Based (Approaches 1 and 2)

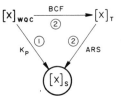

Tissue Threshold Based (Approaches 3 and 4)

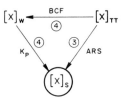

Fig. 23.2
Equilibrium partitioning steps for calculating
a sediment criterion from a water quality
criterion (approaches 1 and 2) or from a
tissue concentration (approaches 3 and 4).

As stated earlier, existing water quality criteria are available for only a few metals and a few persistent pesticides (and PCBs). However, if the soluble portion is the primary bioavailable source of contaminants and if metals and persistent pesticides (and PCBs) are the cause of most sediment-related toxicity problems, the use of sediment bulk analyses with K_p and K_{oc} values should be about as acceptable as current water quality criteria. To reiterate, such an approach should be validated by measurement of interstitial water concentrations of contaminants, an analysis lacking standardized procedures.

The second equilibrium approach suggested by Pavlou and Weston (1) relies on two factors: calculation of a biotic body burden that is equivalent to a chronic water quality criterion and then computation of a sediment criterion as a function of a "site"-observed ARS. Estimation of a body burden equivalent to a chronic threshold concentration is theoretically possible by using the water quality criterion chronic value and multiplying it by the BCF (either known or estimated from K_{ow}) (Table 23.2). It may be reasonable to calculate tissue concentrations that are equivalent to chronic criteria for water by using BCF values. Ths approach is reasonable from the

Table 23.2   Computation of Sediment Criteria Using BCF and Chronic Criteria Values from Current Water Quality Criteria Documents and ARS Values from Pavlou and Weston [1]

Chemical		BCF	Chronic criterion (μg/l)	Calc. chronic threshold conc. in tissue (μg/kg)	Tissue residue limits		Sediment criteria (μg/kg)		
					FDA (μg/kg)	Other (μg/kg)	Chronic threshold ARS[a]	FDA Action level ARS[a]	Other ARS[a]
Chlordane	FW	4,702	0.17	799	300	—	13.1	4.9	—
	SW	4,702	0.0064	30	300	—	0.5	4.9	—
DDT	FW	17,870	0.74[c]	13,220	5,000	150[d]	217	82	2.5
	SW	17,870	no data	—	5,000	150[d]	—	82	2.5
Dieldrin	FW	1,557	0.29	451	300	—	7.4	4.9	—
	SW	1,557	0.084	131	300	—	2.2	4.9	—
Endosulfan	FW	2,500[b]	0.056	140	—	—	2.3	—	—
	SW	2,500[b]	0.0087	22	—	—	0.4	—	—
Endrin	FW	1,324	0.045	60	300	—	1.0	4.9	—
	SW	1,324	0.0093	12	300	—	0.2	4.9	—
Heptachlor	FW	5,222	1.26[b]	6,580	300	—	108	4.9	—
	SW	5,222	1.58[b]	8,251	300	—	135	4.9	—
Lindane (hexachloro-cyclohexane)	FW	486[b]	0.080	39	—	—	0.6	—	—
	SW	—	no data	—	—	—	—	—	—
PCB	FW	45,000	0.2[b]	9,000	5,000	640[c]	117	65	8.3
	SW	45,000	0.1[b]	4,500	5,000	640[c]	58	65	8.3

[a] Using maximal ARS values of 61 for chlorinated pesticides and 77 for PCBs.
[b] Data based on freshwater (FW) or saltwater (SW) criteria.
[c] Based on less than minimally acceptable for establishing water quality criteria.
[d] Based on effects noted in wildlife feeding studies.

standpoint that most chronic toxicity tests are conducted in a manner similar to those from which BCF values are calculated. Indeed, it should be possible to measure chronic toxicity and BCF in the same test. However, where BCF values vary considerably between species of organisms (e.g., fish and worms, clams and shrimp), the extrapolation of data from one taxa to another is a highly presumptive operation.

A similar extrapolation risk is entailed using ARS values. In the examples in Table 23.2, the maximum ARS values reported by Pavlou and Weston [1] were used. The maximum ARS values were sixty-one for chlorinated pesticides and seventy-seven for PCBs, both values for crabs from Puget Sound (table 5 in Pavlou and Weston [1]). The several other taxonomic groups listed characteristically had ARS values of about one-tenth that for crabs.

When the biotic threshold is coupled with an ARS observed in the field, the result is a "site-specific" sediment criterion. However, since some portion of the body burden may be derived from direct water column exposure, the protectiveness of the sediment criterion is questionable. Sites characterized by concentrations of contaminants at or near criterion values in the water should approach the biotic threshold values without any contribution from the sediment. Because water quality criteria based on BCF values are normalized for a 1% body lipid content, all the theoretical sediment criteria in Table 2 would have to be adjusted for the body lipid content of the species in question.

The third approach suggested by Pavlou and Weston uses the ARS quantity to calculate safe sediment levels based on the ratio between observed tissue levels and sediment contaminant concentration at a site. For example, if tissue levels in biota are twice the acceptable values (either acceptable values computed from BCF and chronic threshold values for tissues or FDA limits for edible species), then the sediment criterion would be one-half the existing concentration.

The last approach for establishing a sediment criterion via equilibrium partitioning is similar to the third approach except that the sediment criterion is computed from a biological threshold concentration via BCF and K_p rather than directly through an ARS.

These last two approaches both assume that food/sediment has no direct input to the organisms in question or that apparent equilibrium tissue levels are independent of uptake route, assumptions not currently verified but having strong theoretical support. The second, third, and fourth approaches require the use of biological (tissue) threshold concentrations that might be estimated from chronic water quality criteria and BCF values, but that have not been validated in chronic toxicity tests. Finally, the second and third approaches require measured ARS values that should be obtained for the site in question and can be complicated by direct uptake of water column contaminants not routed through the sediments.

Complications also arise from the tendency of some compounds (e.g., PCBs) to accumulate in tissues without attaining equilibrium maxima. Weininger (8) argued that PCB levels in trout tissue do not follow equilibrium partitioning, but that tissue levels simply reflect uptake and growth, with no appreciable depuration. In this case, food, not sediment or water, was the route of uptake. However, at the lowest levels of the food chain, the sources had to be either water or sediment.

An approach that may help evaluate the adequacy of existing K_p and BCF factors is suggested by the concept of ARS values given by Pavlou and Weston (1). In addition, the approach may provide evidence regarding the relative importance of water and ingestion routes, and hence the credibility of using water quality criteria for interstitial or interfacial waters as sediment criteria.

If only soluble forms of a contaminant are taken up by biota, and no significant portion comes through feeding, then the ARS should be predicted by the ratio of K_p and BCF. This assumes that the biota dwell in water that is in equilibrium with the sediment, and hence the biota actually are exposed to the soluble concentration predicted from K_p. Estimating K_p and BCF from K_{ow} (1), computing theoretical ARS for the water route only, and comparing with observed ARS value from the field, provides a combined measure of the adequacy of the estimates of K_p and BCF, the applicability of water quality to interstitial water, and the measurement of ARS.

For example, hexachlorobenzene has a log K_p of about 3.6 and a log BCF of about 4.2. With a water concentration of 1 μg/L, this should provide an equilibrium sediment concentration of 4 mg/kg and a biotic tissue concentration of about 16 mg/kg. If applicable to the field site and the water exposure of the biota, the ratio of these values ($16 \div 4 = 4$) should equal the ARS if ingestion has a negligible effect on apparent equilibrium values. The ARS values reported for hexachlorobenzene ranged from 4.0 ± 1.0 to 93 ± 55; however, the latter value, for fish liver, should not be used since most BCF values are for whole organisms and because most fish (but not all) do not live in interstitial or sediment-equilibrated interfacial waters. The other ARS values were 4.0 ± 1.0 for clam, 8.1 ± 5.1 for shrimp, 17 ± 19 for worm, and 74 ± 68 for crab. Three points are clear, most confidence limits include the ARS estimate of 3.98 (no uptake via ingestion); most measured mean values exceed 3.98 by from 2 to 20×; and measured ARS values are extremely variable, probably a result of patchy sediment contamination and patchy exposure, especially of motile organisms. If the mean ARS values are used as best estimates, then ingestion need provide very little uptake for the clam, 50% for the shrimp, 75% for the worm and 95% for the crab. If the K_p, BCF and ARS values have credence, this example argues against the direct application (for this chemical) of water quality criteria to interstitial/interfacial waters as a suitable sediment criterion. If these quasi-equilibrium values lack the credence to be applied in the above manner, then they also lack credence for current use in a regulatory manner or as interim criteria.

This exercise of comparing computed ARS values (water, the only source) with observed ARS values would help in the development of sediment criteria. This is essentially the approach used in the laboratory when test organisms are exposed in water alone and in water plus sediments. Although taking many experimental forms, the final analysis of the data compares the uptake or effect under both exposure conditions, and especially looks to detect any impact resulting from sediment contribution in addition to the direct water route. Field ARS values, while more difficult to obtain than experimental ARS values, have the advantage of being based on sediment that was contaminated, aged, and populated by organisms in a natural manner, whereas laboratory tests sediments are spiked (usually once) with chemical, unaged, and the test organisms often stocked and fed in a highly artificial manner.

A final clarification should be made regarding the relative importance of food and water as pollutant sources for bioaccumulation. The significance of ingestion and water as routes of uptake depends upon the relative rates of uptake from the two sources, the degree of independence of apparent equilibrium concentrations in tissues from route of uptake, and the toxicological sensitivity of the organism to uptake via each route. The first two points are illustrated in Figure 23.3 where curve A represents uptake from water only, and curves B and C represent uptake from water and food. Curves A and B each attain eventual equilibrium at the same tissue concentration, which is therefore independent of route of uptake. For short-lived organisms or chemicals with extremely high bioconcentration potential, it is possible that tissue concentrations may never exceed the pattern indicated by case I in Figure 23.3.

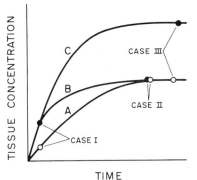

Fig. 23.3
The practical significance of ingestion
as a route of chemical uptake (B and C)
compared to uptake from water only (A)
depends on whether ingestion speeds up
the kinetics (Case I) but reaches the same
equilibrium (Case II) or attains a higher
tissue equilibrium (Case III).

In this instance, the relative importance of food and water as routes in uptake (curves A and B) are dictated by kinetic factors that are independent of the potential equilibrium. However, tissue concentrations never exceed the equilibrium potential from water.

In the case of curve C, the dual source of uptake proceeds to an apparently higher equilibrium tissue concentration than that achievable from water alone (curve A). This assumes that chemicals taken up from water and from food are handled in two independent or very slowly equilibrating compartments within the organism, or that the consumer organism spends most of the time in water with chemical concentrations below sediment equilibrium concentrations.

The significance of the three simple models of chemical uptake illustrated in Figure 23.3 to measured bioaccumulation should be obvious; in cases I and III, ingestion is important to the measured tissue concentration, but in case II route of uptake has no effect on measured uptake. Equilibrium concentration models that accurately predict case II concentrations from water will provide accurate or conservative estimates for all but case III (which may not, in fact, occur). In all three cases, the toxicological significance of tissue concentrations is not known, and is, in fact, likely to be different for different routes (and rates) of uptake.

Biological Response

The fourth class of approaches for establishing sediment criteria (1) involves the empirical observation of adverse effect(s) caused (presumably) by contaminated sediments. Two diverse approaches are recommended, one determining adverse effects by bioassaying sediments from a suspect site, the other determining if environmental damage can be demonstrated at a particular site. Both approaches are highly empirical and cannot lead directly to numerical criteria for application at other sites. These facts are clearly stated by the authors:

"Bioassays, as presently employed, serve as a pass/fail test without identifying the corrective measures required... The greatest difficulty encountered in attempting to establish criteria by this approach (biological field surveys) is in determining the contaminant(s) responsible for an observed impact on the biological community."

In addition, bioassay conditions may considerably increase the amount of soluble contaminants over the concentrations actually present in the field, thereby yielding misleading results. Increased soluble fraction may result from handling (suspending or oxidizing) sediment and from static water-column contamination far in excess of that seen in the field where dilution and transport may be significant operations.

A meaningful array of bioassay proceedures would include static or flow-through core bioassays in which organisms are exposed in a manner similar to that in the field, elutriate bioassays optimizing sediment/water interaction solubility, and bioassays of sediments from various depths in the core. If possible, bioassays should include initial and final measurement of contaminants in the sediment (bulk analysis), the overlying water, the interstitial water, and the test organisms. Preferably, tests should measure acute responses, chronic effects, and bioconcentration. Test organisms should include burrowing or detrital-feeding species. As stated earlier, regardless of the degree of test sophistication, the biological response is a pass/fail test, with no easy identification of the cause(s) or effect. The mortality of test organisms may even have a biological cause such as disease, or predation; attempting to minimize biological causes from predatory or infectious organisms within the sediment sample by screening or disinfection can significantly alter pollutant bioavailability. Thus, interpretation of the simple pass/fail bioassay test can be rather complicated.

Biological field surveys avoid many of the problems of bioassays but are beset by several other inherent complications. One of these was identified by Pavlou and Weston (1): "natural spatial and temporal variability of benthic communities often confounds efforts to ascribe an observed change directly to a pollution impact." It might be added that separating the effects of sediment pollution from water-column pollution can also present significant problems, especially in riverine systems where a possible cause (soluble pollutants) may have been transported downstream for many miles from the affected site and may have occurred weeks past.

The use of bioassays to establish sediment criteria in the manner in which water quality criteria were established has been suggested by Lehnig et al. (2). They point out inherent problems in establishing test methods and simulating an array of sediment conditions in laboratory tests. In addition to these problems, the cost of such testing makes desirable the pre-selection of test chemicals through the identification of the chemicals most likely to be causing sediment contamination. Obvious choices would include metals and highly persistent organics. A screening procedure such as that recently proposed by Gillett (14) could be used and probably should be used to evaluate chemicals found in currently contaminated sediments. His procedure includes use of three factors, K_{ow}, Henry's Law constant, and biodegradability half-life, which influence either persistence or toxicity.

ADDITIONAL CONSIDERATIONS

Tumors

One phenomenon apparently associated with contaminated sediment and not considered in most water quality criteria is the high incidence of tumors and tumor-like growths in fishes from suspect sites. This incidence, while not necessarily the most environmentally damaging aspect of sediment contamination, is probably the one that raises the greatest concern in the general populace largely because of its potential, but unknown, implications for human health. An orderly, analytical study of the etiology of the phenomenon is needed. Unfortunately, such studies are lengthy and expensive. A preliminary study could include test procedures such as the Ames test for sediment extracts (15) and the trout hepatoma test (16-18). Ultimately, mammalian dietary tests with sediment extracts or contaminated fish, might be necessary to evaluate human health risk.

Metals

Of the chemicals certain to accumulate in sediment, the ubiquitous and persistent metals represent a special case. The matrix of natural mineral sediments can contain significant quantities of heavy metals, but these are not usually considered to be biologically available. At the other extreme, free, soluble metals are usually highly bioavailable. In between these two extremes are a multitude of physical-chemical metal forms of highly variable bioavailability and toxicity. A very good review of metal bioavailability was recently prepared by Luoma (19).

Luoma concluded that food can be a significant route of metal uptake in some predator species but that the relative importance of food and water as sources is currently impossible to quantify. Because of complex binding and bioavailability differences in various sediments, metal uptake is not simply a function of feeding rate and metal concentration. Findings for one metal and one predator-prey combination should not be considered as generally applicable to other metals or other predator-prey combinations.

In several important aspects, the problem of metals contamination of sediments is considerably easier than that posed by organic chemicals. First, there are a limited number of metals, and second, water quality criteria already exist for the most troublesome metals. However, because the water quality criteria for metals address the question of metal form in a very simplistic manner, the application of the criteria to sediments could be extremely complicated because of the potential complexity of metal speciation in sediments and interstitial waters.

CRITERIA APPLICATION

If numerical criteria could be established based on any of the procedures already discussed, there would be two possible modes of application. One mode would entail applying the criteria to areas of suspected sediment contamination and identifying those sediments currently causing presumptive environmental damage. The other application would be in controlling ongoing or proposed activities predicted to cause problems of environmental damage by exceeding sediment criteria. In most cases the latter predictive process would require application of sophisticated chemical fate models. The capability of existing fate models is an area with which I am unfamiliar, but satisfactory validation of fate models, at the level of resolution required to provide valid estimates of spatial patterns of sediment concentration, must be accomplished if sediment criteria are to be used to regulate discharges.

Once existing sediment contamination is shown to be causing environmental damage, the issue will be what, if anything, to do about it. Assuming that the sediment cannot be decontaminated in situ, the choices are to bury it by the addition of overlying sediment or dredge it and treat or dispose of it elsewhere. In most cases, treatment will be an unlikely option, so the effect of dredging and transport to a less susceptible disposal site must be weighed against either doing nothing or in situ burial by additional sediment. The options are neither environmentally nor economically attractive. In some instances, the best choice may be to treat ongoing inputs at the source (if possible) and presume that gradual burial with less-contaminated sediment will eventually solve the problem. Many of these options will depend upon decision-making factors already developed in lake restoration, dredging, and ocean dumping programs.

RECOMMENDATIONS

Development of sediment criteria cannot, in the near term, follow the pattern established for water quality criteria, which basically was to develop pollutant-specific numerical criteria to be applied on a nationwide basis. Ultimately, we have realized that to provide the accuracy of effect or risk prediction requisite in today's society, we often need to provide procedures for site-specific and even waste-specific criteria. Given the pressure to produce water quality criteria in the early 1970s, the approach taken was not illogical. However, there is no comparable pressure for the establishment of sediment criteria per se nor a comparable data base on chemical toxicity in sediments. We should not create an atmosphere of pressure within regulatory agencies. However, we also most certainly cannot ignore the potential problem until such time as external pressure for sediment criteria (in whatever form) becomes so intense as to lead to consent decree action (e.g., the priority pollutants).

In my opinion, no method for establishing numerical criteria for sediment contaminants is sufficiently robust to gain the current approval of objective scientists. The closest may be the use of water quality criteria for interstitial waters of sediments. The approach could use either measured concentrations or models of partitioning between the sediment and water. Unsophisticated sampling procedures for interstitial water would negate this approach if sediments were unduly disturbed in the sampling process. Development of adequate sampling procedures for interstitial water should be an agency research goal. Remember that there are full-fledged water quality criteria for only a few chemicals, and while these are usually chemicals of great environmental persistence, there may be many other troublesome chemicals in sediments.

There is only one circumstance that I can interpret as a sediment criterion "violation" based on current knowledge: when in the absence of any significant current input into a system, the sediment serves as a source to the water column of a chemical in excess of its water quality criterion. This is also an instance where cleanup, rather than discharge control, is the regulatory action; depending on the assessment of effect, perhaps no action would be taken.

Any other sediment-related contaminant effects (or violations of any proposed sediment criteria) are on more tenuous scientific ground than are current water quality criteria. The primary scientific problem with numerical sediment criteria is that cause and effect data are neither as convincing nor as plentiful as those for water-column effects. In many instances, even the water column-criteria, or concentration-effect relationships, are correctly questioned as to their exact validity. Even when (for the sake of argument) they are assumed to be exact predictions of effect (at the population level), there is no consensus as to the breadth of their environmental applicability.

There are many areas of research that will contribute knowledge regarding the effects of contaminants in sediments. The agency, probably with good reason, usually concentrates its research on more directly applicable aspects of problems. Given the current state of knowledge regarding sediment contamination, the proposed approaches for sediment criteria should not immediately be implemented as criteria, but they should be considered as hypotheses to be tested in laboratory and field investigations. The balance between laboratory and field investigations should be much more even than was the case with water quality criteria research, which was heavily weighted towards laboratory dose-response bioassays. Sediment-related laboratory tests should be geared more toward mechanisms of effect and routes of effect rather than a compendium of chemical-specific, species-specific bioassay responses.

There are several primary questions to be answered in the next few years. Two questions are of particular importance: (1) what is the threat posed by contaminants when they are deposited in sediments? and (2) geographically, what is the current extent of the problem?

Scientific investigations over the past decade, especially over the last few years, have begun to answer these questions, but often not in a rigorous manner. Agency-wide efforts have not been coordinated towards a well-defined research goal. Perhaps the current push to find a way to establish sediment criteria will catalyze the implementation of a more-focused agency research goal (and effort).

If a more balanced effort between laboratory and field investigations is to be efficient, the choice of field sites must be made from some listing of potential sites. A compendium of sites of probable areas of sediment contamination should be compiled by region, by state, and/or by basin or system. Some idea of probable contaminants would be helpful, as well as knowledge of whether discharge of the contaminant is historic or ongoing. Desirable historic information on the site would include studies of effects in the field, laboratory bioassays, sediment or water-column chemical analysis, tissue residue values from resident biota, and data on sedimentation rates. Sites should be selected based on the availability of data to validate models of fate, both on the macro- and microscale.

Studies that fail to separate sediment/food and water-column effects, or that fail to address the chemical cause of an effect, will be of minimal value in helping establish a foundation for a sediment criterion decision. Not all studies or all sites, perhaps none, will have an ideal set of conditions for investigating all aspects of the problem. A few good studies of the better sites, however, is preferable to studies of more numerous, poorly screened sites.

Before we establish sediment criteria, we need to improve the definition of ecological significance of contaminated sediments and validate or improve existing models of fate and bioavailability. Until we accomplish this, what we propose as criteria are really only hypotheses in need of testing.

REFERENCES

1. **Pavlou, S. P. and D. P. Weston.** 1983. Initial development of alternatives for development of sediment related criteria for toxic contaminants in marine waters (Puget Sound). Phase I: Development of conceptual framework. Prepared by JRB Associates for the U.S. Environmental Protection Agency under EPA Contract No. 68-01-6388.

2. **Legnig, D., G. Hickman and C. Clarkson.** 1983. Sediment criteria. Draft prepared by Camp Dresser and McKee for the U.S. Environmental Protection Agency.

3. **Lyes, M. C.** 1979. Bioavailability of a hydrocarbon from water and sediment to the marine worm *Arenicola marina. Mar. Biol.* **55**:121-127.

4. **Petrocelli, S. R. and J. W. Anderson.** 1976. Distribution and translocation of residues of dieldrin—a chlorinated hydrocarbon insecticide—among water, sediments, and estuarine organisms from San Antonio Bay. In A. H. Buoma, ed., *Shell Dredging and its Influence on Gulf Coast Environments.* Gulf Publishing Co., Houston, Texas, pp. 185-218.

5. **McLeese, D. W. and C. D. Metcalfe.** 1980. Toxicities of eight organochlorine compounds in sediment and seawater to *Crangon septemspinosa. Bull. Environ. Contam. Toxicol.* **25**:921-928.

6. **McLeese, D. W., C. D. Metcalfe and D. S. Pezzack.** 1980. Uptake of PCBs from sediment by *Nereis virens* and *Crangon septemspinosa. Arch. Environ. Contam. Toxicol.* **9**:507-518.

7. **Elder, D. L., S. W. Fowler and G. G. Polikarpov.** 1979. Remobilization of sediment-associated PCBs by the worm *Nereis diversicolor. Bull. Environ. Contam. Toxicol.* **21**:448-452.

8. **Weininger, D.** 1978. Accumulation of PCBs by lake trout in Lake Michigan. Ph.D. thesis, University of Wisconsin, Madison.

9. **Davis, J. C. and J. N. Camerson.** 1971. Water flow and gas exchange at the gills of rainbow trout, *Salmo gairdneri. J. Exp. Biol.* **54**:1-18.

10. **U.S. Environmental Protection Agency.** 1980. Ambient water quality criteria for polynuclear aromatic hydrocarbons. EPA-400/5-80-069. Washington, D.C.

11. **Samoloff, M. R., J. Bell, D. A. Birkholz, G. R. B. Webster, E. G. Arnott, R. Pulak and A. Madrid.** 1983. Combined bioassay-chemical fractionation scheme for the determination and ranking of toxic chemicals in sediments. *Environ. Sci. Technol.* **17**-334.

12. **Pavlou, S. P.** The use of the equilibrium partitioning approach in determining safe levels of contaminants in marine sediments. This volume, pp. 388-412.

13. **U.S. Environmental Protection Agency.** 1980. Ambient water quality criteria for polychlorinated biphenyls. EPA-440/5-80-068. Washington, D.C.

14. **Gillett, J. W.** 1983. A comprehensive prebiological screen for ecotoxicologic effects. *Environ. Toxicol. Chem.* **2**:463-476.

5. **Allen, H. E., K. E. Noll and R. E. Nelson.** 1983. Methodology for assessment of potential mutagenicity of dredged materials. *Environ. Technol. Letters* **4**:101-106.

16. **Wales, J. H., R. O. Sinnhuber, J. D. Hendricks, J. E. Nixon and T. A. Eisele.** 1978. Aflatoxin B₁ induction of hepatocellular carcinoma in the embryos of rainbow trout (*Salmo gairdneri*). *J. Natl. Cancer Inst.* **60**:1133-1139.

17. **Hendricks, J. D., R. A. Scanlan, J. L. Williams, R. O. Sinnhuber and M. P. Grieco.** 1980. Carcinogenicity of N-methyl-N'-nitro-N-nitro-osguinidine to the livers and kidneys of rainbow trout (*Salmo gairdneri*) exposed as embryos. *J. Natl. Cancer Inst.* **6**:1511-1519.

18. **Hendricks, J. D., R. O. Sinnhuber, J. H. Wales, M. E. Stack and D. P. H. Hsieh.** 1980. Hepatocarcinogencity of sterigmato-cystin and versicolorin A to rainbow trout (*Salmo gairdneri*) embryos. *J. Natl. Cancer Inst.* **5**:1503-1509.

19. **Luoma, S. N.** 1983. Bioavailability of trace metals to aquatic organisms—a review. *Science of the Total Environment* **28**:1-22.

Chapter 24

Establishing Sediment Criteria for Chemicals—Industrial Perspectives

Warren J. Lyman

INTRODUCTION

The U.S. Environmental Protection Agency (EPA) has recently taken the first steps of a process that may lead to the establishment of national sediment quality critiera. Although it has not been so stated, it may be presumed that these criteria could be used in a manner similar to EPA's existing water quality criteria. This might include adoption of the criteria directly as standards by the states, use of the criteria to evaluate sediment quality in impacted areas, and as a basis to regulate future activities projected to impact on sediment quality (e.g., point source discharges, control of area run-off, dredging).

What does "industry" think of this prospect? Although relatively few people, including those in industry, are aware of these efforts to establish sediment criteria, this author feels that a significant amount of opposition can be expected unless certain changes are made. In my opinion, the opposition could focus on concerns in three areas:

1. A failure to demonstrate a real need for such criteria; to describe how they are to be used; and to project the benefits that would be obtained from their use.

2. The use of a criteria-setting process that did not obtain adequate scientific peer review or sufficient input from other concerned parties.

3. Scientific weaknesses in the criteria associated with, for example, the use of unjustified assumptions, poor quality data, or over-simplification.

In addition to potential industry opposition, it appears that at this time there exists a diversity of opinions with the federal agencies (EPA and USACE) on the need for and approach to follow in establishing sediment criteria. While the prospect of future opposition and questions of a sound scientific basis for this new policy would appear to spell trouble for any future sediment quality criteria, there is still plenty of time for the concerned parties to work together to develop criteria whose need, basis, and method of use are generally accepted.

RECENT WORK TO DEVELOP SEDIMENT CRITERIA

The primary works of interest related to recent EPA initiatives are those by Pavlou and Weston prepared jointly for EPA's Water Quality Criteria and Standards Division of the Office of Water Programs and for EPA's Region X office (1, 2). This work was designed, in part, to aid evaluation of contaminated sediment in Puget Sound.

Pavlou and Weston's Phase I report (1) and Engler's chapter (3) on the subject provide a brief history of prior work in the area of sediment quality criteria development. Table 24.1 provides a summary of the chemical coverage of several sediment quality criteria proposals.

Table 24.1   Coverage of Chemical Classes by Various Sediment Quality Criteria Proposals

	Number of elements or chemicals for which numerical criteria are given			
Criteria source	Heavy metals	Pesticides	Conventional pollutants[a]	Other organics[b]
1. Jension Criteria, EPA, 1971	3	—	4	1
2. EPA Region V Guidelines for Pollution Classification of Sediments, 1977	11	—	7	1
3. USGS Sediment Alert Levels	12	15	1	1
4. Ontario Ministry of the Environment Dredge Spoil Guidelines, 1976	9	—	7	1
5. EPA Region VI Proposed Guidelines for Sediment Disposal	9	—	—	—
6. EPA Region VI Sediment Quality Indicators, Interstitial or Elutriate Water	10	10	5	2
7. US EPA/USACE Criteria for Evaluating Dredged Material, 1977	No criteria set for allowed concentrations; if exclusionary criteria not met, bioassay required.			
8. Pavlou and Weston Partitioning-Based Criteria, 1984	6	6	—	41

[a] Conventional pollutants include: nitrate, ammonia, phosphorus, $NO_2 + NO_3$, NO_2 as N, cyanide, volatile solids, COD, Kjeldahl nitrogen, oil and grease.
[b] Except for nos. 7 and 8, coverage is only for PCBs or phenol.
Source: Adapted from refs. 1 and 2 (see esp. table 2 in ref. 1, tables 6 and 7 in ref. 2).

Most of these criteria were designed to assist in evaluating sediment from proposed dredging areas and are based on total sediment analysis. Although most also have some focus on heavy metal contamination, coverage of pesticides, conventional pollutants and other (toxic) organics is much less detailed. Technical flaws associated with many of the earlier criteria (e.g., nos. 1-6 in Table 24.1) include the lack of a credible scientific basis, a failure to allow consideration of the geochemical location and background levels of the pollutants, and a failure to address the potential availability of the contaminants to organisms (3). An earlier review by Lee and Plumb (4) also concluded that "There is little or no evidence to support the premise that there is a relationship between bulk-sediment composition and the pollutional tendencies of the sediment."

EPA/Army regulations and criteria for ocean dumping (no. 7 in Table 24.1) (5) replace a total sediment analysis with an elutriate test involving laboratory exposure of test organisms. While the test is sensitive to a variety of factors (e.g., solid:liquid ratio, DO concentration, pH, contact time, particle size), it is considered to be a valuable test (3, 4). Since the chemicals responsible for observed organism effects remain unidentified in this procedure, no insight is gained into appropriate control measures or remedial actions for polluted sediments. The time and costs associated with this elutriate test are of concern but are obviously less than the battery of tests that would be required for complete scientific understanding of all potential environmental effects.

In their preliminary report, Pavlou and Weston (1) evaluated four basic approaches applicable to the development of sediment quality criteria:

1. Background levels of contamination in unpolluted sediments.
2. Adaptation of existing water quality criteria.
3. Equilibrium partitioning.
4. Biological evaluation of sediment toxicity.

In addition, they reviewed three sediment quality assessment protocols including an ecological risk index proposed by Hakanson (6). The latter allows simultaneous consideration of all of the pollutants present and the sensitivity of the receiving water body. All approaches were ranked according to nine criteria: adequacy of the existing data base, development costs, application costs, availability of methodology, cost effectiveness, relative complexity, ability to address synergism, adaptability to new compounds, and utility to management decision making.

In Phase II of their work, Pavlou and Weston developed the equilibrium partitioning approach (including use of existing water quality criteria) and applied it to six trace metals and 47 synthetic organic compounds (2). The basic tenet of this approach is that "... the concentration of each contaminant in sediment should be at or below a level which ensures that its concentration in interstitial water does not exceed the EPA water quality criterion" (2). The pollutant is assumed to be in equilibrium between the solid (sediment) and aqueous phases, and empirical correlations are used to relate sediment concentrations, which would be measured by the criteria user, to the interstitial water concentration. The estimated interstitial water concentrations are then compared with EPA water quality criteria for the protection of aquatic life; chronic criteria are preferred, but in many cases (mostly for organics) only acute criteria are available.

This equilibrium partitioning approach includes a number of assumptions; Table 24.2 lists several discussed by Pavlou and Weston. Other individuals I contacted have identified several other potential problem areas and limitations with the equilibrium approach:

- The approach requires that a water quality criterion be available for a chemical; while there are several based upon acute toxicological endpoints, only a few exist that are based upon more desirable chronic endpoints (e.g., at least 10-d exposures).

- The approach is probably valid for neutral organic chemicals of limited solubility; it may not be applicable to organic acids and bases, to other highly-soluble organic, or to metals.

- The approach will not work for sediments with very low TOC values (e.g., <0.1%).

- There is inadequate consideration given to the kinetics of uptake and release by the sediments.

Underlying some of these comments appears to be a more basic belief that not enough is currently known about the behavior (i.e., environmental chemistry and related bioavailability) of trace pollutants in sediments to allow the development of reliable, defensible sediment quality criteria.

DISCUSSION

Three areas of concern related to the current initiatives in the development of national sediment quality criteria are discussed below. Each involves issues that could be taken up by industry and/or the scientific community in general, possibly resulting in obstacles to the EPA's goals. A failure to address these concerns increases the probability for private and public attacks, delaying tactics, and the well-known rush to the courthouse following any formal promulgation.

Demonstrated Need for Criteria

Has the need for the development of sediment-quality criteria been demonstrated? Does the current program have a clearly stated set of goals and objectives and a plan to achieve these? What will the benefits be? Rather astonishingly, I have not seen these questions fully addressed in any of the

Table 24.2   Assumptions of the Equilibrium Approach and Their Potential for Violation

Assumption	Potential for violation of assumption	Impact of violation of proposed criteria
Sediment-water partitioning of a contaminant is at equilibrium	Medium	Criteria may be too conservative. Violations would be site-specific and source-specific, not compound specific.
Magnitude of partition coefficient dependent upon sediment organic content	Low (organics) High (metals)	At low organic carbon levels the proposed criteria would be too conservative. At high organic carbon levels the effect criteria would be liberal.
Variations in the redox potential and pH will not impact criteria	Low (base-neutral organics); High (metals and some organics)	The effect of redox potential on trace metal partitioning has been largely accounted for by empirical derivation of partition coefficients. In some instances, the redox potential may effect partitioning beyond the level already accounted for. Field verification is needed to determine potential impact on criteria.
Independence of partitioning on temperature	Low	Typical environmental variations in temperature will affect criteria by 10% or less.
Independence of partitioning on salinity (ionic strength)	Medium	For most contaminants of environmental concern variations in salinity typical of marine water will have a negligible impact on the criteria. Variation of salinity between fresh and saltwater could effect criteria by about 25%. An increase in ionic strength of interstitial water would result in a higher criteria value therefore criteria may be too conservative.
Independence of partitioning on dissolved organic matter (DOM)	Medium	Criteria would be lowered as DOM increases. Criteria may be too liberal.
Particle size is not important in determining partitioning other than as a covariate with organic carbon	Low	Evidence to date is very limited but in very sandy sediments criteria may be too liberal.
Entire contaminant burden of sediments is in a bioavailable form	High	If a fraction of the contaminant is not bioavailable, the proposed criteria would be too conservative.

Table 24.2 (Contd.)

Assumption	Potential for violation assumption	Impact of violation of proposed criteria
Water quality criteria are appropriate for benthic organisms	Low	Criteria would be either too liberal or too conservative depending upon toxicant sensitivity of benthic organisms.
Ingestion of sediment-bound contaminants by deposit-feeders does not increase contaminant body burden above levels attained by exposure to the dissolved contaminant fraction	Medium	Criteria would be too liberal if assumption were violated.

Source: Pavlou and Weston (2).

publicly available documents obtained during my information gathering done in preparation for this workshop. Should it not be a fundamental prerequisite of such an undertaking to have a clear statement of need and purpose? The absence of clear objectives leaves questions open as to overall need and utility of the entire sediment criteria issue. Needless to say, the statement of objectives must include sufficient discussion of the scientific issues and regulatory options to clearly defend the chosen program.

At least some needs for, and potential uses of, sediment quality data and criteria are fairly clear (1, 7):

- Sediments are an important part of the aquatic environment providing habitat, feeding and breeding areas for many aquatic biota; we should have means to assess the quality of these sediments and associated contaminant concentrations so that the biota can be protected through appropriate control measures.

- The fact that many pollutants tend to concentrate in sediments makes the above reason especially important. In some cases, pollutants from point source discharges can accumulate to harmful levels in sediments even though the individual discharges do not violate their NPDES permits or water quality criteria, the latter being applicable outside some defined (often ill-defined) mixing zone. Recently-issued criteria have specifically addressed only the "dissolved" or "active" fraction of a pollutant in a sample (8); this raises the possibility that the remaining particulate fraction of the pollutant load may accumulate in sediments, change in the extent of bioavailability, and lead to adverse impacts.

- Because sediments act as natural concentrators of many pollutants, analyses of the sediments can be done with less sensitive analytical methods than would be necessary for water samples. The sediment quality could then potentially be used as a surrogate measure of the quality of the overlying water.

- In areas where sediments tend to accumulate, sediment core data can be used to look at historical pollutant loads in the upstream basin areas. Even where sediments themselves are not accumulating, the pollutant may be; in this case the sediments provide a time and space averaged sample for analyses.

- When sediments become contaminated, they can become a source of pollution to overlying water that may be relatively clean; the ocean or freshwater discharge of dredged material from contaminated areas is a special concern.

Sediment quality data and criteria could be used as a tool to monitor or assess sediment and water quality in both informal and formal actions; as a decision tool that could indicate when more detailed studies are necessary prior to some action; and as an absolute limit that could be invoked directly to require certain restrictions or remedial actions. The need for methods to assess dredged material is clear and, as mentioned above, regulations have been promulgated for the control of dredged material disposal (5).

It should be noted that the use of sediment quality data as a surrogate measure of water quality is, in a sense, the converse of Pavlou and Weston's proposal to use water quality criteria to set sediment quality criteria; both involve the same assumptions and thus all of the caveats associated with the latter proposal must also be associated with the former.

What steps could be taken to evaluate the need for sediment quality criteria? The first step should be a review of sediment pollutant concentrations in a nationwide sampling of marine and freshwater sites; various types of data (e.g., on total sediment, interstitial waters, extracts, particle size differentiation, depth profiles) should be collected so that a better understanding of the pollutant's sediment chemistry and bioavailability can be made. Available data on bioassays with various sediment fractions or elutriates should be collected and evaluated. Data are also needed from biological field surveys in which organism pollutant loadings have been measured and/or histopathological examinations have been conducted. Data from both control and affected areas are required.

Subsequent steps could include a review of the fate (uptake, release, complexation, degradation, etc.) of pollutants in sediments and the development of correlations between subgroups of data, e.g., sediment pollutant loads or "risk index" versus frequency of histopathological abnormalities or other noted effects. All of this would then be followed by an overall assessment of existing sediment quality, future expected trends in sediment quality, available control or remedial measures for impacted sites, and an assessment of the beneficial role that sediment quality criteria could play in the whole process.

A Standards Setting Process without Consensus?

Recent initiatives by the Water Quality Criteria and Standards Division of EPA's Office of Water Programs to develop sediment quality criteria seem to be lacking any real scientific input or support from outside interests. These interests include:

- Federal research laboratories (e.g., the Corps of Engineers Waterways Experiment Station and many of EPA's own research laboratories).
- Other research groups (e.g., Southern California Water Research Project, International Joint Commission, and industrial research groups).
- Industry, municipalities and other groups who, because of their wastewater discharges, could be impacted by any criteria or standards adopted.

- States or other groups (e.g., river basin commissions) who might be asked to adopt federal criteria as local, legally-enforceable standards.

- Other parties, including public interest groups, professional societies, industry associations, or standards-setting organizations.

A fairly recent initiative started in 1982 by the Water Quality Criteria and Standards Division focused on one problem site, Puget Sound. Since this is a recent undertaking, there is time to start a new approach that includes more participation from these other groups.

Industry, municipalities, and other potentially impacted groups may not want to be as much of an active participant in any criteria development process, but they certainly want to be kept informed of progress and directions, given a chance to comment on the work of others; to know the basis for the work being conducted (regulatory need, scientific basis, etc.) and how the criteria and standards would be used. There also exists a need to know how "accurate," or "reliable" the proposed criteria are, how prone they are to misapplication due to a failure to identify properly all assumptions and limitations, and to know how much the new criteria and standards may potentially cost them.

What is apparently occurring here in this new initiative for the development of sediment quality criteria is by no means unique. In a recent article on "Resistance to Standards Development," William Kirchhoff describes the "... conflict between scientific and social imperatives ..." that is often a part of any standards setting process (9). Kirchhoff points out many other pitfalls in such processes and argues for standards to be developed by a consensus approach that includes non-experts as well as experts. He also sees a need for putting a higher emphasis on general agreement between the regulated community and the regulating agency than on obtaining a high degree of precision in the numerical criteria derived or in the data used, for example, in compliance monitoring. In addition, interlaboratory evaluation of interim or final written protocols would be beneficial, as well as the use of the American Society for Testing and Materials (ASTM) to assist in the development of consensus-based standards.

Scientific Weaknesses a Serious Concern?

The second section of this paper ("Recent Work to Develop Sediment Criteria") contained a discussion of several scientific issues that could eventually become serious impediments to any attempt to promulgate criteria containing such weaknesses. It should be assumed that industry, municipalities, and other potentially impacted groups will not let these weaknesses go unnoticed. The recognition of this not only makes the consensus approach more imperative, but also requires that the regulating agency itself carefully identify and evaluate each perceived weakness and clearly state the limitations of each one.

REFERENCES

1. **Pavlou, S. P. and D. P. Weston.** 1983. Initial evaluation of alternatives for development of sediment related criteria for toxic contaminants in marine water (Puget Sound)-Phase I: Development of Conceptual Framework. Final Report, Work Assignment No. 42 of EPA Contract No. 68-01-6388. Water Quality Branch, EPA Region X, Seattle, WA.

2. **Pavlou, S. P. and D. P. Weston.** 1984. Initial evaluation of alternatives for development of sediment related criteria for toxic contaminants in marine water (Puget Sound)-Phase II: Development and test of the sediment-water equilibrium partitioning approach. Final Report, Work Assignment No. 42, EPA Contract No. 68-01-6388. U.S. Environmental Protection Agency, Washington, DC.

3. **Engler, R. M.** 1980. Prediction of pollution potential through geochemical and biological procedures: Development of regulation guidelines and criteria for the discharge of dredged and fill material. In R. A. Baker, ed., *Contaminants and Sediments*, Chapter 7. Ann Arbor Science Publishers, Inc., Ann Arbor, MI., pp. 143-170.

4. **Lee, G. F. and R. H. Plumb, Jr.** 1974. Literature review on research study for the development of dredged material disposal critiera. U.S. Army Engineer Waterways Experiment Station, Vicksburg, MI. (Available as NTIS AD-789 755.)

5. **U.S. Environmental Protection Agency/U.S. Army Corps of Engineers (EPA/USACE).** 1977. The ecological evaluation of proposed discharge of dredged material into ocean waters; Implementation Manual for Section 103 of PL 92-532. Environmental Effects Lab, USACE, Waterways Experiment Station, Vicksburg, MS.

6. **Hakanson, L.** 1980. An ecological risk index for aquatic pollution control: A sedimentological approach. *Water Res.* **14**:975-1001.

7. **Sly, P. G. and H. L. Golterman.** 1981. Significance of sediment information in studies of the aquatic environment. *Water Quality Bulletin* **6**:29.

8. **U.S. Environmental Protection Agency.** 1984. Water quality criteria, request for comments. *Fed. Reg.* **49**:4551.

9. **Kirchhoff, W. H.** 1984. Resistance to standards development, *ASTM Standardization News*, (June) pp. 21-23.

Chapter 25

The Use of the Equilibrium Partitioning Approach in Determining Safe Levels of Contaminants in Marine Sediments

Spyros P. Pavlou

INTRODUCTION

Historically, the development of water quality criteria was the focus for providing some degree of assurance that contaminant concentrations will be within acceptable limits for the protection of aquatic life and human health. However, in Puget Sound, as well as other estuaries, there is increasing evidence of sediment quality degradation in many of the heavily urbanized areas. Macrobenthic communities in the vicinity of point source discharges have demonstrated significant changes in species composition and abundance (1-3). Sediments from urbanized areas have been shown to induce mortality in sensitive benthic species (4), and demersal fishes from heavily polluted areas have been shown to have a higher incidence of histopathological abnormalities than those from reference areas (5; 2).

These observations argue for the need to establish some sort of sediment criteria as a supplement to existing water quality criteria in assessing the magnitude, severity, and significance of the contamination. There are a number of specific reasons why sediment criteria deserve consideration: (1) most toxic compounds are highly insoluble, so the majority of the contaminant is not dissolved in the water but is associated with the organic matrix on sediment particles. For example, sediments in Elliott Bay contain 60,000 times more PCBs than the overlying water (6); (2) sediments serve to integrate contaminant concentrations over time, eliminating the high degree of temporal variability that plagues sampling of toxicants in the water column; (3) sediments serve as sinks for most toxic materials, thus a long-term low level discharge of a contaminant may result in a dangerous buildup in the sediments even though water quality degradation may not be apparent; (4) sediments can serve as reservoirs (sources) of contaminants, which could be reintroduced to unpolluted overlying water; and (5) a large number of organisms, including many of commercial importance, spend most of their lives in or on the sediments. For these species, the contaminant level in the sediments and associated interstitial water may be more critical than that in the overlying water in controlling their bioaccumulation potential.

Criteria Development

In response to the apparent need for developing sediment-related criteria, the Criteria and Standards Division of the U.S. Environmental Protection Agency (EPA) initiated an effort in 1983 to identify and evaluate alternative approaches to the establishment of this type of limit. The current state of knowledge regarding sediment criteria was updated and summarized by Pavlou and Weston [7]. A number of approaches were identified of which three appeared to be of immediate utility in developing contaminant safe levels for Puget Sound. These are briefly summarized below.

Background level approach. The concentration of contaminants in the sediments of relatively unpolluted reference areas is determined and the criteria are then established at some permissible level of enrichment above background levels.

Burden-effect relationships. Observations of adverse biological impacts (alternations in benthic community structures, lethal or sublethal effects observed in bioassays, incidence of pathological disorders) are correlated with contaminant concentration in order to determine those concentrations at which no impacts are evident.

Equilibrium-partitioning approach. A sediment safe level is established at a concentration that ensures that EPA water quality criterion is not violated in the interstitial water.

Since the development of sediment criteria by the background-level approach and burden-effect relationships is an ongoing effort in connection with other work within region X, the equilibrium-partitioning approach was evaluated to test its suitability for providing "first cut" numerical safe level values in Puget Sound. The approach is simple, is immediately adaptable to a large number of chemical contaminants, is based primarily on existing data, has low implementation costs, and can be immediately utilized in management decisions.

This paper summarizes the results and discussion presented in an earlier report submitted to the Environmental Protection Agency [8].

COMPUTATIONAL FRAMEWORK

A brief discussion of the concept of equilibrium partitioning and background references have been presented by Pavlou and Weston (7). As applied to the sediment/water interface, contaminant transport between the solid and aqueous phases occurs via a rapid molecular exchange. This exchange is continuous and therefore maintains the system at chemical equilibrium. The instantaneous concentration of the contaminant in either of the two components can be expressed as a function of its concentration in the other component and an equilibrium constant specific to that contaminant. These constants are commonly referred to as partition coefficients and are expressed mathematically as:

$$K_D = \frac{C_s^x}{C_w^x} \tag{1}$$

K_D is the partition coefficient and C_s^x and C_w^x are the concentrations of contaminant x in the sediment (s) and surrounding water (w), respectively. Contaminant concentrations are expressed as mass of x/dry mass of sediment and mass of x/mass of water.

A schematic representation of the aqueous-solid components in a marine environment where this mechanism is operable is shown in Figure 25.1. Partitioning of a contaminant occurs in both the water column and the sediments. In the former case, the exchange is between suspended particulate matter (SPM) and ambient water; whereas in the latter case, it is between sediment particles and the interstitial water. In applying the equilibrium-partitioning approach to sediments, the zone of interest for developing safe levels is the bioturbation layer where most of the biological activity occurs. The derivation of these quantities is summarized below.

The partition coefficient K_D was adjusted to account for the dependence on organic carbon. The modified partition coefficient is defined as:

$$K_{oc} = \frac{C_{s/oc}^x}{C_{iw}^x} \tag{2}$$

where K_{oc} is the organic carbon normalized partition coefficient and $C_{s/oc}^x$ is now the concentration of contaminant x in the sediment expressed in units of mass of x/mass of organic carbon. The term C_{iw}^x refers specifically to the concentration of x in interstitial water; the units remain the same as defined in eqn (1). Equation (2) can also be expressed as:

$$K_{oc} = \frac{C_s^x}{C_{iw}^x \, \text{TOC}} = \frac{K_D}{\text{TOC}} \tag{3}$$

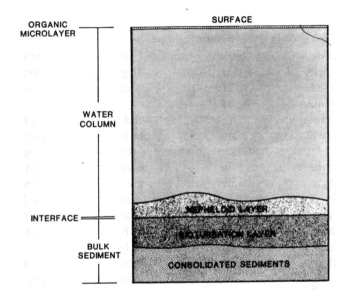

ORGANIC
MICROLAYER

SURFACE

WATER
COLUMN

INTERFACE

BULK
SEDIMENT

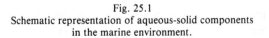

Fig. 25.1
Schematic representation of aqueous-solid components
in the marine environment.

where TOC refers to the total organic carbon content in sediment expressed
as the fractional mean on a dry weight basis. This adjustment was made to
account for the dependence of K_D on organic carbon. This dependence has
been well documented in the literature (9). Modifying eqn (2) by setting C_{iw}^x
as the water quality criterion $C_{w/cr}^x$, the corresponding sediment safe level
for contaminant x, $C_{s/sl/oc}^x$ can be expressed as:

$$C^x_{S/sl/oc} = K_{oc} \, C^x_{w/cr} \tag{4}$$

In order to apply eqn (4) to a specific site, the safe level normalized to organic carbon may be converted to a dry weight normalized quantity if the organic carbon content in the sediment of concern is measured. This modified quantity is expressed as:

$$C^x_{S/sl} = C^x_{S/sl/oc} \, TOC \tag{5}$$

in which TOC is again expressed as the fractional mass on a dry weight basis as in eqn (3). This quantity can be compared directly to the measured contaminant concentration in the sediment of concern.

Based on the above computations and the use of eqn (4), the sediment safe level for a contaminant can be defined as its concentration in the sediment which ensures that the concentration in interstitial water does not exceed the established EPA water quality criterion.

DERIVATION OF SAFE LEVELS FOR DOMINANT CONTAMINANTS IN PUGET SOUND

Sediment safe levels were derived for six priority metals and forty-seven individual priority organic compounds. These contaminants have been measured historically in the sediment of Puget Sound and have shown elevated concentrations within certain subregions and embayments of the main basin. The levels were derived from eqn (4) using the appropriate partition coefficients and correspond to both the acute and chronic limits established for water, when available. For trace metals, the partition coefficients were computed from the literature as an arithmetic mean with an associated standard deviation; whereas for synthetic organic compounds, they were estimated from existing relationships of experimentally derived sediment/water partition coefficients with octanol/water partitioning ratios. Seven regression equations published in the literature for a variety of chemical classes were considered for their applicability to the compounds tested in this study. The optimum equation selected was one derived for nineteen priority organic chemicals.

The applicability of these derived sediment safe levels in Puget Sound was tested with two independent data sets by comparing the measured concentrations in the sediments of various subregions in the sound with the derived values. Exceedences were then estimated as excess factors from the derived acute and/or chronic limit value. The severity of contamination was then evaluated by examining the frequency of exceedences among a number of subregions and embayments. Tables 25.1 and 25.2 summarize the derived safe levels and relevant data for trace metals and synthetic organic compounds, respectively.

Table 25.1   Estimated Sediment Safe Levels for Trace Metals

Metal	EPA water quality criteria ($\mu g/L$)[a]		$(K_{oc} \pm \sigma) \times 10^4$	Sediment criteria	
	Acute	Chronic		$C_{s/cr/oc}(\sigma)$ Acute (σ)	($\mu g/g$-oc) Chronic (σ)
Arsenic (+3)[b]	120	63	1.3 ± 1.2	1,600 (± 400)	820 (± 790)
Cadmium[b]	38	12	6.4 ± 8.6	2,400 ($\pm 3,300$)	700 ($\pm 1,000$)
Copper[b]	3.2	2.0	170 ± 210	5,400 ($\pm 6,700$)	3,440 ($\pm 4,200$)
Lead[b]	220	8.6	38 ± 40	84,000 ($\pm 88,000$)	3,300 ($\pm 3,400$)
Mercury[b] (inorganic)	1.9	0.10	0.8 ± 1.1	15 (± 21)	0.8 (± 1.1)
Zinc	170	58	33 ± 63	56,000 ($\pm 110,000$)	19,000 ($\pm 38,000$)

[a] Acute water quality criteria are the maximum permissible concentrations for protection of saltwater aquatic life as obtained from the Federal Register (10) or, in the case of some trace metals from draft criteria documents. Chronic criteria are 24-hr average concentrations when obtained from the Federal Register or 30-day average when obtained from draft criteria documents.
[b] Draft criteria documents.

Table 25.2   Estimated Sediment Safe Levels for Synthetic Organic Compounds

Compound	EPA water quality criteria ($\mu g/L$)		$\log K_{ow}^a$	$K_{oc}^{b,c}$	Sediment safe level $C_{s/cr/oc}$ ($\mu g/g$-oc)	
	Acute	Chronic			Acute[d]	Chronic[d]
Phenol	2,900[e]		1.46	24	70	
Acenaphthene*	475[e]	355[e]	4.17	4,700	2,300	1,650
Anthracene*	150[e,f]		4.40	7,400	1,100	
Benzo(a)anthracene*	150[e,f]		5.61	37,000	5,500	
Benzo(a)pyrene*	150[e,f]		6.31	300,000	45,000	
Benzo(b)fluoranthene*	150[e,f]		6.57	500,000	75,000	
Benzo(k)fluoranthene*	150[e,f]		6.84	840,000	125,000	
Fluoranthene	20[e]	8[e]	5.33	45,000	900	360
Fluorene*	150[e,f]		4.18	4,800	700	
Naphthalene*	1,175[e]		3.31	890	1,050	
Phenanthrene*	150[e,f]		4.52	9,300	1,400	
Acenaphthalene*	150[e,f]		4.07	3,900	600	
Dibenz(a,h)anthracene	150[e,f]		6.50	160,000	24,000	
Isophorone	6,450[e]		1.67	37	240	
Nitrobenzene	3,340[e]		1.83	50	165	
1,2-dichlorobenzene	80[e]	65[e]	3.40	1,100	90	70
1,4-dichlorobenzene	80[e]	65[e]	3.37	1,000	80	65
2,6-dinitrotoluene	290[e]		2.05	77	22	
Benzo(g,h,i)perylene	150[e,f]		7.23	1,800,000	270,000	
Chrysene*	150[e,f]		5.61	77,000	11,500	
Indenopyrene	150[e,f]		7.66	4,100,000	600,000	
Pyrene*	150[e,f]		5.18	33,000	4,950	
Butylbenzyl phthalate	1,472[e,f]		4.05	3,700	5,500	
Di-n-butyl phthalate	1,472[e,f]		5.20	35,000	50,000	
Di-octyl phthalate	1,472[e,f]		9.20	82,000,000	120,000,000	
Diethyl phthalate	1,472[e,f]		1.40	22	32	
Dimethyl phthalate	1,472[e,f]		1.61	33	49	
Hexachlorobutadiene	16[e]		3.74	2,000	32	
α-BHC	0.17[e]		3.81	2,300	0.49	
Lindane*	0.16[g]		3.72	1,950	0.31	
DDD*	1.8[e]		6.03	180,000	325	
DDE*	7[e]		5.74	99,000	700	
DDT*	0.13[g]	0.001[g]	5.98	160,000	21	0.16
Aldrin*	1.3[g]		—	400[h]	0.52	
2-PCB*		0.014[f,g]	4.81	4,600		0.064
3-PCB*		0.014[f,g]	5.58	73,000		1.0
4-PCB*		0.014[f,g]	5.75	100,000		1.4
5-PCB*		0.014[f,g]	6.30	370,000		5.2
6-PCB*		0.014[f,g]	6.57	500,000		7

Table 25.2 (Contd.)

Compound	EPA water quality criteria (μg/L)		$\log K_{ow}$[a]	K_{oc}[b,c]	Sediment safe level $C_{s/cr/oc}$ (μg/g-oc)	
	Acute	Chronic			Acute[d]	Chronic[d]
Benzene	2,550[e]	350[e]	2.12	97	245	34
Ethylbenzene	265[e]		3.15	650	140	
Methyl chloride	6,000[e]	3,200[e]	0.91	8	48	
Methylene chloride	6,000[e]	3,200[e]	1.25	16	95	50
Tetrachloroethylene	5,100[e]	225[e]	2.53	200	1,000	45
Toluene	3,150[e]	2,500[e]	2.21	100	315	250
Trichloroethylene	1,000[e]		2.42	160	160	
1,2-Dichloropropane	5,150[e]	1,520[e]	2.28	120	600	180

* Criteria for these compounds tested against measured concentrations in Puget Sound.

[a] K_{ow} values from Callahan et al. (11), Dexter (12); Kenaga and Goring (9); Veith et al. (13)

[b] Derived from regression 8 in Table 3.

[c] Standard error of K_{oc} values $\times$ 3, 95% confidence interval $\times$ 10.

[d] Standard error of sediment criteria $\times$ 3, 95% confidence interval $\times$ 10.

[e] Not naturally adopted water quality criteria, but rather one-half the lowest concentration at which toxic effects have been noted as reported in the Federal Register (10). Actual criteria, when established, are likely to be different.

[f] Based on the class criteria for polynuclear aromatic hydrocarbons (300 μg/l), phthalate esters (2,944 μg/l), or polychlorinated biphenyls (0.014 μg/L).

[g] EPA Water Quality Criteria. Acute criteria are the maximum permissible concentrations for protection of saltwater aquatic life as obtained from the Federal Register (10). Chronic criteria are 24-h average concentrations from the same document.

[h] K_{oc} obtained from Karickhoff (14). No literature value of K_{ow} is available to permit derivation of K_{oc} by the regression equation.

Trace Metals

It can be seen that for the trace metals, the large uncertainty associated with the derived safe levels is attributed to the error in the calculated partition coefficients introduced by (1) normalization of the coefficients to organic carbon and (2) diversity of sites and variation in environmental covariates. A detailed discussion regarding these sources of error was presented by Pavlou and Weston (8). A brief summary follows.

The environmental variables and physical/chemical processes responsible for mediating the transport of trace metals in marine sediments are neither well understood nor adequately quantifiable to allow a theoretical computation of sediment/water partition coefficients. Therefore, these coefficients were derived empirically by using measurements of trace metal concentrations in the interstitial water and bulk sediments from a wide variety of substrate types. Empirical K_{oc} values were then calculated for each substrate type, and a mean of the individual K_{oc} values was computed to determine a partition coefficient most representative of that particular trace metal. The derivation of the K quantities rely solely on the data obtained by Brannon et al. (14). This publication contains the most comprehensive information available in the recent literature. In the study of Brannon et al., samples were taken from a wide variety of areas throughout the country, primarily from marine sediments, but including a few freshwater sites as well. From each site, measurements were made of trace metal concentrations in the bulk sediment and in the interstitial water. The K_D values calculated from this data base were converted to K_{oc} values by dividing by the fractional mass of the sediment organic carbon measured at each location.

Although there is clear justification of normalizing the partitioning coefficient of synthetic organics to sediment organic content, for trace metals the issue is more complicated and requires further examination. Of the six metals evaluated, a statistically significant relationship between K_D and organic content was evident for Cu, Cd, and Pb. For these three metals, it was clear that K_{oc} would be a better estimate of the sediment/water partitioning than would K_D. For the remaining three metals (Zn, As, Hg), the apparent independence of partitioning on sediment organic content was not taken as definitive evidence against the use of K_{oc}. Historical studies do not negate this procedure. For example, Lindberg and Harris (16) found comparisons of mercury concentrations among sites to be more meaningful after the concentrations were first normalized to sediment organic content. Furthermore, there is widespread evidence in the scientific literature that the trace metal content of marine and freshwater sediments is highly correlated with the concentration of organic carbon in the sediment. Crecelius et al. (17) demonstrated this relationship for arsenic, antimony, and mercury in the sediments of Puget Sound. This correlation is a result of the fact that organic material serves as one of the major trace metal sinks in marine sediments. Organic substances and iron plus manganese oxides dominate the trace metal sorptive properties of sediment (18, 19). Thus, sediments with a high organic content can generally be expected to exhibit a greater affinity for trace metals, and therefore have a higher K_D, than sediments with a low organic content. Normalization of trace metal concentrations to organic content, but use of K_{oc}, can be expected to provide a better estimate of partitioning, in most cases, than use of K_D alone. There is some limited evidence (20) that such an approach may not be appropriate for zinc and cobalt, but pending further clarification of the anomalous behavior of these two metals, K_{oc} was adopted as the proper partition coefficient for the sediment safe level determinations.

Synthetic Organic Compounds

The derived sediment safe levels for the forty-seven priority organic compounds measured in sediments of the central basic in Puget Sound (21) were based on the K_{oc} values estimated from the K_{oc}/K_{ow} regression (eqn (8) shown in Table 25.3). Again the associated standard error reflects the uncertainty in the K_{oc} values. Though both acute and chronic safe levels are

Table 25.3   Regression Equations from the Literature for the Estimation of K_{oc} from K_{ow} [a]

No.	Regression	Number of compounds considered	r [b]	Chemical class	Reference
1	$\log K_{oc} = 0.544 \log K_{ow} + 1.377$	45	0.86	Pesticides	[8]
2	$\log K_{oc} = 0.937 \log K_{ow} - 0.006$	19	0.97	Aromatics, polynuclear aromatics, triazines, dinitroaniline herbicides	Brown et al. in prep.
3	$\log K_{oc} = 1.00 \log K_{ow} - 0.21$	10	1.00	Mostly aromatic and poly-nuclear aromatics	[22]
4	$\log K_{oc} = 0.94 \log K_{ow} + 0.02$	9	—	Triazines and dinitro-aniline herbicides	[23]
5	$\log K_{oc} = 1.029 \log K_{ow} - 0.18$	13	0.95	Variety of herbicides, insecticides, and fungicides	[24]
6	$\log K_{oc} = 0.524 \log K_{ow} + 0.855$	30	0.92	Substituted phenyl ureas and alkyl N-phenylcarbamates	[25]
7	$\log K_{oc} = 0.989 \log K_{ow} - 0.346$	5	1.00	Aromatic and polynuclear aromatic hydrocarbons	[14]
8	$\log K_{oc} = 0.843 \log K_{ow} + 0.158$	19	0.96	Priority pollutants	[8]

[a] Modified from Lyman et al. [23].
[b] Correlation coefficient.

FSBC—N*

presented, the chronic value is recommended in order to ensure adequate protection of marine life. Specifically, chronic safe levels are more appropriate than acute values since (1) sediment contaminant concentrations reflect long-term conditions and do not demonstrate the extreme temporal variability of water column contaminant concentrations; and (2) benthic organisms often lack the mobility required to escape a contaminated environment and therefore are susceptible to impacts resulting from long-term chronic exposure. However, no chronic water quality criteria are currently available for the majority of organic compounds, and in these cases, only acute sediment safe levels have been presented in Table 25.2. It may be possible to establish an estimated chronic value by one of two methods: (1) for those compounds having a freshwater chronic criterion, this may be used as an estimate of the saltwater criterion; and (2) the sediment chronic level can be estimated from the acute level using the general "rule of thumb" that the acute:chronic ratio is 100:1. Both approaches warrant further consideration, but a review of the toxicological literature is necessary to determine if one or both approaches would be suitable for establishment of interim sediment safe level guidelines.

It is important to note that for many compounds EPA has not established a water quality criterion but only identifies the lowest concentration at which adverse biological effects have been reported. In these cases, a sediment safe level has been determined on the basis of a water quality "criterion" established at one-half the lowest concentration causing adverse biological effects. The procedure closely parallels the protocol followed in development of water quality criteria in which a final acute value (FAV) designed to protect 95% of a diverse group of species is determined, and the criteria is established at one-half of the FAV. This approach has been adopted as an interim attempt to estimate the concentration at which a water quality criterion may eventually be established. However, it is important to recognize that for some compounds the lowest concentration causing adverse effects is based on a very limited toxicological data base. In these cases, the estimated water quality "criterion" used in this report could be substantially different from the value eventually established. Therefore, these "criteria" may be inadequate to ensure protection of marine life, and therefore any violation of these "criteria" could be of serious consequence.

The EPA water quality criteria for the PCBs, phthalates, and polynuclear aromatic hydrocarbons are class criteria based on the cumulative concentration of all members of the class. In the derivation of the sediment safe levels, it has been necessary to apply the class level to each member of the class individually, since each has a unique K_{oc}. In environments where one class member comprises the majority of the sediment burden of the class, this approach should prove adequate. However, if numerous class constituents are significantly enriched, a safe threshold for the class as a whole may be exceeded even though no individual constituent violates its predicted safe level.

LIMITATIONS AND ASSUMPTIONS

Although the advantages of using the equilibrium-partitioning approach were presented earlier, there are a number of limitations and assumptions inherent in this method that must be considered as well. These have been discussed extensively together with their implications by Pavlou and Weston (8). The presentation that follows highlights these discussions.

Limitations

Lack of Comprehensive Water Quality Criteria. For the majority of synthetic organic contaminants, no water quality criteria are available. This problem was addressed by using one-half the lowest concentration causing adverse effects in cases where definitive water criteria were not available.

Synergism and Antagonism. Synergistic or antagonistic interactions among contaminants can result in a mixture of contaminants having either a greater or lesser toxicity than would be expected simply on the basis of the toxicities of the individual contaminants. Although all individual contaminants may be at concentrations below their predicted safe level, synergistic interactions may still induce adverse biological impacts.

Level of Uncertainty in the Derived Sediment Safe Levels. The sediment safe levels estimated in this study are presented as a mean with a specified level of uncertainty. This uncertainty is a result of our inability, based upon the current state of knowledge, to predict accurately the degree of partitioning of a contaminant between sediment and water. The water quality criteria is considered a fixed value; therefore, the level of uncertainty associated with the derived sediment safe level is entirely a consequence of the uncertainty in the K_{oc} value.

Assumptions

Since the equilibrium-partitioning concept represents a new and unique approach to the development of sediment safe levels, the analyses and evaluations performed by Pavlou and Weston (8), as summarized in this paper, have focused on the identification of all assumptions inherent in the approach. None of the assumptions are considered significant enough to invalidate the approach, but it was deemed essential to assess their potential impact. They have been discussed at considerable length in the author's earlier publication, both to ensure against misuse of the approach and to direct further efforts of refinement.

These assumptions are summarized in Table 25.4 and are cross-referenced with the section in which they are discussed in the Pavlou and Weston report (8).

Table 25.4   Assumptions of the Equilibrium Approach and
Their Potential for Violation

Assumption	Pavlou and Weston [8] reference section	Potential for violation of assumption	Impact of violation of proposed safe levels
Sediment-water partitioning of a contaminant is at equilibrium	3.5.1	Medium	Safe levels may be too conservative. Violations would be site-specific and source-specific, not compound-specific.
Magnitude of partition coefficient dependent upon sediment organic content	3.5.2	Low (organic) High (metals)	At low organic carbon levels, the proposed safe levels would be too conservative. At high organic carbon values, the levels would be liberal.
Variations in the redox potential and pH will not impact safe levels	3.5.3	Low (base-neutral organics); High (metals and some organics)	The effect of redox potential on trace metal partitioning has been largely accounted for by empirical derivation of partition coefficients. In some instances, the redox potential may affect partitioning beyond the level already accounted for. Field verification is needed to determine potential impact on safe levels.
Independence of partitioning on temperature	3.5.3	Low	Typical environmental variations in temperature will affect safe levels by 10% or less.
Independent of partitioning on salinity (ionic strength)	3.5.3	Medium	For most contaminants of environmental concern, variations in salinity typical of marine waters will have a negligible impact on safe levels. Variation in salinity between fresh and saltwater would affect levels by about 25%. An increase in ionic strength of interstitial water would result in a higher safe level value, therefore these limits may be too conservative.
Independence of partitioning on dissolved organic matter (DOM)	3.5.3	Medium	Safe levels would be lowered as DOM increases. Values may be too liberal.
Particle size is not important in determining partitioning other than as a covariate with organic carbon	3.5.3	Low	Evidence to date is very limited, but in very sandy sediments safe levels may be too liberal.
Entire contaminant burden of sediments is in a bioavailable form	3.5.3	High	If a fraction of the contaminant is not bioavailable, the proposed safe levels would be too conservative.
Water quality criteria are appropriate for benthic organisms	3.5.5	Low	Safe levels would be either too liberal or too conservative depending upon toxicant sensitivity of benthic organisms.
Ingestion of sediment-bound contaminants by deposit-feeders does not increase contaminant body burden above levels attained by exposure to the dissolved contaminant fraction	3.5.5	Medium	Safe levels would be too liberal if assumption was violated.

For each assumption, a potential for violation has been given that represents the likelihood (high, medium, or low) that the assumption may prove invalid by further investigation.

Many of the assumptions are unavoidable because of the inadequacy of our current state of knowledge. In these cases, the potential for violation represents a "best guess" based on the limited information available. Should any of the assumptions be violated, the potential impact on the proposed safe level is also given. The potential impact is quantified in some cases (e.g., temperature and salinity) although for most assumptions, it is impossible to go beyond a qualitative statement of potential impact at this time.

FEASIBILITY TESTING IN PUGET SOUND

To test the applicability of the derived safe levels to actual measurements, two independent sediment residue data sets obtained in Puget Sound were selected. These data had sufficient ancillary information to allow direct comparisons of contaminant sediment concentrations to the numerical safe level values computed by the equilibrium-partitioning approach.

The first set of test data used was that published by Malins et al. (5), encompassing six embayments in Puget Sound (Commencement Bay, Elliot Bay, Sinclair Inlet, Port Madison, Case Inlet, and Budd Inlet). These sub-regions are known to vary in degree of contamination and therefore provided a wide enough spectrum of contaminant concentrations in sediments to determine the extent of safe level exceedences.

Safe levels were also tested by comparison with the concentration of contaminants in sediments measured in Elliott Bay and within the vicinity of the West Point outfall as published in the Metro Toxicant Pretreatment Planning Study (TPPS) Draft Report (21). These measurements constitute the most recent information on toxicant accumulation in sediments in this urban environment.

COMPARISON OF MEASURED CONCENTRATIONS WITH
DERIVED SAFE LEVELS

The sediment contaminant concentration in the trial data set of Malins et al. (5) and Metro/TPPS were compared to the safe levels as follows. The data from each test station were illustrated as a scatterplot on a graph of sediment contaminant concentration versus percent organic carbon. The acute and chronic safe level graphs were then superimposed on these plots, and the position of the points about these curves was examined. Typical plots for selected trace metals and synthetic organic compounds are shown in Figures 25.2 through 25.4. Examination of all plots generated in this study revealed the following:

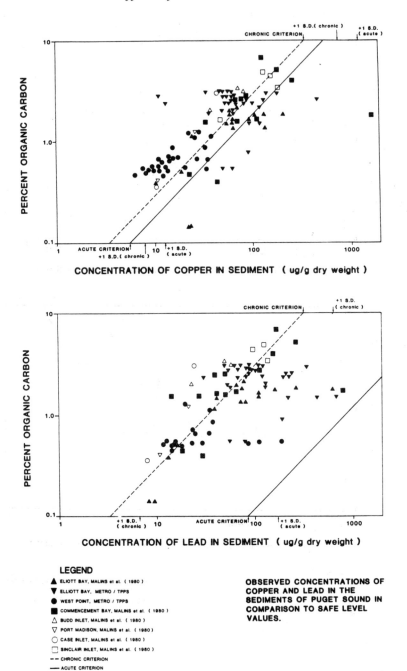

Fig. 25.2
Concentration of lead in sediment (μg/g dry weight).

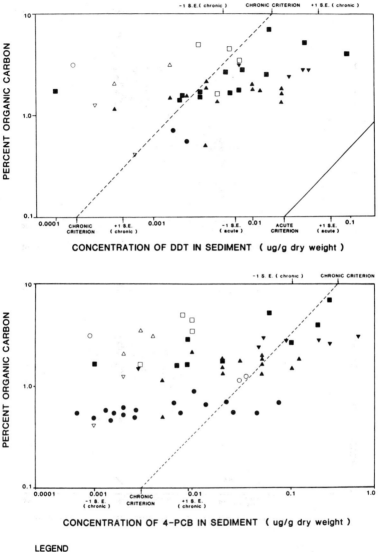

CONCENTRATION OF DDT IN SEDIMENT (ug/g dry weight)

CONCENTRATION OF 4-PCB IN SEDIMENT (ug/g dry weight)

LEGEND

▲ ELIOTT BAY, MALINS et al. (1980)

▼ ELLIOTT BAY, METRO / TPPS

● WEST POINT, METRO / TPPS

■ COMMENCEMENT BAY, MALINS et al. (1980)

△ BUDD INLET, MALINS et al. (1980)

▽ PORT MADISON, MALINS et al. (1980)

⊓ CASE INLET, MALINS et al. (1980)

⊏ SINCLAIR INLET, MALINS et al. (1980)

— — CHRONIC CRITERION

—— ACUTE CRITERION

OBSERVED CONCENTRATIONS OF
DDT AND 4-PCB IN THE
SEDIMENTS OF PUGET SOUND IN
COMPARISON TO SAFE LEVEL
VALUES.

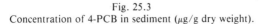

Fig. 25.3
Concentration of 4-PCB in sediment (μg/g dry weight).

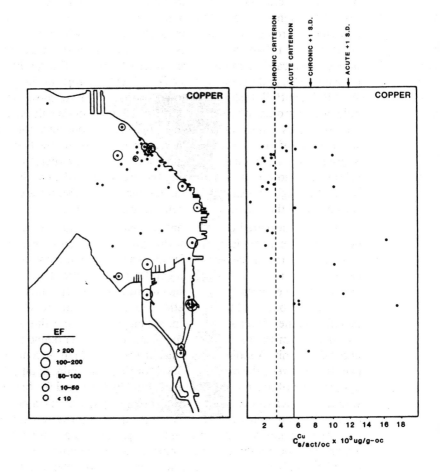

Fig. 25.4
Spatial representation of excess factors (EF) for chronic safe levels computed
for Cu and corresponding sediment concentrations measured at specific sites
in Elliott Bay, Washington.

- For the trace metal examined, the chronic safe levels are exceded in the order: Hg > Pb > Cu > As > Zn > Cd. Acute values are exceeded primarily by the metals Hg > Cu > As; no exceedences are apparent for Cd, Zn, and Pb. Among the subregions examined, proposed safe levels are exceeded most frequently in Elliott Bay and Commencement Bay. Chronic levels for Hg are exceeded in all areas, suggesting that the predicted levels for Hg may be unrealistically low.

- For the synthetic organic compounds tested, the absence of chronic water quality criteria for most contaminants prevented an evaluation of the significance of the observed levels. However, for PCBs and DDT, chronic values are obviously exceeded within the most contaminated urban embayments. Acute values for the remainder of the chemicals examined were exceeded only by fluoranthene, phenanthrene, and pyrene within a fairly localized area north of the West Point outfall. It should be noted, however, that for most of the polynuclear aromatic hydrocarbons (PAHs), the criteria values used are both class criteria and one-half the lowest concentrations at which toxic effects have been noted as reported in the Federal Register (10). Individual compound safe levels may therefore be substantially lower, thus increasing the probability of exceedences within the regions examined. Similar limitations should also be considered in the use of class chronic levels for PCBs.

- Most of the measured contaminant concentrations that exceed the derived sediment safe levels fall within plus one standard deviation or standard error of the mean safe values as computed from the uncertainty assigned to the partition-coefficient quantities (K_{oc}). This may have some implication on the enforceability of the numerical action levels computed by this method. One possible approach is to establish an alert level at the computed mean safe level value and a maximum permissible sediment contaminant at the upper bound of the uncertainty limit. A sediment in which the concentration of contaminant falls between the mean safe level for that contaminant and the upper bound of the uncertainty range could be identified for further studies in order to demonstrate the absence of any environmental danger. For those sites at which a contaminant concentration exceeds the upper bound of the uncertainty range, immediate corrective action may be required. This upper uncertainty limit could be established either at one standard deviation from the mean or at the 95% confidence limit, dependent upon the degree of conservatism desired.

Spatial Comparisons

The frequency with which safe levels are exceeded and the relative severity of these occurrences were used to evaluate the extent of degradation within various subregions of Puget Sound. Comparisons between subregions were made numerically and graphically. These methods are summarized below.

Numerical Approach. The frequency with which predicted safe levels are exceeded for individual contaminants can be defined as:

$$f_v^x = \frac{\text{number of measurements exceeding safe level} \times 100}{\text{total number of measurements}} \quad (6)$$

where f_v^x is defined as the frequency of safe level exceedence and refers to the specific contaminant measured. Values of f_v^x were calculated for both acute and chronic sediment safe levels for six trace levels as shown in Table 25.5. It is possible to sum the individual f_v^x values to obtain a frequency of exceedence for a given subregion integrated for a class of contaminants. This quantity is defined as:

$$F_v = \frac{\sum_n f_v^x}{n} \quad (7)$$

where F_v is the class integrated frequency of safe level exceeded and n is the number of contaminants in the class. Values of F_v are illustrated in the last columns of Table 25.5. Although the sampling frequency among the subregions is quite variable, it is clear that Elliott Bay, Commencement Bay, Sinclair Inlet, and the area around the West Point outfall exhibit the highest frequency of exceedences. This, of course, is not surprising since these areas have historically been known to receive high contaminant inputs originating from land-based point and nonpoint sources.

Graphical Approach. Excess factors, representing the percent enrichment of a contaminant above safe levels, can be calculated at each sampling site as:

$$EF = \frac{C_{s/act/oc}^x - C_{s/sl/oc}^x}{C_{s/sl/oc}^x} \quad (8)$$

where $C_{s/act/oc}^x$ equals the actual measured contaminant concentration normalized to organic carbon content. Computation of excess factors allows for site specificity and definition of areas where remedial and/or control-enforcement action may be necessary. It is possible to obtain this site specificity by spatial representation of excess factors. In this manner, a gradient

Table 25.5   Percent Frequency of Safe Level Exceedences (f_v^x) for Six Trace Metals in Several Subregions of Puget Sound[a]

Subregion	As		Cd		Cu		Pb		Hg		Zn		F_V[b]	
	Acute	Chronic	Acute	Chronic	Acute	Chronic	Acute	Chronic	Acute	Chronic	Acute	Chronic	Acute	Chronic
Elliott Bay	5.5 (2/36)[b]	25 (9/36)	0 (0/36)	0 (0/36)	25 (13/53)	42 (22/53)	0 (0/53)	74 (39/53)	59 (31/53)	100 (53/53)	4 (2/53)	13 (7/53)	22 (46/212)	57 (121/212)
Commencement Bay	—	—	—	—	21 (3/14)	50 (7/14)	0 (0/14)	50 (9/14)	43 (6/14)	100 (14/14)	7 (1/14)	7 (1/14)	18 (10/56)	52 (29/56)
Sinclair Inlet	—	—	—	—	0 (0/4)	50 (2/4)	0 (0/4)	25 (1/4)	100 (4/4)	100 (4/4)	0 (0/4)	0 (0/4)	25 (4/16)	44 (7/16)
Budd Inlet	—	—	—	—	0 (0/3)	0 (0/3)	0 (0/3)	0 (0/3)	0 (0/3)	100 (3/3)	0 (0/3)	0 (0/3)	0 (0/12)	25 (3/12)
Case Inlet	—	—	—	—	0 (0/2)	0 (0/2)	0 (0/2)	0 (0/2)	0 (0/2)	100 (2/2)	0 (0/2)	0 (0/2)	0 (0/8)	25 (2/8)
Port Madison	—	—	—	—	0 (0/2)	0 (0/2)	0 (0/2)	0 (0/2)	0 (0/2)	100 (2/2)	0 (0/2)	0 (0/2)	0 (0/8)	25 (2/8)
West Point Outfall Region	10 (3/31)	36 (11/31)	0 (0/31)	0 (0/31)	3 (1/31)	16 (5/31)	0 (0/31)	21 (9/31)	52 (16/31)	100 (31/31)	3 (1/31)	16 (5/31)	15 (18/124)	40 (50/124)
Total	7 (5/67)	30 (20/67)	0 (0/67)	0 (0/67)	16 (17/109)	33 (36/109)	0 (0/109)	51 (56/109)	52 (57/109)	100 (109/109)	4 (4/109)	11 (13/109)		

[a] Numbers in parentheses are the number of stations which exceed criteria over the total number of stations tested.
[b] Excludes As and Cd.

in the degree of exceedence may be established that could lead to specific-source control action. A typical example of this method is depicted in Figure 25.4. It is apparent that chronic safe levels for Cu are exceeded within the immediate proximity of the city's waterfront, the industrialized area around Harbor Island, and a very short radius from shore at the Denny Way combined sewer overflow (CSO) site. This representation points to the severity of contamination being highly localized and therefore amenable to remedial action. Similar plots could be constructed for other contaminants to improve site specificity of safe level exceedences.

Correlations With Other Sediment Threshold Limits

Pavlou and Weston (7) discussed the history of sediment criteria development and represented a variety of proposed threshold limits, many of which had been developed from the Great Lakes. The sediment contaminant concentrations permitted by these earlier limits were established in a rather subjective manner, but were generally based upon correlations of sediment contaminant burden with observations of adverse biological impact. It should also be pointed out here that the effect of the environmental variable discussed earlier was not considered in the development of these limits.

A comparison of these earlier levels with the values proposed in this study are presented in Table 25.6.

Table 25.6   Comparison of Historical Sediment Threshold Limits with
Safe Level Predicted by the Equilibrium Partitioning Approach[a]

Contaminant	Historical limits	Safe levels (this study)[b]
As	3-8	16
Cd	1-6	15
Cu	25-50	68
Pb	40-50	66
Hg	0.3-1	0.006
Zn	75-100	380
PCB	0.05-10	0.06[c]

[a] Concentrations are given in mg/kg dry weight.
[b] Based on sediments containing 2% organic carbon.
[c] Based on an average criterion from two to six chlorinated biphenyls.

The limits from the literature represent the range of concentrations encompassed by the EPA Region V, EPA Region VI, and Ontario Ministry of the Environment guidelines as reported by Pavlou and Weston (7). The safe levels derived by the equilibrium approach are calculated assuming a 2% organic carbon content. This value is intermediate in the range of organic content typical of sediments and represents material consisting of 50% fines (silt and clay).

The safe levels derived by the equilibrium approach agree remarkably well with the limits developed for Great Lakes sediments. With the exception of Hg, all safe levels estimated by the partitioning approach are within one order of magnitude of the literature limits, and the difference is generally much less. The fact that the proposed Hg safe level is far below the literature value suggests that the estimates from this study may be overconservative. Though the proposed PCB safe level agrees well with the literature values, it should be noted that it represents the mean value collected for 2-6 chlorinated biphenyls. The appropriate safe level can best be determined only after consideration of the relative proportions of all PCB homologs.

SUMMARY AND CONCLUSIONS

With the increased use of our nation's coastal and inland waters, regulatory agencies are frequently confronted with difficult decisions in resolving conflicts between alternative uses of these waters, while at the same time striving to protect overall environmental quality. A significant number of management decisions facing these agencies concern the definition of permissible levels of contaminants in marine sediments. The development of criteria that define environmentally safe levels of contaminants in sediment would be invaluable tools in environmental management decisions. However, given the current state of knowledge, regulatory agencies have been forced to adopt short-term, interim decision criteria until scientifically sound and legally defensible limits can ultimately be established.

This paper presents one of the approaches currently being evaluated as a method for developing safe levels of contaminants in sediments. The equilibrium-partitioning approach is a method that holds much promise for establishing these types of limits. The approach provides a relatively simple mechanism whereby the large toxicological data base incorporated in the EPA water quality criteria can be adapted to determine permissible contaminant concentrations in marine sediments, which would ensure protection of benthic biota and possibly higher trophic level organisms. The basic tenet of the equilibrium-partitioning approach is that the concentration of each contaminant in sediment should be at or below a level that ensures that its concentration in interstitial water does not exceed the EPA water quality criterion.

The key to this approach is the determination of the sediment/water partition coefficient for each compound of interest. This coefficient was determined empirically for trace metals based on measurements of trace metal concentrations in the interstitial water and bulk sediments from a wide variety of substrates. The sediment/water partition coefficients for synthetic organics were estimated from octanol/water partition coefficients. Since the partitioning of a contaminant between sediment and water is strongly dependent upon the organic carbon content of the sediment, the partition coefficient was normalized to this parameter.

Using this approach, safe levels in sediments were derived for six trace metals and forty-seven synthetic organic compounds of concern in Puget Sound. The derived limits were tested against existing data on sediment contaminant concentrations in a variety of areas in Puget Sound, including Elliott Bay, Commencement Bay, Sinclair Inlet, Budd Inlet, Case Inlet, Port Madison, and the West Point area. In most cases, the predicted safe levels were exceeded only in areas that historically have received high inputs of contaminants from point and nonpoint sources. Among the trace metals, the chronic sediment limits were exceeded most frequently for Hg, Pb, Cu, and As in Elliott Bay and Commencement Bay. Among the synthetic organics, only PCBs and DDT consistently exceeded the predicted values, primarily in the urban embayments noted above. The polynuclear aromatic hydrocarbons were generally well below the acute safe levels, except for a few isolated sites in the vicinity of a municipal outfall. Both numerical and graphical procedures were developed to relate the frequency and magnitude of safe level exceedences to specific sites.

During development of the equilibrium-partitioning approach, a number of shortcomings in our current knowledge of pollutant interactions between sediment, water, and biota were identified. The most significant of these shortcomings include: (1) inability to consider toxicological synergistic or antagonistic interactions among contaminants; (2) poor understanding of the influence of environmental variables (e.g., organic carbon, pH) on the chemical behavior of the contaminants; and (3) inability to differentiate the bioavailable fraction of a contaminant from the total sediment burden of the contaminant.

Until some of the issues identified above are resolved and a field verification of this method is performed, the sediment safe levels derived in this study are not recommended for adoption by regulatory agencies. It is important to recognize that although the numerical values suggested as permissible levels of contaminants in marine sediments are referred to as "safe levels," they are only "first cut" estimates. The predicted safe levels are reported with the expectation that they will be refined through an interactive process involving input from both the scientific community and environmental managers. However, as an immediate step in a long evaluation process, the equilibrium-partitioning approach appears very promising in its ability to establish acceptable limits for a wide diversity of hydrophobic synthetic organic chemicals and possibly some trace metals. On a long-term basis, these limits may ultimately be refined to the point where they find broad regulatory application.

REFERENCES

1. **Armstrong, J. W., R. M. Thom, K. K. Chew, B. Arpke, R. Bohn, J. Glock, R. Hieronymus, E. Hurlburt, K. Johnson, B. Mayer, B. Stevens, S. Tettelback and P. Waterstrat.** 1978. The Impact of the Denny Way Combined Sewer Overflow on the Adjacent Flora and Fauna in Elliott Bay, Puget Sound, Washington. College of Fisheries, University of Washington, Seattle.

2. **Malins, D. C., B. B. McCain, D. W. Brown, A. K. Sparks, H. O. Hodgins and S. L. Chan.** 1982. Chemical contaminants and abnormalities in fish and invertebrates from Puget Sound. NOAA Technical Memo OMPA-19, National Oceanic and Atmospheric Administration, Boulder, CO.

3. **Comiskey, C. E., T. A. Farmer and C. C. Brandt.** 1983. Dynamics and biological impacts of toxicants in the main basin of Puget Sound and Lake Washington, Vol. II—Assessment of the impacts of sewage discharge from the West Point Treatment Plant and the Denny Way CSO on the structure of macroinfaunal communities in the central basin of Puget Sound. Submitted to the Municipality of Metropolitan Seattle, WA.

4. **Swartz, R. C., W. A. DeBen, K. A. Sercu and J. O. Lamberson.** 1982. Sediment toxicity and the distribution of amphipods in Commencement Bay, Washington, USA. *Mar. Pollut. Bull.* **13**:359-364.

5. **Malins, D. C., B. B. McCain, D. W. Brown, A. K. Sparks and H. O. Hodgins.** 1980. Chemical contaminants and biological abnormalities in central and southern Puget Sound. NOAA Technical Memo OMPA-2, National Oceanic and Atmospheric Administration, Boulder, CO.

6. **Pavlou, S. P. and R. N. Dexter.** 1979. Distribution of polychlorinated biphenyls (PCBs) in estuarine ecosystems. Testing and concept of equilibrium partitioning in the marine environment. *Environ. Sci. Technol.* **13**:65-71.

7. **Pavlou, S. P. and D. P. Weston.** 1983. Initial evaluation of alternatives for development of sediment related criteria for toxic contaminants in marine waters (Puget Sound). Phase I: Development of conceptual framework. Prepared by JRB Associates for the U.S. Environmental Protection Agency under EPA Contract No. 68-01-6388.

8. **Pavlou, S. P. and D. P. Weston.** 1984. Initial evaluation of alternatives for development of sediment related criteria for toxic contaminants in marine waters (Puget Sound). Phase II: Development and testing of the sediment/water equilibrium-partitioning approach. Prepared by JRB Associates for the U.S. Environmental Protection Agency under EPA Contract No. 68-01-6388.

9. **Kenaga, E. E. and C. A. I. Goring.** 1980. Relationship between water solubility soil sorption, octanol/water partitioning, and concentration of chemicals in biota. In J. G. Eaton, P. R. Parrish, and A. C. Hendricks, eds., *Aquatic Toxicology.* ATM STP 707. American Society for Testing and Materials. Philadelphia, PA, pp. 78-115.

10. **Federal Register.** 1980. Vol. 45, No. 231, November 28.

11. **Callahan, M., M. Slimak, N. Gabel, I. May, C. Fowler, R. Freed, P. Jennings, R. Durfee, F. Whitmore, B. Maestri, W. Mabey, B. Holt, and C. Gould.** 1979. Water-related environmental fate of 129 priority pollutants, Vol. 1. Prepared by Versar, Inc. for the U.S. Environmental Protection Agency under EPA Contract No. 68-01-3852 and 68-01-3867.

12. **Dexter, R. N.** 1976. An application of equilibrium adsorption theory to the chemical dynamics of organic compounds in marine ecosystems. Ph.D. thesis, University of Washington, Seattle.

13. **Veith, G. D., K. J. Macek, S. R. Petrocelli and J. Carroll.** 1980. An evaluation of using partition coefficients and water solubility to estimate bioconcentration factors for organic chemicals in fish. In J. G. Eaton, P. R. Parrish and A. C. Hendricks, eds., *Aquatic Toxicology.* ASTM STP 707. American Society for Testing and Materials, Philadelphia, PA, pp. 116-129.

14. **Karickhoff, S. W.** 1981. Semi-empirical estimation of sorption of hydrophobic pollutants on natural sediments and soils. *Chemosphere* **10**:833-846.

15. **Brannon, J. M., R. H. Plumb, Jr. and I. Smith, Jr.** 1980. Long-term release of heavy metals from sediments. In R. A. Baker, ed., *Contaminants and Sediments*, Vol. 2. Ann Arbor Science Publishers, Ann Arbor, MI, pp. 221-226.

16. **Lindberg, S. E. and R. C. Harris.** 1974. Mercury-organic matter associations in estuarine sediments and interstitial water. *Environ. Sci. Technol.* **8**:459-462.

17. **Crecelius, E. A., M. H. Bothner and R. Carpenter.** 1975. Geochemistries of arsenic, antimony, mercury, and related elements in sediments of Puget Sound. *Environ. Sci. Technol.* **9**:325-33.

18. **Jenne, E. A.** 1977. Trace element sorption by sediments and soils—Sites and processes. In W. R. Chappell and K. K. Peterson, eds., *Molybdenum in the Environment*, Vol. 2. Marcel Dekker, New York, pp. 425-553.

19. **Jenne, E. A. and S. N. Luoma.** 1975. Forms of trace metals in soils, sediments, and associated waters: An overview of their determination and biological availability. In *Biological Implications of Metals in the Environment*, Proceedings 15th Annual Hanford Life Sciences Symposium at Richland, WA. ERDA Symposium Series, No. 42, pp. 110-343.

20. **Luoma, S. N. and E. A. Jenne.** 1975. The availability of sediment-bound cobalt, silver, and zinc to a deposit-feeding clam. In *Biological Implications of Metals in the Environment*, Proceedings 15th Annual Hanford Life Sciences Symposium at Richland, WA. ERDA Symposium Series, No. 42, p. 2.

21. **Romberg, G. P., S. P. Pavlou, R. F. Shokes, W. Hom, E. A. Crecelius, P. Hamilton, J. T. Gunn, R. D. Meunch and J. Vinelli.** 1984. Toxicant Pretreatment Planning Study Technical Report C1: Presence, Distribution, and Fate of Toxicants in Puget Sound Municipality of Metropolitan Seattle, Seattle, WA.

22. **Karickhoff, S., D. S. Brown, and T. A. Scott.** 1979. Sorption of hydrophobic pollutants on natural sediments. *Water Res.* **13**:241-248.

23. **Lyman, W. J., W. F. Reehl and D. H. Rosenblatt.** 1982. *Handbook of Chemical Property Estimation Methods: Environmental Behavior of Organic Compounds.* McGraw-Hill Book Co., New York.

24. **Rao, P. S. C. and J. M. Davidson.** 1980. Estimation of pesticide retention and transformation parameters required in nonpoint source pollution models. In M. R. Overcash and J. M. Davidson, eds., *Environmental Impact of Nonpoint Source Pollution.* Ann Arbor Science Publishers, Ann Arbor, MI.

25. **Briggs, G. G.** 1973. A simple relationship between soil adsorption of organic chemicals and their octanol/water partition coefficients. Proceedings 7th British Insecticide and Fungicide Conf., Vol. 1. The Boots Company Ltd., Nottingham, GB.

Chapter 26
Synopsis of Discussion Session 5:

Regulatory Implications of Contaminants Associated with Sediments

G. Chapman (*Chairman*), W. Adams, H. Lee, W. Lyman, S. Pavlou and R. Wilhelm

INTRODUCTION

General agreement exists that sediments decrease the bioavailability of chemicals in water, effectively reducing chemical exposure to organisms dwelling in the water column. Even though water-column exposure to many chemicals is reduced by the presence of suspended and settleable solids, the potential exposure of benthic organisms to chemicals in sediment interstitial water and sorbed to sediments may not be similarly reduced. Some sediment-associated contaminants almost certainly present a direct hazard to benthic organisms, and the weight of scientific evidence to date indicates that chemicals associated with sediments cannot be arbitrarily considered as nonbioavailable and nontoxic.

The use of bulk measurements of sediment chemical concentrations is of limited utility in assessing the hazard of sediment-associated chemicals, especially for metals and polar organics. There are supporting data to indicate that bulk measurements of neutral lipophilic organic chemicals may be used in hazard assessment by normalizing the contaminant concentration to the organic carbon content of the sediment. These normalized values may be used to calculate theoretical equilibrium tissue concentrations in benthic organisms by normalizing tissue concentrations to tissue lipid content. However, similar normalization procedures are lacking for metals and polar organics.

Where FDA action levels exist (e.g., PCBs, Hg) there is evidence indicating that tissue levels in fish feeding upon contaminated benthic organisms can exceed the FDA limits. This evidence exists for highly chlorinated organic chemicals and methyl mercury, chemicals that are not easily metabolized and that also have a high log P (i.e., 5 to 7). Chemicals with these characteristics represent about 10% of the approximately 19,000 chemicals currently used in U.S. commerce (G. D. Veith, USEPA Duluth, MN, personal communication). FDA action levels are used as regulatory criteria based upon tissue levels but no other regulatory procedures exist that are triggered by bioaccumulation of chemicals. The relationship between tissue levels and toxic effects in the exposed organism is not well known and may be complex. This means that there are no available data other than a few FDA action levels that specify bioaccumulation levels representing harm either directly to the bioaccumulating organisms or indirectly to its consumer.

NEED FOR SEDIMENT QUALITY ASSESSMENT

There is a consensus that there is a need for sediment quality assessment or sediment hazard assessment of some type. However, the development of simple numerical criteria for individual chemicals in sediments is considered to be currently impractical and speculative. A number of procedures for evaluating sediment hazard have been considered by people variously concerned with the fate of chemicals in water, dredging effects, or the protection of fish and benthic populations. These procedures have been outlined by Pavlou and Weston [1] and discussed critically by Chapman [2]. The most promising method for establishing numerical sediment criteria is the equilibrium partitioning approach discussed by Pavlou [3]. Consideration of the above cited papers, however, makes it clear that sediment quality evaluations can be difficult, expensive, and inconclusive.

The sediment assessment procedures depend upon: (1) an evaluation of the extent to which contaminated sediments are currently harming aquatic ecosystems and perhaps threatening human health; (2) an assessment of the statutory requirements of federal laws applicable to sediments; and (3) the significance of sediments to aquatic biota including those consumed by man.

On the first point, an assessment of the current quality of sediments in U.S. water, the workshop digested a significant amount of numerical and anecdotal data from a few sites with sediment contamination. These data provided little insight into the extent of the sediment contamination problem. Some type of national sediment quality survey would help define the extent of the problem. It is clear that there are several severely contaminated sediment sites where resultant impacts on biota have been demonstrated or are strongly suspected. Some of the severely contaminated sites appear to have been contaminated by relatively uncontrolled discharges of chemicals prior to the NPDES permit system implementation. In addition, there may be numerous small foci of sediment contamination in systems currently receiving pollutant loads from human activities.

On the second point, relating to the needs of various state and federal program offices charged with implementing federal laws, a number of specific needs were identified. Table 26.1 provides a listing of the federal acts that, directly or indirectly, require an assessment of actual or potential sediment quality for contaminated sediments.

Table 26.1   Statutory Needs for Sediment Quality Assessment

Law*	Area of need
TSCA	— Section 5: Premanufacture notice reviews
	— Section 6: Reviews for existing chemicals
NEPA	— Preparation of environmental impact statements for projects with surface water discharges
CWA	— NPDES permitting, especially under BAT in water-quality-limited water
	— Section 403(c) criteria for ocean discharges; mandatory additional requirements to protect marine environment
	— Section 301(g) waivers for POTWs discharging to marine waters
	— Section 404 permits for dredge and fill activities (administered by the Corps of Engineers)
MPRSA	— Permits for ocean dumping
CERCLA	— Assess need for remedial action with contaminated sediments; assess degree of clean-up required, disposition of sediments
RCRA	— Assess suitability (and permit) on-land disposal or beneficial use of contaminated sediments considered "hazardous"
EPA/ORD	— Evaluation of sediment quality for research or screening purposes

*TSCA   = Toxic Substances Control Act.
 NEPA   = National Environmental Policy Act.
 CWA   = Clean Water Act.
 MPRSA = Marine Protection, Resources and Sanctuary Act.
 CERCLA = Comprehensive Environmental Response, Compensation and
                Liability Act ("Superfund").
 RCRA   = Resources Conservation and Recovery Act.
 EPA/ORD = (U.S.) Environmental Protection Agency, Office of
                Research and Development.

It is important to note that where contamination of sediments is a possible outcome, TSCA, CWA/NPDES-permitting, and NEPA require assessments of potential sediment quality degradation in advance of any actual discharge or action. Thus, in these cases, there are no actually contaminated sediments with which to conduct pollutant analyses or bioassays. Such acts may require a substantially different assessment protocol from those protocols that may be developed for other acts dealing with existing sediment problems.

With regard to the third point, the need to conduct future research on sediments, the answer seems clear. Given that aquatic sediments are an important (often vital) component of aquatic ecosystems, providing habitat for benthic organisms and feeding and hatching areas for water-column dwellers, and that these sediments frequently concentrate pollutants to very high levels, it becomes imperative to understand the fate, transport, and bio-availability of pollutants in the sediments. Future research programs may need to look for easy-to-monitor indicators of sediment quality to eliminate the need for an exhaustive series of surveys and bioassays for each assessment.

Considering all of the above, there was a strong consensus among the workshop attendees that some form of sediment quality assessment is desirable. This consensus was achieved only after expressions of concern over the form such criteria might take, the scientific basis for their establishment, and the criteria that could be derived with currently existing knowledge and data.

The attendees of the workshop were highly skeptical that criteria expressed as chemical concentrations in sediment (i.e., mg chemical/kg sediment) could be established as national criteria and hence national standards to regulate the discharge of chemicals. The science is not well enough developed at this time to allow for the calculation of such numbers. Furthermore, what was strongly favored is a tiered assessment approach for estimating the hazard of chemicals associated with sediments.

The question was debated whether it is possible for sediment quality criteria (possibly developed from the water quality standard-equilibrium partitioning approach) to be exceeded if all discharges to the water meet the applicable water quality criteria. There was no agreement among the work-shop attendees on this question, which is more than academic because of the potential application of sediment quality criteria to NPDES-permitted discharges and vice versa.

Finally, we would point out that by agreeing to the need for sediment quality criteria, we do not intend to imply that sediment quality can or should be regulated as a separate environmental compartment. Sediments are directly connected with surface and groundwaters, and indirectly connected with natural soils (which are displaced into the aquatic environment and become sediments) and the atmosphere (which can be a source of pollutants for water). The interconnection between sediments and aquatic biota is intimate and complex. Any attempt to protect or improve sediment quality should be done with a simultaneous consideration of the impacts on the other environmental compartments.

SCIENTIFIC APPROACHES TO SEDIMENT QUALITY ASSESSMENT

A number of potential approaches to assessing sediment contamination are being used and others have been suggested. Some approaches generate numerical criteria based on sediment chemistry, whereas others involve biological responses. Tiered approaches may involve multiple approaches, a single approach in sequentially increasing detail, or a combination of these procedures.

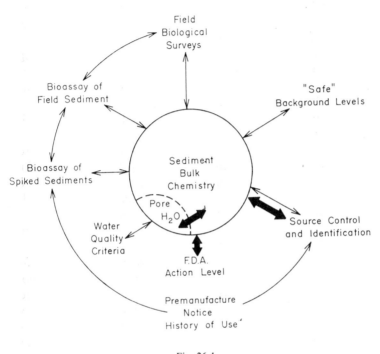

Fig. 26.1
Diagram of the various procedures used for estimating the
hazard of contaminants associated with sediments.
Dark arrows represent modeling applications.

Possible approaches to assessing sediment hazard are shown in Figure 26.1 and are listed below:

- Background Levels—compare site sediment chemistry with that of other sites determined to be unimpacted (numerical pass/fail).
- Field Biological Surveys—conduct on-site studies of biota to evaluate possible impact at site (biological response pass/fail).
- Bioassay of Field Sediment—conduct laboratory (or in situ) toxicity or bioaccumulation bioassay of sediment to evaluate possible impact at site (biological response pass/fail).
- Bioassay of Spiked Sediment—estimate effect/no effect sediment concentrations for a specific chemical (numerical criterion). May be desired in clearance of new chemicals.

- Tissue Action Level—link sediment concentration to safe tissue concentration (e.g., FDA action level or body burden-response data) through application of equilibrium or kinetic models (numerical criterion).
- Aqueous Toxicity Data—apply toxicity data from typical water-column bioassays to sediments through direct measurement of interstitial (pore) water concentration or estimations of pore water from sediment concentrations through application of equilibrium models. May be desired in evaluation of new chemicals.
- Source Control and Identification—identify by waste loading and fate models the source of sediment contamination and the reduction in loading required for reducing impact.

These approaches are discussed in several of the papers in this workshop. No single approach is best for all applications. Depending on the nature of each sediment contamination problem, the method or methodological sequence of choice may vary, i.e., the approaches selected from the assessment scheme in Figure 26.1 depend on site specificity, available data, type of problem, resource availability, etc. Each approach has several advantages and disadvantages and these can be listed based upon whether the approach uses biological assessment procedures or numerical criteria.

Numerical Criteria

Advantages

1. Easy to apply.
2. Require no biological expertise.
3. Amenable to modeling approaches.

Disadvantages

1. May be costly to develop for all chemicals of concern.
2. If criteria exist for some chemicals of concern but not for others, a bioassay test may still be required for most sediments.
3. Analysis requires chemical expertise.
4. Normalization procedure may be required for each chemical (or class of chemicals).
5. Some equilibrium modeling approaches will fail if tissue concentration and toxicity can be independent.

Biological Criteria

Advantages

1. Integrate effects of multiple factors including sediment characteristics and complex or unknown wastes.

2.   Require little chemical expertise.
3.   Field surveys are site-specific, require minimum extrapolation.

Disadvantages

1.   Difficult to establish cause of observed effect.
2.   Field surveys are costly.
3.   Bioassay organisms may not represent sensitivity of the natural species assemblage.
4.   Spatial variability and organism motility may confound field surveys.
5.   Require biological expertise.

One of the first steps in developing a sediment quality assessment methodology should be deciding what environmental and regulatory problems need to be addressed. Different goals dictate different research and regulatory approaches. For example, if our major concern is to protect benthic communities, per se, one possible approach is to develop concentration-effect relations for sediment-associated contaminants using acute and chronic benthic bioassays. However, if our major concern is to regulate the amount of contaminants in edible fish and shellfish, then possible approaches are to use partitioning or kinetic bioaccumulation models and bioaccumulation tests. The more explicitly the potential uses of a sediment assessment are defined, the more rapidly the required research can be completed and regulatory strategies developed.

For most cases of sediment-associated contaminants, assessing specific sediment quality directly gives a more meaningful measurement of environmental quality, appears to be more sensitive, and gives a better temporal-spatial evaluation of contamination than indirect generic approaches. Although direct sediment assessment is preferred in evaluating problems associated with sediment contamination, a combination of direct and indirect approaches may be required in some cases.

Two problems arise with the proposals to apply the present water quality criteria to sediment pore water or with various equilibrium partitioning approaches using these criteria values: (1) there are only water quality criteria values for eight organics and for a few metals, and (2) the compounds causing the problems may not be any of the priority pollutants. These two problems limit the immediate application of a sediment quality approach based on the water quality criteria. However, the development and validation of a sediment quality approach for even a limited set of chemicals (e.g., PCBs, Cd) would be a major achievement in understanding sediment contamination. Furthermore, it is important to recognize that this limitation is a data gap correctable by a directed research effort and not a limitation of these methods of sediment assessment, per se.

Experience with the application of regulatory water quality criteria shows that there is an inverse relationship between the ability of the criterion to define effect thresholds for specific species and sites and the simplicity of the criterion. In adopting a numerical criterion as a pass/fail trigger, to reasonably assure protection, it is common to use a worst-case approach. Thus, it is common to protect most species and sites by using criteria for the protection of a sensitive species and site. As a result, simple numerical criteria, which should be protective in situations where sensitive species are present at sensitive sites, must be overprotective to some degree in other situations.

As seen in Figure 26.2, single criteria for the protection of all sites (lower bound) can be unnecessarily low for the protection of many specific sites, and simple tier I assessments are less accurate than more complex tier II or subsequent assessments. In some instances, more is gained in going site-specific immediately, while in other cases, greater accuracy is obtained from subsequent tiers.

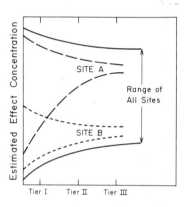

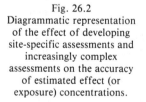

Fig. 26.2
Diagrammatic representation
of the effect of developing
site-specific assessments and
increasingly complex
assessments on the accuracy
of estimated effect (or
exposure) concentrations.

In Figure 26.2, "all sites" may represent any inclusive area (e.g., an estuary, a stream segment, or the world), and sites A and B, any subdivisions of the inclusive area. The sites may be scaled up or down without altering the validity of the comparisons.

Providing proper protection with a single numerical criterion based only on the average condition means that protection will be less than desired in about half the cases. For example, the average estimated effect concentration for all sites in Figure 26.2 would be protective at site A but not at site B. The only way to do better is to provide a more complex regulatory criterion. This more complex approach may entail site-specific criteria to allow for effects of different water and sediment types and different species or may include a tiered approach where the first tier provides basic assessment data upon which decisions for further assessment are determined.

Sediment Hazard Assessment

Previous Pellston conferences dealing with hazard assessment of chemicals (4) and effluents (5) have stressed the importance of using a step-wise tiered approach to assessing the hazard of chemicals and effluents. This also appears desirable for sediments. This approach incorporates the concept that the assessment of hazard is based on a comparison of effect levels with exposure levels (i.e., a calculation of a margin of safety). A small margin of safety signifies potential hazard and indicates the need for additional testing and movement to the next tier. The initial tier may consist of estimated or easily obtained effect and exposure data, and latter tiers will require more extensive effects testing and refined or measured exposure concentrations. This provides a cost-effective hazard assessment approach.

It does not appear possible at this time to design a tiered approach with specific tests for sediment hazard assessment. However, a generic tiered approach for assessing the hazard of neutral lipophilic organics sorbed to sediments for benthic organisms can be described. Different tiered approaches for sediment-chemical hazard assessment might be used for assessment of existing chemicals, for premanufacture review of new chemicals, or for assessment of existing contaminated sediments.

The initial tier for new and existing chemicals might consist of compiling basic physical/chemical and toxicological properties for the chemical of interest from the literature or from estimation techniques. These data could be used to provide an acute effect level and an estimated sediment interstitial water concentration for an initial calculation of a safety factor. This calculation employs an equilibrium partitioning approach (6) and has received only limited testing (7). This is presented as one approach that may prove to be useful. To use this approach, the chemical's K_{oc} and acute effect level have to be estimated, as well as the organic carbon level and chemical concentration of the sediment. The magnitude of the margin of safety indicates the need for futher assessment of the chemical. A large margin of safety would indicate little need for additional assessment. A small margin of safety indicates a need to make sure the margin of safety is reliable and accurate. An additional tier I approach could be a model estimation of maximum body residues from thermodynamic data.

The next tier of assessment would be designed to provide better estimates of the chemical concentration in sediment interstitial water and the toxicity of the chemical to benthic organisms. This could be accomplished by measuring the chemical's K_{oc}, the organic carbon content of the sediment of interest, and the acute toxicity of the chemical to one or more representative benthic organisms. The margin of safety could be calculated again and used to determine the need for further refinement of the assessment of the chemical. If the margin of safety is small and additional testing is indicated, additional steps can be taken to increase the confidence that exposure levels do not exceed effects levels. This testing might include chronic toxicity tests with benthic organisms and bioaccumulation tests with dosed sediment for premanufacture assessment; for existing chemicals, chronic and bioaccumulation tests with spiked sediments or sediment from an area known to be contaminated with the chemical of interest could be performed. Exposure estimates could be refined by measuring the interstitial water concentration of contaminated sediments for existing chemicals.

The goal is to derive in the most cost-effective manner a reliable calculation of the concentrations of a chemical sorbed to sediments and present in the sediment interstitial water that will not impact the benthic environment. This may ultimately require extensive chronic testing or field testing, including ecosystem survey of the impact of an existing chemical. New chemicals might ultimately have to be field tested on a small scale or monitored as the chemical is introduced.

This same assessment approach can be used for sites that are known to be contaminated. Tier I estimation of margins of safety may not be appropriate, and the tier systems could be entered at tier II or higher. This tiered approach is intended to be neither the ultimate nor singular method to assess sediments, but is presented for illustrative purposes and to stimulate further discussion, thinking, and testing.

INSTITUTIONAL CONSIDERATIONS

Should the EPA or other federal agencies develop and promulgate sediment quality criteria (or even allow informal action levels to become widespread), there would certainly be institutional impacts on industries and municipalities discharging to surface waters. Among the institutional impacts would be added costs, and possibly time delays, associated with required data generation (e.g., bioassays) and agency review.

Industry, and other impacted groups, thus have cause to follow and evaluate any program to develop and promulgate sediment quality criteria. They would want to see (1) a demonstrated need for the generation of sediment quality criteria; (2) the use of a criteria-setting process that obtained the input of affected parties, and/or use of a consensus-based approach; (3) criteria that were free of scientific weaknesses that significantly limited the use or reliability of the criteria; and (4) criteria that were easy to use and included, for example, a tiered approach.

RESEARCH NEEDS

Considerably more data are needed before sediment criteria or hazard assessment obtains the level of acceptance of current procedures for evaluating water quality. Although not considered research by some, it is paramount that an evaluation of the extent of the sediment contamination problem be made. What bodies of water have contaminated sediment; what chemicals appear to be most troublesome; what effects can be ascribed to these contaminated sediments; and what are the sources of the contaminants? Such information is needed to evaluate the significance of the problem in relation to other environmental problems also requiring research resources. In addition, a listing of sites and chemicals of concern could focus research onto areas of pertinent chemistry and biology.

Understanding the biological impacts of sediment contamination will require information on the exposure routes for various taxa of biota to chemicals associated with sediments and toxicological data on the relative sensitivity of many more sediment-dwelling organisms than have currently been tested. The etiology of the various fish diseases and histopathological phenomena associated with contaminated sediments needs to be investigated, and the significance of various genotoxic and mutagenic test indices needs to be evaluated.

Further recommendations for research are separated into two categories: short-term needs and long-term needs. The short-term needs address specifically the Equilibrium Partitioning Theory (EPT). They respond to the consensus reached among workshop participants that the EPT approach is useful as a first tier in establishing sediment criteria for neutral organic compounds. We are identifying the research needs of the EPT approach because of the probable near-term application. These needs are intended to provide a chemical and biological verification of the approach to confirm its potential utility and address three major limitations:

- EPT does not account for toxic interactions such as additivity, synergism, or antagonism.
- The influence of environmental variables and processes at the sediment/water interface and the bioturbation layer (active layer of sediment) on the partitioning process is unknown.
- EPT may not differentiate the potentially bioavailable fraction from total sediment burden.

The long-term needs represent research efforts needed to expand the current state of knowledge on the transfer of contaminants between sediment, water, and biota. In this regard, they are important, not only in the development of the EPT approach, but in providing the scientific basis in support of any sediment criteria eventually adopted. It is recommended that the needs identified below be pursued prior to adoption of any type of sediment criterion for regulatory action. Within each category, the needs are presented in order of priority based on their relative importance in expediting near-term approaches and cost effectiveness.

Short-term Needs

- Verify partition coefficients predicted from empirical relationships. This effort should involve laboratory studies, as well as field studies as designated sites of contamination spanning a range of contaminant levels. Measurements should include contaminant concentrations in interstitial water and sediment, as well as ancillary chemical variables that might affect partitioning. As part of this study, existing techniques in sampling interstitial water should be reviewed and, if necessary, the appropriate methodology for obtaining large quantities of water at the active sediment layer should be developed.

- Verify with sediment bioassays the criteria predicted by various numerical approaches. This study should be performed simultaneously with the chemical verification described above, at the same sites, to determine whether the sediment criteria are indeed below concentrations causing adverse biological effects. These bioassay tests should employ appropriately sensitive aquatic species of common occurrence to the specific region in question.

- Relate occurrence of criteria exceedence to occurrence of biological effects for areas where exceedence is predicted. A biological effect can be represented by effect parameters such as: percent mortality obtained in a bioassay test; chronic effects; a change in benthic community structure; percent incidence of disease; percent incidence of physiological abnormality; immune system suppression; enzyme inhibition; and others. This effort should be based primarily on existing information.

Long-term Needs

- Determine the bioavailability of contaminants in the biotic zone of sediments. The factors affecting bioavailability of both trace metal and synthetic organic chemicals should be quantified. These studies should include an evaluation of the importance of ingestion of particulate-bound contaminants in determining the ultimate contaminant body burden of the resident benthic biota. Regarding the relevance of this study to the applicability of the numerical approach, there is a need to examine whether the ingestion of contaminated particles will increase the body burden above that which is attained via the interstitial water.

- Perform studies to determine the sediment-water partitioning and speciation of trace metals as influenced by environmental conditions (organic carbon, pH, redox conditions, iron, manganese, dissolved organic matter, temperature, salinity, or conductivity). These studies should include (1) a review of pertinent literature, (2) field measurements, and (3) controlled experiments with natural and spiked sediments.

● Develop acute and chronic toxicity data for additional chemicals associated with sediments.

● Develop a computerized data base management system for sediment-related chemical and biological data. This system for handling, processing, and accessing these data by the regulatory body and other potential users should be flexible and should be updated as more information becomes available.

REFERENCES

1. **Pavlou, S. P. and D. P. Weston.** 1983. Initial evaluation of alternatives for development of sediment related criteria for toxic contaminants in marine sound (Puget Sound). Phase I: Development of Conceptual Framework. Prepared by JRB Associates for the U.S. Environmental Protection Agency under EPA Contract No. 68-01-6388.

2. **Chapman, G. A.** Establishing sediment criteria for chemicals-regulatory perspective. This volume, pp. 355-377.

3. **Pavlou, S. P.** The use of the equilibrium partitioning approach in determining safe levels of contaminants in marine sediments. This volume, pp. 388-412.

4. **Cairns, J., K. L. Dickson and A. Maki.** 1978. *Estimating the Hazard of Chemical Substances to Aquatic Life.* ASTM STP 657. American Society for Testing and Materials, Philadelphia, PA.

5. **Bergman, H., R. Kimerle and A. Maki.** *Hazard Asssessment for Complex Effluents.* Special publication. Society of Environmental Toxicology and Chemistry (in press).

6. **Pavlou, S. P. and D. P. Weston.** 1983. Initial evaluation of alternatives for development of sediment related criteria for toxic contaminants in marine water (Puget Sound). Phase II: Development and testing of the sediment water equilibrium partitioning approach. Prepared by JRB Associates for the U.S. Environmental Protection Agency under EPA Contract No. 68-01-6388.

7. **Adams, W. J., R. A. Kimerle and R. G. Mosher.** 1985. Aquatic safety assessment of chemicals sorbed to sediments. In R. D. Cardwell, R. Purdy and R. C. Bahner, eds. *Aquatic Toxicology and Hazard Assessment: Seventh Symposium.* ASTM STP 854. American Society for Testing and Materials, Philadelphia, PA, pp. 429-453.

Workshop Summary

Chapter 27

Workshop Summary, Conclusions and Recommendations

J. Fava (*Chairperson*), J. Gilford, R. Kimerle, R. Parrish, T. Podoll, H. Pritchard and N. Rubinstein

INTRODUCTION

Many compounds (especially organics) that enter aquatic sysems have an affinity for particulate material and, in time, become associated with sediments. Sediment-associated contaminants, under certain conditions, have been shown to be biologically available to a wide variety of aquatic organisms. Understanding physical, chemical, and biological factors that influence the bioavailability of contaminants in sediments is essential for rational management of aquatic habitats and resources.

The purpose of the Sixth Pellston Workshop was to assess the role of sediments in influencing the fate and effects of chemicals in aquatic systems. The intent of viewing this topic from the perspective of different interest groups (scientists, regulators, industries, and others) was to establish relationships between the state of the science and the formulation of predictive and regulatory strategies. A combined workshop effort to recognize, describe, and quantify the environmental problems posed by contaminated sediments stimulated ideas and concepts for further research, methods development, and hazard assessment. The resulting discussions and exchange of information focused timely attention on the magnitude of the problem and identified applicable research needs.

This final chapter integrates the conclusions and recommendations identified during the workshop by each session into a single set of workshop conclusions and recommendations and discusses the extent to which the workshop objectives were met.

DYNAMICS OF WATER/SEDIMENT INTERACTIONS

The difficulty in assessing whether sediment-associated contaminants will cause toxicity or will bioaccumulate is determining the bioavailability (or exposure) of these chemical constituents to aquatic organisms. Figure 27.1 shows the possible sediment exposure pathways discussed and evaluated at this workshop. While it is recognized that not all possible pathways may have been identified, we believe that this schematic illustrates the holistic nature of various environmental compartments that determine the bioavailability of sediment-associated chemicals to aquatic organisms. These dynamic interactions are theoretically reversible, although various environmental factors control the extent to which thermodynamic equilibria are obtained. Some of the major factors affecting these relationships are included in this concept. Thermodynamic considerations of the exposure pathways have led to the development of predictive models that can be tested readily in laboratory and field studies and, when verified, can be used to assess the hazard of chemicals bound to sediments.

The importance of sediments in controlling concentrations of toxic chemicals in the water column is also shown in Figure 27.1. Clearly, sediments act as both sinks and sources. Aquatic sediments (top 10 to 20 cm) represent a relatively small volume of the entire ecosystem, and thus high concentrations of sediment-bound chemicals tend to be found on a relatively localized basis. Once a chemical enters the water column, it remains in the dissolved state or it partitions to suspended particulate matter. In either state, the chemical may be available to pelagic and benthic organisms. Sorbed chemicals are exchanged between sediments and animal tissue depending on the residence time of the particulate material in the organism, the equilibrium partition of the chemical, the dissolved organic carbon associated with the particles, and possibly other surface characteristics. Concentrations of dissolved and sorbed chemicals may be significantly affected by various fate processes, particularly photolysis and biodegradation.

Transport of contaminants (in solution or sorbed) to the sediment bed is strongly influenced by system hydrodynamics. Dissolved chemical concentrations in the sediment may be significantly reduced by a variety of processes, including, for example, sorption. Chemical concentrations in pore water will depend on the surface characteristics of the particulate material and chemical gradients.

Accumulation by benthic biota depends on the distribution of chemical gradients among sediments, pore water, and animal tissue. Animal tissue, as a potential sink for chemicals, should be initially defined as total lipid or lipid per organ, until a clearer understanding of active and passive pharmacokinetics

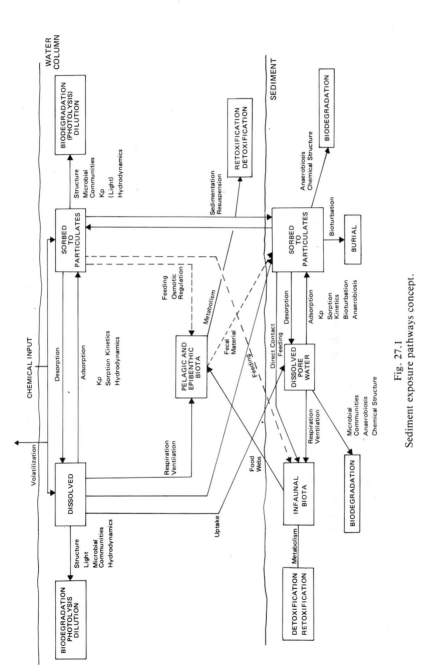

Fig. 27.1
Sediment exposure pathways concept.

is obtained. Adverse effects to the organisms may occur as a result of a particular, as yet undefined, body burden.

Distribution of dissolved and sorbed chemicals in the sediment bed, including release back into the water column, is determined by sediment resuspension, biological activity (bioturbation) and interstitial mixing, sediment reworking (e.g., tides, currents, storms), and burial. Chemical concentrations, either dissolved or sorbed, are additionally affected by biodegradation and the factors that control biodegradation rates. Biodegradation and bioturbation processes may influence chemical speciation of some sediment-associated contaminants.

EVALUATION OF OVERALL WORKSHOP OBJECTIVES

The organizers of the workshop developed six objectives as part of preparing the participants for an active exchange of information and data:

1. Enhance communications and provide a forum for transfer of information between researchers and decision makers from the regulatory, industrial, and academic communities.
2. Assess the state of the science regarding the role of suspended and bed sediments in controlling the fate and effects of chemicals in aquatic systems.
3. Identify existing, consensus methodology for predictive tests of sediment impact on environmental fate and effects of chemicals in the laboratory and field.
4. Identify individual chemical properties and major chemical, physical, and biological properties in the aquatic environment that key the need for sediment/sorption testing in a predictive hazard assessment process.
5. Identify the water quality significance of sediment interaction/ bioavailability, including
    —regulatory implications,
    —sediment action concentrations, and
    —consequences to beneficial users.
6. Identify specific research needs for sediment interactions/bioavailability assessment, and document the workshop findings for general distribution.

The following is our evaluation of the discussions, syntheses, and conclusions reached during the workshop, and their relationship to the workshop's overall objectives.

An important objective was to enhance communication on selected environmental and technical subjects. This was accomplished by inviting key technical experts from regulatory, industrial, and academic communities to participate in the deliberations. The areas of expertise represented included toxicology, aquatic biology, chemistry, physical chemistry, environmental fate, and modeling. The conference atmosphere and format for the sessions were conducive to open communication of thoughts, ideas, and the wealth of experience that the participants brought to the workshop.

By bringing together these experts, it was possible to assess the state of the science on the role of suspended and bed sediments in affecting the exposure and effects of chemicals in aquatic ecosystems. Although this workshop also advanced the science to some extent, it was more successful in bringing together a current broad base of experience and viewpoints to unify future directions in science and regulatory issues related to sediment-sorbed chemicals. A conclusion of the workshop was that chemicals sorbed to sediments are an important environmental issue that must be carefully understood in order to manage our aquatic safety programs and to protect all environmental compartments. Workshop participants recognized that the current status for evaluating the environmental impact of chemicals in sediments was in an early state of development. Therefore, specific answers to numerous technical questions on exposure effects and data interpretation were not addressed. Instead, the discussions and recommendations that resulted should stimulate interest and guide research in the science of sediment-sorbed chemicals.

The workshop participants discussed numerous methods used (1) to estimate exposure of aquatic organisms to organic chemicals and metals in and on sediments, and (2) to assess the bioaccumulation and toxic effects of chemical exposure to both marine and freshwater organisms. Currently, only a few of the methods in use have been "standardized," as discussed in the Session 3 Summary. The use of these methods, however, has produced data that demonstrate the importance of understanding the role of sediment-sorbed chemicals. The workshop also identified mechanisms for improving the utility of the available methods and indicated areas where new methods need to be developed.

One of the most important objectives of the workshop was to identify the methods for incorporating the data on physical, chemical, and biological properties of sediment-sorbed chemicals into a hazard assessment process. Such a process has worked successfully in other aquatic safety programs, and it was concluded that a hazard assessment procedure should and could be adopted to collect data more effectively and to make decisions about the potential hazard of chemicals sorbed to sediments. The chemical exposure (Session 2) and biological effects (Session 3) groups presented guidelines on how data might be collected for tiered evaluation processes. These guidelines provide an opportunity to identify the chemicals most likely to create an environmental problem because of sorption to sediments.

Without a clear understanding of how to predict and measure the biologically available exposure concentrations and biological effects of chemicals associated with solids, the scientific and regulatory communities cannot effectively manage environmental problems. As a result, the desired beneficial uses of our aquatic resources may be impaired. This workshop made significant progress toward understanding the relationships between effects and exposure of sediment-associated chemicals. It is anticipated that the material presented here will provide a foundation on which others can build in the future.

ANALYSIS OF WORKSHOP EXPECTATIONS

During the first session entitled "Background and Perspectives," a number of important research questions about the availability and toxicity of chemicals sorbed on sediments was identified. The charge to the workshop participants was to examine these questions critically in an attempt to document the breadth of our current data base, the need for additional data and information, and the research required to fill the existing knowledge gap.

The workshop summary group examined those questions raised by Session 1 and developed expectations for the workshop. The expectations were classified into four basic categories:

1. Scope of the Problem.
2. Environmental Exposure.
3. Test Methods/Field Studies.
4. Regulatory Control Implications.

Based on these expectations, an attempt was made to document the general concerns of the workshop participants as to what we know, what we need to know, and to present conclusions and recommendations for future work. The results of this analysis are shown in Table 27.1, which summarizes the many pages of discussion from each of the five summary sessions. Based on an assessment of the table, the expectations of the workshop were met. The major conclusions that can be drawn from this table are summarized below.

MAJOR CONCLUSIONS

While each session developed specific conclusions, as discussed in their respective summary sections and highlighted in Table 27.1, twelve overall major conclusions were drawn by the workshop summary session. In some cases, these conclusions resulted from a synthesis of results from several sessions. The following section identifies and discusses each of these major conclusions.

Table 27.1   Summary of Workshop Expectations, Conclusions, and Recommendations

Expectation	Scope of the Problem		Recommendations
	What is known	Need to know	
Develop useful definition of sediment.	Sediments are hydraulically transported organic and inorganic particles.	—	Consider whether sewage sludge, dredge materials, and drilling fluids should be classified as sediments.
Determine sorption dependence on sediment characteristics.	Organic carbon content is the dominant characteristic of sediments that determines sorption of neutral hydrophobic chemicals	Dependence of metals and charged organics sorption on sediment composition, charge distribution structure, and particle size.	Develop quantitative relationships between sorption of metals and charged organics and determine influencing factors.
Define distribution of contaminated sediments in U.S. and their adverse effects to biota.	There are localized areas sufficiently contaminated to warrant concern.	Extent of contaminated areas and potential toxicity.	Define the extent of problem through field monitoring and toxicity testing.
Determine problem chemicals in sediments.	Sorbed hydrophobic chemicals with high log P and sorbed heavy metals may be bioavailable.	Bioavailability.	Verify mathematical models and structure-activity relationships through field and laboratory studies.
	High molecular weight polymers that irreversibly sorb are not problems.	Biodegradability in or on sediments.	

Table 27.1 (Contd.)

	Scope of the Problem (Continued)		
Expectation	What is known	Need to know	Recommendations
Determine a predictive framework for assessing bioavailability and toxicity of sediment-bound chemicals.	Bioavailability and toxicity of neutral organics can be modeled.	Approaches to predict bioavailability and toxicity of metals and polar and ionizable chemicals.	Quantitatively describe the relationship between sediment composition and bioavailability of the sediment-bound metals.
		Validity of models.	
		Relationship of body residue and effects.	Develop experimental designs and models to test relationship between body residues and effect.
			Develop Tier I method for bioaccumulation and toxicity of effects.
			Develop field database.
		Methods for measuring available fractions for divalent metals.	Develop models for predicting available fractions of metals.
		Sorption/desorption kinetics.	Develop sediment transport models and validate.
		Hydrodynamic and bioturbation effects on exposure.	Examine effects of sorption on biodegradation kinetics.
		Biodegradation of available fractions.	Examine biodegradation capabilities of sediment-associated bacteria.

Table 27.1 (Contd.)

Expectation	What is known	Environmental Exposure Need to know	Recommendations
Define relationship between chemical concentration in the water column, interstitial water, and sediment.	Chemical loading into sediments and partition coefficients determine distribution.	Quantitative effects of hydro-dynamics and bioturbation on distribution. Methods for collecting interstitial water in the field.	Develop models and validate. Develop standardized methods for interstitial water sampling in the field.
Define routes of exposure to infauna, epibenthic, and pelagic organisms.	Ingestion, sorption, and respiration are the primary routes of exposure.	Relative contribution of each of the three routes under different conditions for different species.	Design and conduct studies to determine exposure routes for representative species and contaminants of concern.

Table 27.1 (Contd.)

Expectation	What is known	Need to know	Recommendations
		Test Methods/Field Studies	
Laboratory			
Identify methods for determining toxic effects of contaminated sediments.	Methods exist.	Cost-effectiveness and variability of existing methods.	Update and improve existing methods. Conduct intra- and inter-laboratory validation tests for existing methods. Develop new methods as appropriate and validate.
Assess adequacy of sorption/desorption tests for hazard assessment.	Methods exist but are not used universally.	Usefulness of standardized tests. Errors associated with non-standardized tests.	Update the tests and standardize.
Field			
Determine need for field studies to assess impacts and to validate laboratory studies.	Validation is essential. Field study methods exist for impact assessment.	Validity of methods. Environmental relevancy of laboratory studies.	Perform validations. Develop criteria for uncertainty analysis.
Sample			
Describe procedures for collecting, handling, and storing representative sediment sample.	Sediment collection, handling, and storage affect sediment characteristics.	Relative importance of handling and storage variability to hazard assessment. Most appropriate methods for handling and storing sediments.	Design experimental program to assess quantitatively effects of handling and storage.

Table 27.1 (Contd.)

	Regulatory Control Implications		
Expectation	What is known	Need to know	Recommendations
Evaluate need for sediment criteria.	The sediment criteria concept is controversial.	Methods to measure response of aquatic systems to sorbed chemicals.	Develop response-based assessment procedures.
	Available data are not sufficient to set numerical criteria.		
Develop process to assess sediment hazard.	Problems with new chemicals will occur.	Predictive methods for bioavailability and toxicity of problem chemicals.	Develop methodology.
	Principles of hazard assessment exist (tiers, decision criteria, and minimum databases) and can be applied to sediment assessment.		Develop tiered system.
	Levels of uncertainty exist.		

1.   *The public should have a more realistic understanding of the impact of contaminated sediments on aquatic systems and public health.*

Sediments contaminated with toxic chemicals have been found in many areas of the United States. The impact of these contaminated sediments on the aquatic system in which they occur, however, is uncertain; adverse effects on aquatic systems have been associated with some contaminated sediment, whereas in other contaminated areas, adverse impacts have not appeared. The existing information on the occurrence of contaminated sediments and the impact or lack of impact on aquatic systems should be assembled and used to bring the current public perspective and its influence on regulatory actions to a more realistic understanding of the subject.

2.   *Toxic chemicals that sorb to sediment are a potential hazard to aquatic systems.*

Toxic chemicals introduced into aquatic systems can sorb to sediments. Sorption is affected by physical and chemical properties of the chemical, sediment, and water phase, and the sediment/water mixing characteristics. Unless a sorbed chemical becomes bioavailable, it does not manifest a toxic effect on the system into which it is introduced. However, if a strong thermodynamic potential exists in the aquatic system for sorbed toxic chemicals to partition into aquatic organisms, the sorbed chemicals can produce adverse effects. Even though all toxic chemicals that sorb to sediments do not become bioavailable, those that do are potential hazards to aquatic systems and should be assessed in that light.

3.   *Determining exposure concentrations of sediment-associated chemicals is essential for assessing the impact of contaminated sediments in the aquatic environment.*

Bulk chemical analysis of sediments cannot be used directly to predict biological response (i.e., accumulation in tissues and acute and/or chronic effects) of aquatic organisms. Determining the extent and duration of exposure of biota to biologically available concentrations of sediment-associated chemicals requires additional information on sediment characteristics and chemical gradients. Although additional validation is necessary current evidence indicates that for neutral hydrophobic chemicals (i.e., $\log P = 3.5$ to 7), exposure concentration or bioavailability to aquatic life can be predicted simply on the basis of equilibrium partition coefficients, total chemical concentrations in sediments, and the organic carbon content of the sediment and the organism. Prediction of bioavailable heavy metal concentration appears to be more complex, and appropriate normalizing factors must be determined. Until predictive methods for determining bioavailability of contaminants in sediments can be validated, empirical measurements of body burden and effects as determined by the toxicity tests and field monitoring provide the most direct approach for evaluating the impact of contaminated sediment in the aquatic environment.

4. *Equilibrium partitioning models provide an estimate of the maximum amount of sorbed material that is bioavailable.*

Understanding sorption equilibria and kinetics is critical in assessing the bioavailability of sorbed chemicals. If the amount of chemical on the sediment is known, then knowledge of sorption equilibrium partitioning gives the maximum amount of sorbed chemical available for bioaccumulation. Sorption processes that are slow relative to exposure times will lead to bioaccumulation that is only a fraction of this maximum equilibrium amount.

Sorption equilibrium depends on chemical type and sediment characteristics. Because of the large number of chemicals of interest and the diversity of sediment types, it is critical to develop sorption models in the laboratory that will simplify the prediction of sorption partitioning in the aquatic environment. As a first step, it is important to realize that hydrophobic organics, metals, and polar (or ionizable) organics will interact very differently with sediments, and therefore separate sorption models are required for these three classes of chemicals.

5. *A hydrophobic sorption model: Bioavailability of hydrophobic solutes is dependent on organic carbon content.*

The sorption of hydrophobic organics on sediments is dominated by nonspecific interactions of these solutes with the organic fraction of the sediment. Thus, partition coefficients for hydrophobic solutes normalized for organic carbon can be used to predict the partitioning of those solutes in or on an aquatic sediment if the organic carbon content of the sediment is known. Furthermore, because partitioning of hydrophobic solutes into biota also can be normalized for lipid content, the maximum amount of sorbed hydrophobic chemical partitioning into biota can be predicted.

The sorption of polar or ionizable organic solutes is more difficult to model than the sorption of hydrophobic organics because of specific interactions of these solutes with sediment surface functional groupings. Any sorption model for these organics will depend on the organic carbon content of the sediment and one or more measures of charge distribution on the sediment, such as pH or cation exchange capacity. Conversely, the activity of these polar or ionizable organics on biota is expected to be influenced primarily by organic carbon.

6. *A means of modeling metal sorption to sediment has been proposed.*

The bioavailability of metals from sediment is more difficult to model because metal/sediment interactions are more complex. During this workshop, a sorption model for metals was proposed that is somewhat more complex but analogous to the hydrophobic sorption model. Sorption equilibria of the divalent metal cations in aerobic sediments should be dominated by interactions of metals with the iron and manganese oxides and the organic carbon in the sediment. If sorption of a given metal divalent cation species can be normalized with respect to these oxide and organic fractions, the sorption partitioning of the divalent cations in an aquatic environment can be predicted reasonably well for characterized sediments.

7. *Sorption models must be fully assessed in the laboratory before field testing.*

The hydrophobic sorption model has been verified in laboratory studies. Initial experiments have shown the usefulness of the sediment/biota equilibrium partitioning model for hydrophobic solutes. The proposed sorption model for metals has not been tested in the lab, and a great deal of experimental research and model development is needed for sorption of polar and ionizable organics. It is critical that the applicability and limitations of these equilibrium models be addressed more fully in the laboratory before they are universally applied to field situations. Moreover, studies of sorption and bioaccumulation kinetics should be performed to refine these equilibrium models.

8. *Test methods and appropriate organisms should be recommended for assessing the toxicity of chemicals that sorb to sediments.*

Methods and species to be used in testing the toxicity of contaminated sediment need to be identified in order to encourage the development of data that adequately assess toxicity or other adverse effects and that are suitable reference data for other investigators. The use of recommended methods and species also will make it possible, in time, to pool data from different studies to create a larger data base on the toxicity of contaminated sediments. The Biological Effects Synthesis Committee (Session 3) identified a number of acute and chronic tests and species that can be used for assessing toxicity of sediment-associated chemicals. From this review of other methods and species, a list of recommended methods and species should be developed and presented.

9. *The relationship between fate, distribution, and bioavailability of sediment-associated chemicals, and the oxidized state of sediment should be thoroughly investigated.*

The physical integrity, pore-water hydrodynamics, and biological activity of sediments allow for the formation of microenvironments with oxidation-reduction states that vary widely. As a consequence, unique interactions between the sediment biota and their physical environment regulate the cycling of organic and inorganic materials. Investigations into the degradative and remineralization activities of microbial communities and the reaeration and sediment reworking activities (bioturbation) of benthic invertebrates have clearly indicated the complexity of these interactions. These activities strongly influence aerobic and anaerobic conditions and related oxidation-reduction states. Since pollutants contributed by man are processed by the biological systems as if they were natural materials, a thorough knowledge of the effect of aerobic and anaerobic environments on the physical and biological fate of these organic and inorganic chemicals must be developed. Research must be conducted in carefully designed laboratory and field experiments that preserve the integrity of the sediment bed, its micro- and macro-biotic community, and its associated oxidized states.

10. *Bioaccumulation of sediment-associated contaminants by aquatic organisms is not presently an effective measure of adverse ecological impact because of the paucity of residue-effects correlations.*

Bioaccumulation of sediment-associated contaminants has been reported for a variety of aquatic organisms. There is currently no available guidance for interpreting the biological consequence of elevated body residues. The few Food and Drug Administration (FDA) action levels that are available are for protection of man from consumption of contaminated finfish or shellfish, and do not provide guidance for interpreting the biological consequence of elevated body residues. If bioaccumulation is to be used as an effective measure of adverse ecological impact resulting from exposure to contaminated sediments, then the appropriate residue-effects data base must be developed for representative aquatic species and contaminants of concern. A model that may provide a focus for future research incorporates an equilibrium approach for predicting the distribution of contaminants within the organisms and thereby indirectly relates chemical concentration in non-target organs or whole-body residues to observed effects.

11. *At this time, it is not feasible to develop numerical sediment quality criteria.*

The participants at the workshop concluded that it was not feasible at this time to develop a technically defensible approach for the establishment of numerical sediment quality criteria. This conclusion was based on the belief that there are too many uncertainties in current methods, and the incompleteness of validation of theories. However, there was agreement on the need to utilize a hazard assessment approach for collecting data and for making decisions on the safety of chemicals sorbed to sediments. Because of the unique physical and chemical characteristics of each sediment that affect exposure, it was concluded that judgements should be made on a sediment-specific basis.

12. *An approach to assess the hazard of sediment-bound chemicals is needed.*

A recurring concern through the workshop was the lack of a unified approach to assess the potential hazard of chemicals bound to sediments.

This need arose from two basic concerns: historical problems with contaminated sediments, and the potential of contaminating sediments with new chemicals. Historical sediment problems generally resulted from past disposal and dumping practices that occurred before much of our current environmental legislation was enacted. Many of these disposals resulted in areas of sediment contamination with significant environmental adverse effects. These areas are potential concerns of the U.S. Army Corps of Engineers because disposal of dredged materials may impair beneficial uses of receiving waters. The second major concern was the release of existing chemicals into the environment through either point or non-point discharges or the evaluation of new chemicals as required under the Toxic Substances Control Act.

Both concerns require cost-effective approaches for determining unacceptability of contaminated sediments or the approval basis for new chemicals for production. Three of the previous five Pellston Workshops identified and described approaches to assess the hazard of chemicals to aquatic life. These hazard assessment strategies illustrated the need to relate exposure estimates to no-observed-effect concentrations, the need to include decision criteria (including the concept of the margin of safety) to determine the need for additional data and information, and the need for a logical sequence of testing. Session 5 in this workshop (Regulatory Control Implications) also supported the concept of a tiered hazard assessment scheme for evaluating the potential hazard of chemicals in sediments. Session 2 (Environmental Fate and Compartmentalization) and Session 3 (Biological Effects) specifically identified and discussed techniques for evaluating the potential exposure of aquatic organisms to chemicals in sediments and the potential effects of residues in aquatic organisms.

Although a specific, tiered hazard assessment procedure was not developed in these sessions, it was concluded by the workshop participants that a tiered hazard assessment procedure should be developed, tested, and verified.

An important point that should be re-emphasized is that contaminated sediments are only a portion of the aquatic environmental quality problem, as illustrated in Figure 27.1. As such, hazard assessment schemes for evaluating the role of chemicals in sediments should eventually be integrated into holistic hazard assessment approaches.

MAJOR RESEARCH NEEDS

1. Establish the relationship between real-world impacts on aquatic ecosystems (mortality, growth, reproduction, species composition and abundance, behavior, disease, and physiological abnormalities) and the predicted or perceived safe concentration of organic chemicals and metals sorbed to sediments.

2. Establish the validity of partition coefficients predicted from empirical relationships.

3. Verify predictions of chemical exposure based on equilibrium partitioning theory by comparing laboratory- and field-derived data.

4. Determine the factors that affect the bioavailability of contaminants in sediments.

5. Determine the effect of metal speciation on sediment/water partitioning, and describe and quantify the environmental conditions that affect metal speciation.

6. Examine the relationship between bioconcentration and other biological effect responses.

7. Develop and standardize sediment toxicity testing methods for use with freshwater and marine species.

8. Evaluate methods for collecting, handling, and testing sediments for chemical concentration and toxic effects.

9. Determine relative contribution of water column particulate-bound contaminants and interstitial water contaminants to sediment toxicity.

10. Develop and verify the aerobic-sediment heavy-metal activity prediction method and verify that toxicity to benthic organisms is caused by chemical activity.

11. Conduct the research necessary to evaluate the utility of a tiered hazard assessment process in evaluating the impact of sediment-associated contaminants on aquatic ecosystems.

12. Characterize the metabolic capabilities and biodegradation kinetics of microbial populations on sediment surfaces and examine the environmental parameters that control these biodegradation processes.

13. Conduct the research necessary to evaluate the hazard assessment process as an alternative to numerical sediment quality criteria.

14. Characterize the extent of sediment quality in the United States (Annual Report of the President's Council on Environmental Quality might be used as a vehicle for this annual analysis).

15. Conduct research to improve quantification of the role of hydrodynamics on the release of chemicals from sediments.

Index